ASTEROIDS
Astronomical and Geological Bodies

Asteroid science is a fundamental topic in planetary science and key to furthering our understanding of planetary formation and the evolution of the Solar System. Ground-based observations and missions have provided a wealth of new data in recent years, and forthcoming missions promise further exciting results. This accessible book presents a comprehensive introduction to asteroid science, summarizing the astronomical and geological characteristics of asteroids. The interdisciplinary nature of asteroid science is reflected in the broad range of topics covered, including asteroid and meteorite classification, chemical and physical properties of asteroids, observational techniques, cratering, the discovery of asteroids, and how they are named. Other chapters discuss past, present, and future space missions and the threat that these bodies pose for Earth. Based on an upper-level course on asteroids and meteorites taught by the author, this book is ideal for students, researchers, and professional scientists looking for an overview of asteroid science.

THOMAS H. BURBINE is Director of Williston Observatory at Mount Holyoke College and has a Ph.D. in Planetary Sciences from MIT. Asteroid (5159) Burbine is named in his honor.

Cambridge Planetary Science

Series Editors: Fran Bagenal, David Jewitt, Carl Murray, Jim Bell, Ralph Lorenz, Francis Nimmo, Sara Russell

Books in the series

[†] Reissued as a paperback

ASTEROIDS

Astronomical and Geological Bodies

THOMAS H. BURBINE

Mount Holyoke College

CAMBRIDGE
UNIVERSITY PRESS

University Printing House, Cambridge CB2 8BS, United Kingdom

One Liberty Plaza, 20th Floor, New York, NY 10006, USA

477 Williamstown Road, Port Melbourne, VIC 3207, Australia

314-321, 3rd Floor, Plot 3, Splendor Forum, Jasola District Centre, New Delhi - 110025, India

79 Anson Road, #06-04/06, Singapore 079906

Cambridge University Press is part of the University of Cambridge.

It furthers the University's mission by disseminating knowledge in the pursuit of education, learning and research at the highest international levels of excellence.

www.cambridge.org
Information on this title: www.cambridge.org/9781107096844
10.1017/9781316156582

First published 2017

A catalogue record for this publication is available from the British Library

Library of Congress Cataloging in Publication data
Names: Burbine, Thomas H., 1966–
Title: Asteroids : astronomical and geological bodies /
Thomas H. Burbine, Mount Holyoke College.
Description: Cambridge : Cambridge University Press, 2017. |
Series: Cambridge planetary science |
Includes bibliographical references and index.
Identifiers: LCCN 2016028019 | ISBN 9781107096844 (hardback)
Subjects: LCSH: Asteroids. | Space flight to asteroids.
Classification: LCC QB651.B86 2017 | DDC 523.44–dc23
LC record available at https://lccn.loc.gov/2016028019

ISBN 978-1-107-09684-4 Hardback

Dedicated to the People Who Inspire Me Everyday

My Parents

Ahlay and Shahla Hussain

Adam Carolla and Drew Pinsky

Contents

Color plate section is between pages 166 and 167.

Preface

This time period may be the Golden Age of asteroid research. A mission (Dawn) has just mapped the third largest body [(4) Vesta] in the main asteroid belt and is currently mapping the largest [(1) Ceres]. Missions (e.g., Rosetta) routinely image asteroids on their voyages to their target bodies. The Hayabusa mission has returned fragments of an asteroid to Earth. Two asteroid missions (Hayabusa2 and OSIRIS-REx) have just been launched. Both will return samples from near-Earth asteroids back to Earth. Other asteroid missions are in development. News stories on asteroids seem to occur almost every week when these objects make a close approach to the Earth. A number of movies (*Deep Impact*, *Armageddon*, *Seeking a Friend for the End of the World*) have been made about objects potentially striking the Earth.

However, "asteroid" is not a term that is officially defined by the IAU (International Astronomical Union), the internationally recognized body for assigning designations to celestial objects. The IAU uses the term "small Solar System bodies," which refers to minor planets and comets that are not considered dwarf planets (like Pluto). However, scientists and the general public generally refer to small bodies (that are not comets) in the Solar System as asteroids. The word "asteroids" is derived from the Greek word *asteroeides* (star-like) due to their point-like appearance in the sky.

Why are asteroids so important to study? Richard Binzel often gives a talk entitled "Asteroids: Friends or Foes." Asteroids can be considered "friends" because these objects could potentially be mined for important resources such as platinum and palladium (Kargel, 1994) for Earth-based use or water and oxygen for space-based purposes. Asteroids also can be considered "foes" because an impacting body could potentially wipe out our civilization. This book will discuss all characteristics of asteroids to allow us to make our own decision on whether these bodies are "friends" or "foes." Or maybe they are both?

For the longest time (~170 years), asteroids were thought of as just astronomical objects that could only be seen as points of light in a telescope. Only their magnitudes (brightnesses) could be determined and their orbits calculated. Edmund Weiss (1837–1917), the Director of the Vienna Observatory, referred to these bodies as "those vermin of the sky" (Seares, 1930) because asteroid trails would ruin photographic images of "more important" astronomical objects.

Asteroids are now known to be geological bodies. Our impressions of asteroids started to first change to a geological point of view when spectrophotometry (and then spectroscopy) of these bodies in the visible and near-infrared began to become relatively easy to do in the 1970s. Reflected light from these bodies at a variety of wavelengths became relatively easy to measure. Many minerals have characteristic absorption bands in the same wavelength range, and the presence of these minerals could be identified on asteroid surfaces. These spectral observations of asteroids showed that these bodies have a wide variety of surface mineralogies and experienced a wide range of heating.

Starting in the 1990s, spacecraft images showed that asteroids were covered with geological features such as craters, grooves, scarps (cliffs), and boulders. In the 2000s, geochemical analyses of asteroids started to be done remotely by spacecraft, which allowed us to better understand the geological processes occurring on these bodies.

Concurrent with these spectroscopic and geochemical studies of asteroids were high-resolution laboratory analyses of fragments of asteroids (meteorites). Meteorites were found to range from those that melted (e.g., irons, stony-irons, achondrites) to those that experienced minimal heating (e.g., ordinary chondrites, carbonaceous chondrites, enstatite chondrites). Laboratory studies showed that there was a wide variety of geological and chemical processes that occurred during the formation and subsequent alteration of asteroids due to heating and shock events (impacts). Technological advances have led to very precise determinations of when these bodies formed and when different types of alteration occurred.

There also now appears to be a continuum of compositions between all types of small bodies (minor planets and comets) in the Solar System. The distinction between asteroids and comets has blurred in the last few years with some asteroids displaying cometary activity and some comets losing their activity and looking asteroid-like. Many volatile-rich objects have also been discovered past the orbit of Neptune. Because of these new discoveries, Pluto is now classified as a dwarf planet and has been given a minor planet number. All of these small Solar System bodies are thought to be the remnants of the planetesimals that did not form the terrestrial and giant planets at the beginning of our Solar System.

Objects labeled as asteroids will be the primary focus of this book. However because of this continuum between all small bodies, any object labeled as a small

Solar System body or dwarf planet will be discussed. The terms "asteroid" and "minor planet" will be used relatively interchangeably. Meteorites will also be discussed in detail since they are almost all fragments of asteroids.

This book is written as a textbook for an undergraduate class on asteroids or as a reference book for anybody wanting to learn more about small bodies. The content of this book most closely resembles *Introduction to Asteroids* (Cunningham, 1988), a book that covered all aspects of asteroid studies. However, that book was published long before the first spacecraft encounters with asteroids and is almost 30 years out of date. *Asteroids* (Gehrels, 1979), *Asteroids II* (Binzel et al., 1989), *Asteroids III* (Bottke et al., 2002), and *Asteroids IV* (Michel et al., 2015) from the University of Arizona Space Science Series are written more for a graduate-level audience with some previous knowledge of asteroids. A few other books have recently been published. However, *Asteroids: Relics of Ancient Time* (Shepard, 2015) was written for a general audience while *Asteroids, Meteorites, and Comets* (Elkins-Tanton, 2010) was written for a high school audience.

The problem with learning about asteroids is that a background in physics, mathematics, chemistry, geology, and astronomy is really necessary to understand topics specific to asteroids. This book will assume a college-level knowledge of these subjects. But the book is written to give as much introductory information as possible so more complicated topics can be understood.

References are given for topics that cannot be completely covered in this book. For example, sections in many chapters on subjects such as CCD detectors (Howell, 2006), orbits (Curtis, 2014), discovery of Ceres (Cunningham, 2016), meteorites (Hutchison, 2004; Grady et al., 2014), cosmochemistry (Lewis, 2004; McSween and Huss, 2010), isotope geology (Dickin, 2005), crystal field theory (Burns, 1993), radiative transfer modeling (Hapke, 2012b), light curves (Warner, 2006), comets (Swamy, 2010), and cratering (Melosh, 1989, 2011) are covered in much more detail in their own specialized books.

The focus of this book is to give enough background information so results from spacecraft missions to asteroids can be understood. Why were these targets chosen? How were these bodies named? What instruments have been used to study these bodies? How well can we link these bodies to particular meteorite groups? What can be learned from returned samples? Will the spacecraft results allow us to protect the Earth from potential impactors?

The first chapter of this book covers light and how it is detected, since all asteroids are observed using light. The second chapter covers orbits and asteroid discoveries, since all these bodies have different orbits and different designations. The third chapter covers meteorites, minerals, and isotopes, since fragments of asteroids (meteorites) are composed of minerals and contain both radioactive and stable isotopes. The first three chapters cover the three most basic characteristics of minor planets. We see these objects because they reflect or emit light, they all have distinct orbits around the Sun, and they are all composed primarily of minerals that can be studied on Earth using meteorites.

The remaining chapters expand on these basic characteristics to discuss how we study minor planets. The fourth chapter covers reflectance spectroscopy and asteroid taxonomy, since the mineralogies of asteroids are primarily determined by how they reflect light. The fifth chapter covers asteroid families and the physical properties (e.g., diameters, masses, densities, distributions) of asteroids. The sixth chapter primarily covers outer Solar System bodies such as comets, Trojan asteroids, and trans-Neptunian objects and the spacecraft missions to these bodies. The seventh chapter covers near-Earth asteroids, cratering, and the impact threat. The eighth chapter covers past, current, and future spacecraft missions to asteroids and how these missions help us understand the geological histories of these bodies.

Equations are written so that the units balance so as to make them more understandable to new researchers not previously familiar with the formulas. However, this means that the equations may be written in a slightly different way than how they are usually given. As much as possible, different symbols are used for different quantities.

At the end of each chapter, practice questions are given. These questions will reinforce topics that were learned from the chapter. The questions range from short answer to calculations.

The book was written for a 13-week class. The course should cover approximately a different chapter every week and a half. During the last few weeks of the class, the students should do short presentations on *Nature* or *Science* articles on minor planets, cratering, or meteorites to apply what they learned throughout the semester. Papers in these journals are chosen because these articles are relatively short and are usually very scientifically important discoveries.

The book is ambitious because it tries to cover as much as possible about asteroids. I always feel it is better to shoot for the Moon with any project than to keep both feet on the ground. I have tried to make it as understandable as possible to an audience unfamiliar with asteroids. I have learned so much writing this book and I hope the reader will too.

Acknowledgments

I would like to thank all the students in my fall 2015 "Asteroids and Comets" seminar at Mount Holyoke College that read and commented on many of the chapters. These students were Sarah Brady, Laura Breitenfeld, Elizabeth Capiro, Patricia Chaffey, Olivia Chen, Qingdong Hu, Laura Hunter, Alyssa Jones, Lydia Koropeckyj-Cox, Laura Larson, Clarissa Leight, Kathryn Morrison, Louisa Rader, Daniel Rono, Isoke Samuel, Claire Schwartz, Ranjana Sundaram, Pa Chia Thao, Helena Valvur, and Erica Watts. They all suffered through some very incompletely written chapters and gave so many helpful comments. I would like to thank my teaching assistant Isha Raut for grading during the class.

I also have had the opportunity to work and study with a number of planetary science legends: Mike Gaffey, Bruce Hapke, Bill Cassidy, Brian Marsden (1937–2010), Roger Burns (1937–1994), Rick Binzel, Jim Elliot (1943–2011), Tim McCoy, Jack Trombka (1930–2016), and Darby Dyar. I learned so much from each of them.

I would like to thank my editor Emma Kiddle who was nice enough to listen to me when I asked if there was any interest in writing such a book on asteroids and for encouraging me to put together a book proposal. I would like to thank my content manager Zoë Pruce for giving me numerous extensions on the book and not getting too annoyed by the numerous versions of the book that I sent her. I would like to eternally thank Aimée Feenan for making all the edits to the proof.

I have made numerous friends and colleagues in planetary science and have had numerous conversations with them on all sorts of subjects. These people include Neyda Abreu- Schienke, Carl Agee, Conel Alexander, Robert Anderson, Tomoko Arai, Matt Balme, Olivier Barnouin, Gerbs Bauer, Jeff Bell, Gretchen Benedix-Bland, Rebecca Blackhurst, Phil Bland, David Blewett, Lars Borg, Oliver Botta, Bill Bottke, Emma Bowden Sadrpanah, Adrian Brearley, Dan Britt, Paul Buchanan, Emma Bullock, Brian Burt, Bobby Bus, Ben Bussey, Joshua Cahill, Bill Cassidy, Nancy Chabot, John Chambers, Queenie Chan, Clark Chapman, Lysa Chizmadia, Fred Ciesla, Ed Cloutis, Barbara Cohen, Harold Connolly Jr., Cari Corrigan, Gordon Cressey, Kate Crombie, Andy Davis, Jemma Davidson, Paul DeCarli

(1930–2013), Francesca DeMeo, Deborah Domingue Lorin, Jason Dworkin, Tasha Dunn, Denton Ebel, Lindy Elkins-Tanton, Tim Fagan, Caleb Fassett, Kelly Fast, Sherry Fieber-Beyer, Luigi Folco, Ian Franchi, Jon Friedrich, Marc Fries, Matthew Genge, Jeffrey Gillis-Davis, Daniel Glavin, Tim Glotch, Monica Grady, James Granahan, Richard Greenwood, Jennifer Grier, Jeff Grossman, Tim Grove, Paul Hardersen, Ralph Harvey, B. Ray Hawke (1946–2015), Chris Herd, Karl Hibbitts, Takahiro Hiroi, Ellen Howell, Marina Ivanova, Eugene Jarosewich (1926–2007), Diane Johnson, Natasha Johnson, John Jones, Jim Karner, Klaus Keil, Mike Kelley, Rachel Klima, Mutsumi Komatsu, Sasha Krot, Dante Lauretta, Samuel Lawrence, Lucille Le Corre, Laurie Leshin, Joanna Levine, Lucy Lim, Paul Lucey, Glenn MacPherson, Amy Mainzer, Zita Martins, Joe Masiero, Molly McCanta, Tim McClanahan, Francis McCubbin, Lucy McFadden, Kevin McKeegan, Hap McSween, Anders Meibom, Keiko Messenger, Scott Messenger, Michelle Minitti, Dave Mittlefehldt, Nick Moskovitz, Smail Mostefaoui, Hanna Nekvasil, Larry Nittler, Sarah Noble, Mike Nolan, Joe Nuth, Daniel Ostrowski, Jisun Park, Misha Petaev, Chris Peterson, Noah Petro, Carlé Pieters, David Polishook, Louise Prockter, Vishnu Reddy, Christina Richey, Andrew Rivkin, Vishnu Reddy, Rachel Roberts, Alan Rubin, Sara Russell, Ian Sanders, Devin Schrader, Eli Sklute, Ed Scott, Derek Sears, Mark Sephton, Tom Sharp, Caroline Smith, Tim Spahr, Jordan Steckloff, Karen Stockstill Cahill, Rhonda Stroud, Jessica Sunshine, Tim Swindle, Mark Sykes, Steve Symes, Jeff Taylor, Bradley Thomson, Pierre Vernazza, Meena Wadhwa, Kevin Walsh, Kees Welten, Linda Welzenbach Fries, John Wasson, Ben Weiss, Gareth Williams, Ian Wright, Shui Xu, Akira Yamaguchi, Qing-zhu Yin, Aileen Yingst, Tom Zega, Nicolle Zellner, and Mike Zolensky.

I would also like to thank numerous relatives and friends who were especially supportive during my studies and the writing of this book. I especially like to thank my mother, sister, and brother-in-law who were always encouraging. I really wish my father (1941–1995) was alive to see the book. My next-door neighbors Franz (1922–2015) and Yolanda (LaLa) Kameka (1925–2016) were like grandparents to me. My good friends include Bill Blackwell, Brian Bowers, Chong-Ren Chien, Amy Chung, Doug Crawford, Holly Crawford, Mario DeCaro, Victoria De la Torre, Lillian Fu, Annakay Johnson, Megan Hepler Blackwell, Jesse Hong, Ahlay Hussain, Shahla Hussain, Fyllio Katsavounidou, Candace Kita, Nitu Kitchloo, Dhaya Lakshminarayanan, Kathy Liu, Hsiang-Wei Lu, John Matz, Karen Matz, Mandy Mobley, Lisa Nagatoshi, Baochi Nguyen, Matt Noblett, Wendy Noblett, Daileen O'Brien, Kevin O'Brien, and Benjie Sun.

I would like to thank the Mount Holyoke College Astronomy Department Academic Department Coordinator Sarah Byrne for her support during the writing. I would like to thank Mount Holyoke College for all their support during this project. I would like to specifically thank the Mount Holyoke College library staff that helped me find books when I needed them and were very helpful with any questions I asked. I wrote in the hallway outside my office where I could stretch out and was luckily only asked once by the Mount Holyoke Campus Police whether I was homeless.

I would also like to thank the Remote, In Situ, and Synchrotron Studies for Science and Exploration (RIS4E) Solar System Exploration Research Virtual Institute (SSERVI) for support during my writing. Part of the asteroid spectral data utilized in this publication was obtained and made available by the MIT-UH-IRTF Joint Campaign for NEO Reconnaissance. The IRTF (Infrared Telescope Facility) is operated by the University of Hawaii (UH) under Cooperative Agreement number NCC 5–538 with the National Aeronautics and Space Administration, Office of Space Science, Planetary Astronomy Program. The MIT (Massachusetts Institute of Technology) component of the IRTF observations is supported by NASA grant 09-NEOO009-0001 and by the National Science Foundation under Grant Numbers 0506716 and 0907766. This research has made considerable use of NASA's Astrophysics Data System.

Extensive use was made of NASA, public domain, and Creative Commons (CC) licensed images. The American Astronomical Society, Audrey Bouvier, Robert Buchheim, Francesca DeMeo, Caleb Fassett, Richard Greenwood, Ralph Harvey, Japan Aerospace Exploration Agency, David Polishook, John Wiley and Sons, Noriko Kita, Rachel Klima, Randy Korotev, Katharina Lodders, Nature Publishing Group, Jisun Park, Vishnu Reddy, Steven Soter, Ted Stryk, Nancy Todd, David Vokrouhlický, Kevin Walsh, and Brian Warner were all so gracious in freely supplying images and figures. Caroline Smith helped me get an image from the Natural History Museum database. I will give them all free copies of the book.

I must have listened to thousands of hours of podcasts (e.g., *The Adam Carolla Show*, *Classic Loveline*, *Loveline*, *Gilbert Gottfried's Amazing Colossal Podcast* and *The Howard Stern Show*) while writing this book. Listening to these shows got me through every day. And when I really wanted to relax, I watched *Game of Thrones* and *The Walking Dead*.

Variables, Constants, and Unit Abbreviations

a	largest axis of triaxial ellipsoid
a	semi-major axis
A	Bond albedo
$Å$	angstrom
A_1	constant used for scattering function 1
A_2	constant used for scattering function 2
a_J	semi-major axis of Jupiter (5.2 AU)
a_p	proper semi-major axis
AU	astronomical unit
A_v	visual Bond albedo
b	constant in phase function
b	intermediate axis of triaxial ellipsoid
b	power law exponent
b	y-intercept
B_1	constant used for scattering function 1
B_2	constant used for scattering function 2
BAR	Band Area Ratio
$B(g)$	backscatter function
BI Center	Band I center
BII Center	Band II center
B_0	amplitude of opposition effect
$B(\lambda,T)$	Planck function for isotropically emitted radiation
$B_\lambda(\lambda,T)$	Planck function per unit solid angle
c	constant in phase function
c	number density of craters when the crater diameter is 1 km
c	smallest axis of triaxial ellipsoid
c	speed of light (3×10^8 m/s)

C	specific heat capacity
C_i	counts in the ith pixel
cm	centimeter
C_{sky}	average count in a background sky pixel
d	distance between two bodies
d	distance metric
D	diameter
d_1	distance of Body 1 to the Sun
d_2	distance of Body 2 to the Sun
d_{cut}	cutoff distance
dF_a	differential absorbed flux
dF_i	differential incident flux
D_{frag}	diameter of a sphere with a volume equivalent to all the fragments
D_i	number of atoms of a nonradiogenic stable isotope of the daughter atom
d_j	average effective particle size for mineral
D_{LM}	diameter of largest member
D_0	original amount of daughter atoms
D_{PB}	diameter of original parent body
dS	differential surface area
$D(t)$	amount of daughter atoms at time t
$\dfrac{dx}{dt}$	change in x divided by change in time
$\dfrac{dy}{dt}$	change in y divided by change in time
$\dfrac{dz}{dt}$	change in z divided by change in time
e	eccentricity
e	emission angle
E	energy
E_1	energy of body 1
E_2	energy of body 2
En	enstatite content
e_p	proper eccentricity
F	flux
F_a	fayalite content
F_a	absorbed flux
f_B	background frequency
F_c	centripetal force

F_g	gravitational force
F_e	emitted flux
F_{ir}	infrared flux
Fo	forsterite content
F_{ref}	reference flux
Fs	ferrosilite content
f_{0,λ_i}	observed flux at a particular wavelength scaled to a phase angle of $0°$
f_{λ_i}	observed flux at a particular wavelength
F_{λ_i}	theoretical flux at a particular wavelength
F_λ (object)	flux at a particular wavelength for an object
F_λ (star)	flux at a particular wavelength for a star
F_λ (Sun)	flux at a particular wavelength for the Sun
g	phase angle
G	gravitational constant (6.67384×10^{-11} m^3 kg^{-1} s^{-2})
G	slope parameter
Ga	billion years
$g(x)$	Gaussian distribution for variable x
h	Planck's constant (6.626×10^{-34} J s)
H	absolute magnitude
hr	hour
$H(x)$	Chandrasekhar H function
Hz	hertz
$H(\alpha)$	reduced visual magnitude at phase angle α
i	incident angle
i	inclination
i	counting variable
I	radiance
I	observed intensity
I_o	original intensity
i_p	proper inclination
j	counting variable
J	irradiance
J	joule
k	Boltzmann constant
k	extinction coefficient
K	kelvin
K	K filter
km	kilometer

kT	kiloton of TNT
m	meter
m	magnitude
m	mass
m	slope
M	mass
M	mean anomaly at epoch
m_1	mass of the primary body
m_1	magnitude of Body 1
m_2	mass of the secondary body
m_2	magnitude of Body 2
magnitude	measurement of brightness
m_{inst}	instrumental magnitude
M_j	mass fraction for each mineral
m_{ref}	reference magnitude
Ma	million years
mm	millimeter
MT	Megaton of TNT
n	mean motion
n	neutron
n	number
n	real part of the complex index of refraction
$\underline{n}$	complex index of refraction
$N_{cum}(\geq D)$	cumulative number of craters equal to or larger than a particular diameter
N_i	number of molecules for each molecular species
nm	nanometer
N_o	original amount of parent atoms
$N(t)$	number of parent atoms at time t
offset	arbitrary constant
ol	olivine content
opx	orthopyroxene content
p	geometric albedo
p	differential slope index
p	orbital period
p	proton
P	Palermo Scale value
P	power
P_A	absorbed power

P_E	extincted power
$p(g)$	single particle phase function
p_i	impact probability
P_S	scattered power
p_v	visual geometric albedo
pyx	pyroxene content
q	perihelion
q	phase integral
Q	aphelion
Q_A	absorption efficiency
QE	quantum efficiency
Q_E	extinction efficiency
Q_i	production rate for each molecular species
Q_S	scattering efficiency
r	distance
r	heliocentric distance in AU
r	orbital distance of the center of a moon to the center of the mass of the system
R	radius
R	reflectance
R_b	reflectance at band center
R_c	reflectance of the continuum at the band center wavelength
r_0	variable in Chandrasekhar H function
R_λ	reflectance at a particular wavelength
$r_\lambda(i,e,g)$	bidirectional reflectance
s	second
s	strength of band
S_0	solar constant (1366 W/m^2)
sr	steradian
t	time
T	orbital period
T	rotation period
T	temperature
$t_{1/2}$	half-life
T_J	Tisserand parameter with respect to Jupiter
T_{SS}	subsolar temperature
$T(\theta)$	surface temperature distribution versus angle
v	velocity
V	volume

V_b	bulk volume
v_e	escape velocity
V_g	grain volume
$V_{obs}(\alpha)$	observed visual magnitude at phase angle α
w	single scattering albedo
W	watt
Wo	wollastonite content
x	x position
y	y position
yrs	years
z	z position
α	absorption coefficient
α	phase angle
α	right ascension
β	mean phase coefficient
γ	variable in Chandrasekhar H function
Γ	thermal inertia
δ	declination
Δ	geocentric distance in AU
δa_p	change in proper semi-major axis
ΔE	change in energy
Δm	amplitude
Δt	change in time
$\Delta\lambda$	change in wavelength
$\delta^{17}O$	deviation in parts per thousand of ^{17}O
$\delta^{18}O$	deviation in parts per thousand of ^{18}O
$\Delta^{17}O$	offset from terrestrial fractionation line
ε	emissivity
η	beaming factor
θ	angle between the incident flux and the normal direction of the surface element
κ	thermal conductivity
λ	wavelength
λ_d	decay constant
λ_{EC}	electron capture decay constant for ^{40}Ar
λ_i	individual wavelength
λ_{max}	maximum wavelength
λ_0	original wavelength
λ_T	total decay constant for ^{40}K

λ_{87}	decay constant for ^{87}Rb
λ_{235}	decay constant for ^{235}U
λ_{238}	decay constant for ^{238}U
μ	center of the band in energy
μ	cosine of emission angle
μ	planetary discriminant
μ_0	cosine of incident angle
μm	micron
ν	frequency
π	pi
ρ	density
ρ_b	bulk density
ρ_g	grain density
ρ_j	single particle density for each mineral
σ	geometrical cross section
σ	standard deviation
σ	width of the band given as a standard deviation
σ	Stefan–Boltzmann constant (5.67×10^{-8} W m^{-2} K^{-4})
σ_A	absorption cross section
σ_E	extinction cross section
σ_i	standard deviation of the ith measurement
σ_S	scattering cross section
τ_i	mean lifetime of the molecular species
$\Phi_1(\alpha)$	function 1 that describes scattering off a surface
$\Phi_2(\alpha)$	function 2 that describes scattering off a surface
υ	true anomaly
χ^2	chi-square value
ω	argument of periapsis
Ω	longitude of the ascending node
$'$	arcminute
$''$	arcsecond
$\circ$	degrees

1

Light and Magnitude

1.1 Light

Except in extremely rare circumstances, minor planets are too faint to be seen with the naked eye. So to study a minor planet, you need a telescope. A telescope gathers and focuses light from the body, allowing objects too faint to be seen with the eye to be observed and studied. The telescope needs a detector to capture the light.

1.1.1 Electromagnetic Spectrum

Light usually refers to the visible portion of the electromagnetic spectrum that can be seen with the naked eye. All parts of the electromagnetic spectrum have been used to study asteroids; however, the visible is the one that is most often used since the radiation from the Sun peaks and the atmosphere is relatively transparent in this wavelength region. The light that allows us to see asteroids originated from the Sun. Light from the Sun is reflected by the asteroid and then detected by the observer.

The regions (Table 1.1 and Figure 1.1) of the electromagnetic spectrum (in order of decreasing wavelength) are radio, microwave, infrared, visible, ultraviolet, X-rays, and gamma rays. Visible light is usually broken up (in order of decreasing wavelength) into red, orange, yellow, green, blue, indigo, and violet. This sequence is often remembered as ROYGBIV. In a vacuum, light travels at a speed (c) of $299\,792\,458$ m/s (usually written as 3×10^8 m/s).

Light can be characterized as both a wave and a particle. Particles of light are called photons and are massless with no charge. Since light also acts like a wave, its properties can also be characterized by both its frequency (v) and wavelength (λ). The units of frequency are usually given as hertz (Hz) (cycles/second) while the units of wavelength vary [e.g., meters (m), microns (μm) (1×10^{-6} m), nanometers (nm) (1×10^{-9} m), angstroms (Å) (1×10^{-10} m)]. The frequency and wavelength of

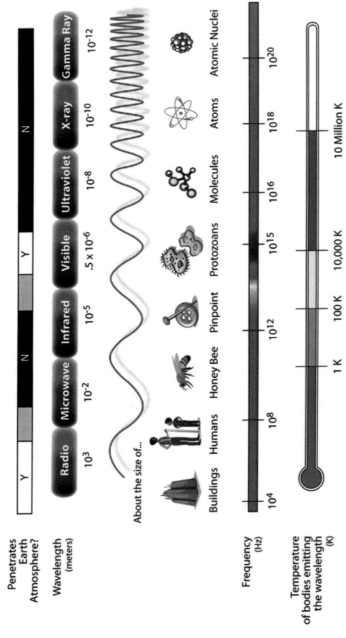

Figure 1.1 The wavelengths (m) and frequencies (Hz) of different regions of the electromagnetic spectrum. The wavelength regions that penetrate the Earth's atmosphere, a list of different bodies with similar sizes to the wavelengths of each region, and the temperatures of different bodies emitting black body radiation primarily at that wavelength. Credit: NASA. (A black and white version of this figure will appear in some formats. For the color version, please refer to the plate section.)

Table 1.1 *The wavelength regions for different parts of the electromagnetic spectrum*

Region	Wavelength range (meters)
radio	> 1
microwave	$1 \times 10^{-3} - 1$
infrared	$7.0 \times 10^{-7} - 1 \times 10^{-3}$
visible	$4.0 \times 10^{-7} - 7.0 \times 10^{-7}$
ultraviolet	$1 \times 10^{-8} - 4.0 \times 10^{-7}$
X-ray	$1 \times 10^{-11} - 1 \times 10^{-8}$
gamma ray	$< 1 \times 10^{-11}$

a photon are inversely correlated and obey the formula $c = \nu\lambda$ in a perfect vacuum. The energy of a photon is $h\nu$ where h is Planck's constant (6.626×10^{-34} J s). Radio waves have extremely long wavelengths, small frequencies, and small energies while gamma-ray photons have extremely small wavelengths, large frequencies, and large energies.

The wavelength, frequency, and energy of a photon are all interrelated, and knowing one of these values allows you to calculate the other two quantities. If you know the wavelength of a photon, you can calculate its frequency and energy. If you know its frequency, you can calculate its wavelength and energy. And if you know its energy, you can calculate its wavelength and frequency. So knowing the value of either the wavelength, frequency, or energy of the light used to study an asteroid, you are able to uniquely define its characteristics.

1.1.2 Atmosphere

All radiation does not pass equally through the Earth's atmosphere. Visible light, short wavelength radio waves (including microwaves), and some infrared wavelengths pass relatively unimpeded through the atmosphere while X-rays, gamma rays, and some ultraviolet wavelengths are absorbed (Figure 1.2).

There is a large atmospheric window (Figure 1.2) in the far ultraviolet, visible, and near-infrared (~0.3 to ~2.4 μm) where most wavelengths of light can pass relatively easily through the atmosphere (Gupta, 2003). From ~2.4 to ~3.5 μm, H_2O, which is a relatively small constituent of the atmosphere, has a number of strong absorption bands that severely affect the transmission of these wavelengths of light through the atmosphere. Ultraviolet photons shortward of ~0.3 μm are significantly absorbed by the atmosphere (particularly by the ozone layer).

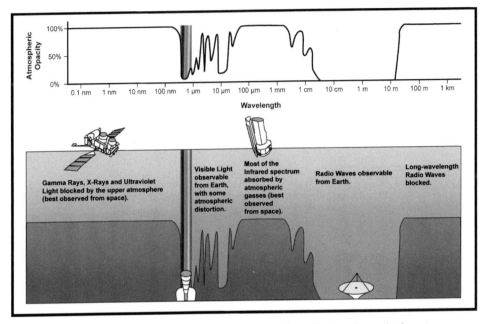

Figure 1.2 The transmittance of electromagnetic radiation through the atmosphere. Microwaves are included as part of the short wavelength radio waves. Atmospheric opacity is the amount of light absorbed by the atmosphere with 100% opacity indicating total absorption and 0% opacity indicating total transmission. Credit: NASA. (A black and white version of this figure will appear in some formats. For the color version, please refer to the plate section.)

1.2 Black Bodies

The Sun, the source of the asteroid's reflected light, is assumed to act like a black body. A black body is a theoretical object that absorbs all radiation that strikes it and emits radiation at all wavelengths. This emission is due to the thermal motion of charged particles in the material and is often called black body or thermal radiation.

Radiation from a star can be modeled as if the radiation was emitted from a black body. However, stars are not "perfect" black bodies. The temperature of a star will vary across its surface, so different regions will have different black body emission so the black body curve will not be the ideal black body curve for a body at one temperature. Also, atoms or ions in the atmosphere of the star absorb radiation being emitted. These absorption lines are apparent in measured stellar spectra (including the Sun) due to the absorption of light by different atoms.

Plots of the emitted intensity of radiation versus wavelength for theoretical black bodies have characteristic shapes that are just a function of temperature and wavelength (Figure 1.3). Using classical physics, the Rayleigh–Jeans law, proposed by Lord Rayleigh (1842–1919) and James Jeans (1877–1946), modeled black body

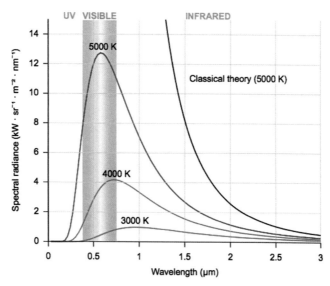

Figure 1.3 Black body curves for bodies at temperatures at 5000 K, 4000 K, and 3000 K plus the classical theory prediction for a body at 5000 K. Credit: Dark Kule. (A black and white version of this figure will appear in some formats. For the color version, please refer to the plate section.)

radiation as proportional to T/λ^4; however, this law inaccurately predicted that emitted radiation would continually increase with decreasing wavelength. The Planck function, proposed by Max Planck (1858–1947) using quantum theory, accurately predicts the spectral distribution of electromagnetic radiation from a black body and can be written as

$$B_\lambda(\lambda,T) = \frac{2hc^2}{\lambda^5} \frac{1}{e^{hc/\lambda kT} - 1} \tag{1.1}$$

where the emitted intensity is given in terms of $W\,m^{-3}\,sr^{-1}$ if the temperature is in kelvin (K) and the wavelength is in meters. The Boltzmann constant (k) is $1.381 \times 10^{-23}\,J\,K^{-1}$. The Boltzmann constant relates the kinetic energy and temperature of a molecule in an ideal gas. A solid angle is a measure of the fraction of the surface of a sphere that a body covers as seen from the center of the sphere. A solid angle is given in steradians (sr) with a sphere subtending 4π steradians.

If the radiation is emitted isotropically (independent of direction), the Planck function can also be written (Delbó and Harris, 2002) as

$$B(\lambda,T) = \frac{2\pi hc^2}{\lambda^5} \frac{1}{e^{hc/\lambda kT} - 1} \tag{1.2}$$

which has units of $W\,m^{-3}$. This formula gives the emitted flux from a black body. A flux is the flow of some quantity per unit area.

The radiant flux (F_e) (or total power per unit area) emitted by the black body is a function of the fourth power of the temperature (T). This equation is called the Stefan–Boltzmann law and can be written as

$$\frac{F_e}{4\pi R^2} = \sigma T^4. \tag{1.3}$$

The $4\pi R^2$ term is the total surface area of the body with a radius R. The Stefan–Boltzmann constant (σ) ($5.67 \times 10^{-8}\ W\,m^{-2}\,K^{-4}$) is a constant of proportionality that relates the intensity of the emitted radiation per unit area to the fourth power of the temperature. As the temperature increases, the amount of energy emitted per second will dramatically increase. If you double the surface temperature of a body, the amount of energy emitted per second by the body will increase by a factor of 16 times.

Asteroids reflect and also emit radiation. However, gray bodies, such as asteroids, do not emit all the energy that strikes them. The Stefan–Boltzmann Law for gray bodies is written as

$$\frac{F_e}{4\pi R^2} = \varepsilon \sigma T^4. \tag{1.4}$$

The emissivity(ε) term is the efficiency with which a body radiates thermal radiation. The emissivity is 1 for a "perfect" black body but will be less than 1 for any type of gray body. For asteroids, the emissivity is often assumed to be 0.9.

The peak of the black body curve (where the object emits most of its light) is inversely proportional to temperature with cooler black bodies peaking at longer wavelengths than hotter bodies (Figure 1.1), which is called Wien's law. Wien's law is given by the equation

$$\lambda_{max} = 2.898 \times 10^{-3}\, \frac{m\,K}{T}. \tag{1.5}$$

The calculated wavelength is in meters and the temperature is in kelvin. Radiation from the Sun peaks in the visible wavelength region while radiation from asteroids peaks in the infrared since they have much cooler surface temperatures.

Radiation detected from an asteroid is usually dominated by reflected light at wavelengths less than 2.5 μm while emitted thermal radiation usually dominates past 5 μm (Kim et al., 2003). The temperature of an asteroid will be primarily be a function of its distance from the Sun and how the dark the surface is. Objects further from the Sun are heated by less solar radiation than those that are closer and will have cooler surface temperatures. The flux of solar radiation striking an asteroid follows an inverse square law where the flux is proportional to the inverse of the distance squared from the Sun. Darker surfaces also tend to absorb more radiation than lighter surfaces. So darker asteroids will tend to be hotter than brighter surfaces at the same distance from the Sun.

For asteroids, the transition between reflected and thermal radiation usually occurs between 2.5 and 5 µm and is a function of the surface temperature of the body. The exception are dark near-Earth asteroids, which have "thermal tails" between ~2 and ~2.5 µm. These thermal tails are due to these bodies having a measurable blackbody flux in the ~2–2.5 µm wavelength region due to their relatively high surface temperatures.

1.3 Albedo

Albedo is a quantity that defines how light or how dark a surface is. Values for albedos tend to range from 0 (perfectly absorbing) to 1 (perfectly reflecting). Albedo is the fraction of radiation reflected from a surface. However, there are many different ways to define the albedo of a surface.

Bond albedo (A) is the fraction of all the total radiation at all wavelengths and at all solar phase angles that is scattered from a surface. The solar phase angle (Figure 1.4) is the angle between the light incident on a body from the Sun and the light reflected from the body and detected by an observer. Since the Bond albedo accounts for all scattered light, it varies from 0 to 1. The visual Bond albedo (A_v) only accounts for radiation scattered in the visible. The Bond albedo and the visual Bond albedo are often assumed to be similar in value since the Sun's flux peaks in the visible.

The geometric albedo (p) is the ratio of the brightness of a body at zero phase angle relative to a theoretical flat and fully reflecting disk with the same cross section that reflects light diffusely (scatters light equally in all directions). Objects with diffuse reflectance are said to have a Lambertian reflectance. The geometric albedo can be greater than 1 if light from the surface is preferentially reflected backwards towards the observer. The visual geometric albedo (p_v) is the geometric albedo that accounts for only visible light.

1.4 Temperature

The shape and intensity of the emitted radiation from an asteroid will be a function of its average surface temperature. This temperature is often called the effective temperature. The effective temperature of an asteroid can be estimated by assuming that it is a black body and then equating the emitted power and the absorbed power. The absorbed flux (F_a) in watts will be

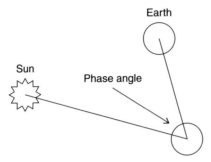

Figure 1.4 Illustration of the phase angle for an observer on the Earth observing a planet or a minor planet.

$$F_a = (1-A)\pi R^2 \left[\frac{S_0}{(r^2 / AU^2)} \right] \tag{1.6}$$

where A is the Bond albedo, S_0 is the solar constant (~1366 W/m²), and r is the heliocentric distance in AU. The solar constant is the average solar flux at the top of the Earth's atmosphere and is not actually a constant because it slightly varies over time. The $\frac{S_0}{(r^2 / AU^2)}$ term adjusts the solar flux for the asteroid's distance from the Sun according to the inverse square law. The πR^2 term is the cross-sectional area of the asteroid that is absorbing the radiation. One astronomical unit (AU) is approximately the mean of the average distance between the Earth and the Sun and is now defined by the IAU as 149 597 870 700 meters.

If you equate the emitted flux with the absorbed flux, the resulting formula will be

$$4\pi R^2 \varepsilon \sigma T^4 = (1-A)\pi R^2 \frac{S_0}{\left[r^2 / (AU)^2 \right]}. \tag{1.7}$$

Solving for T^4 produces the formula

$$T^4 = \frac{(1-A)S_0}{4\varepsilon\sigma\left[r^2 / (AU)^2 \right]}. \tag{1.8}$$

Then solving for the effective temperature produces the formula

$$T = \left(\frac{(1-A)S_0}{4\varepsilon\sigma\left[r^2 / (AU)^2 \right]} \right)^{1/4} \tag{1.9}$$

for the average temperature. So by increasing the albedo so there is less radiation absorbed, the average surface temperature of a body will decrease. Also if the distance from the Sun is increased for a body, its average surface temperature will become lower.

Example 1.1

A near-Earth asteroid has an effective surface temperature of ~270 K and was observed at a distance from the Sun of 1 AU. What is this object's Bond albedo?

Solving for the Bond albedo using *Equation 1.8* produces the formula

$$A = 1 - \frac{4\varepsilon\sigma T^4 \left[r^2 / \left(\text{AU}\right)^2 \right]}{S_0}. \tag{1.10}$$

The emissivity (ε) is assumed to be 0.9. Substituting the values in the equation then produces the formula

$$A = 1 - \frac{4(0.9)(5.67 \times 10^{-8})(270)^4}{(1366)} \tag{1.11}$$

with all the units cancelling. The calculated Bond albedo will then be 0.21.

1.5 Telescopes

Telescopes gather light from astronomical bodies and focus it to produce an image. Telescopes that just use lenses to produce an image are called refracting telescopes, while those that use mirrors are called reflecting telescopes. The first telescopes were refractors and were pioneered by Galileo Galilei (1564–1642) starting in 1609. The Galilean telescope has a convex primary lens that collects the parallel rays of light from an object and focuses it and a concave eyepice that intercepts the light from the lens and renders the rays parallel again. This design was updated by Johannes Kepler (1571–1630) who used a convex lens as the eyepiece. This design was called the Keplerian telescope. Refracting telescopes were the primary research telescopes until the early twentieth century.

The size of a telescope is given by its aperture size, which is the diameter of its main optical component (lens or mirror). The larger the aperture, the brighter and sharper the image tends to be since more photons tend to be collected and focused. However as the size of the lens increases for refracting telescopes, the focal length also tends to increase, which increases the size of the telescope tube. The weight of the lens also increases, which increases the weight of the counterweight needed to balance the telescope.

Telescopes that are used today for research are almost all reflecting telescopes. Reflecting telescopes use a curved mirror (called the primary mirror) or a series of curved mirrors to focus an image. Isaac Newton (1642–1727) built the first reflecting telescope in 1668. Besides having smaller sizes, reflecting telescopes have a number of other advantages compared to refracting telescopes. It is easier to make a high-quality mirror than a lens. Large lenses are much heavier than large mirrors. Lenses are also affected by chromatic aberration where all wavelengths of light cannot be focused to the same point reflecting telescopes are not affected by chromatic aberration. There are a number of different types of designs for reflecting telescopes.

Mirrors can be spherical and have a curved shape that is either convex or concave. However unless corrected for, simple spherical mirrors suffer from spherical aberration where light reflected from the edge of the mirror is focused at a slightly different point than light reflecting from the center of the mirror. Mirrors can be made with hyperbolic or parabolic shapes to eliminate spherical aberration since they focus all light to a single point. Hyperbolic mirrors tend to be more difficult and expensive to make than parabolic mirrors. However, the disadvantage of parabolic mirrors is that they can distort point sources, which is called a coma since it resembles the tail of a comet.

The commonest type of reflecting telescope is the Cassegrain reflector. Invented in the 1600s, a Cassegrain reflector uses two mirrors: a primary concave mirror and a secondary convex mirror. There is an opening in the primary that allows light to pass to the eyepiece. The classical Cassegrain design has a parabolic primary mirror and a hyperbolic secondary mirror.

The most commonly used type of Cassegrain reflector for research is the Ritchey–Chrétien Telescope (RCT), which was invented in the early twentieth century by George Ritchey (1864–1945) and Henri Chrétien (1879–1956). A Ritchey–Chrétien reflector has a hyperbolic primary mirror and a hyperbolic secondary mirror. The advantage of using hyperbolic mirrors is that the images are relatively free of coma and spherical aberration. The Hubble Space Telescopes and the Keck Telescope all are Ritchey–Chrétien telescopes.

Catadioptric telescopes use both refraction (lens) and reflection (mirror) in their optical systems. Schmidt telescopes (often called Schmidt cameras) have a spherical primary mirror and a aspherical correcting lens. This type of telescope was invented by Bernhard Schmidt (1879–1935) in 1930. The main advantage of a Schmidt telescope is that it has a relatively wide field of view.

1.6 Detectors

Detectors are used to record an image produced by a telescope. An image of an astronomical body allows you to determine the position of a body relative to the

Table 1.2 *Quantum efficiencies for a number of detectors*

Detector	Quantum efficiency	Reference
human eye	~10%	Rose and Weimer (1989)
photographic plate	~1–5%	Irwin (1997)
photomultiplier tube	~20%	Duerbeck (2009)
charge-coupled device (CCD)	~90%	Howell (2006)
infrared detector	>70–80%	Rieke (2007)

stars. A series of these images allows you to determine whether the body is moving relative to the stars. This movement is called proper motion. If it is moving, the body lies within our Solar System. To identify a body as a planetary body, the same region of the sky must be observed over a period of time. Any object that moves relative to the stars over a time period of minutes, days, or hours is not a star but a "planet" or comet. The faster the body moves, the closer the object is to Earth. Besides discovering and determining the orbits of planetary bodies, images produced by a detector also can be used to determine how bright an object is in the sky and how this brightness (magnitude) changes over time.

An important quantity for judging how well a detector works is its quantum efficiency (QE). Quantum efficiency is how efficient a detector is at converting photons to measurable electrons. The higher the QE, the better the detector. The approximate QEs for a number of detectors are given in Table 1.2. As can be seen in the table, charge-coupled devices (CCDs) are the most efficient detectors that an astronomer can use.

1.6.1 Eyes

For the longest time, astronomical observations were only recorded by drawing on paper what you saw through the telescope. This changed in the middle of the nineteenth century when photographic plates started to be used. The first photograph of an astronomical object was of the Moon in 1840. The first asteroid photographed was (80) Sappho in 1896 by Isaac Roberts (1829–1904), while the first asteroid discovered by astrophotography was (323) Brucia in 1891 by Max Wolf (1863–1932). The first asteroid discovery survey is also attributed to Max Wolf who discovered over 200 objects using astrophotography.

1.6.2 Photographic Plates

Astrophotography works by exposing a photographic plate to light. Photographic plates are glass plates that are coated with a mixture of silver halide (e.g., AgBr) crystals in a thin layer of gelatin, which is called an emulsion. The gelatin keeps the grains well separated and fixed in place (Smith, 1995). Plates are used since they do not wrinkle like film. Photography works due to a photochemical reaction between the incident light and the emulsion that the photons strike. When a photon strikes a silver halide grain, an electron is released. These electrons can meet some relatively rare silver ions that are also migrating through the crystal lattice of the grain (Smith, 1995). So when these silver halide grains are exposed to light, free silver atoms can be produced. One produced silver atom will become ionized again; however, if several silver atoms come together, they can become a stable silver cluster (Appenzeller, 2012). These free silver atoms are much smaller the silver halide grains.

A chemical reducing agent is then used to develop the image. This developing agent uses these small silver ions as catalysts to convert the entire silver halide grain to silver. The developing needs to be stopped before all grains are turned into silver, which will ultimately happen at a much slower rate for silver halide grains that do not contain free silver. A fixing agent is then used to removed the unexposed silver halide. The image that is ultimately produced is a negative with the areas that appear dark being the regions on the plate that have absorbed photons. The brightest objects will tend to have the largest images on the photographic plate.

One advantage of plates over other types of detectors is that plates can have extremely large fields of views. Schmidt telescopes use photographic plates that could cover areas of ~6° by ~6° in the sky. One disadvantage is that most of the released electrons do not survive long enough to produce a silver atom since most are absorbed by either a positive hole or a halogen atom (Smith, 1995). Therefore, photographic plates are not very efficient in detecting photons. Another disadvantage is that the strengths of images on a plate are not all linearly related to the number of photons that strike those particular areas.

1.6.3 Photomultiplier Tubes

In the 1940s, photomultiplier tubes started to be used to measure the brightnesses of celstial objects. Photons strike a photocathode (negatively charged electrode), which causes an electron to be dislodged due to the photelectric effect. The photoelectric effect is the ejection of electrons from a metal by the absorption of photons. These electrons are then accelerated down the tube using a seried of positively charged electrodes in a vacuum (usually called dynodes), which are the electron multiplier. The positive voltage increases the energy of the electrons and when the electrons strike a dynode, more electrons are released. Each dynode has a higher voltage. More and more electrons are produced during this process, which produces a current that is measured by the final electrode. Photomultipliers have a linear response since the signal is directly proportional to the measured flux of photons. Photomultiplier tubes have a much higher quantum efficiency than photographic plates. However, a single photomultplier tube can not be used to image a region of the sky since it can only measure the intensity of light from a single point source. To image a large region of the sky, an array of photomultiplier tubes must be used.

1.6.4 CCDs

In the 1980s, CCDs (charge-coupled devices) started to be used and became the primary way to record astronomical observations. CCDs are electronic detectors (Figure 1.5) that are broken up into millions of light-sensitive elements called pixels. Each pixel is composed of an epitaxial layer of silicon doped with a number of elements over a substrate of silicon. When photons strike the epitaxial layer, free electrons are produced through the photoelectric effect. For each pixel, a positively charged gate keeps the electrons from returning to the newly created electron holes. The number of electrons released is directly proportional to the number of photons that strike the pixel.

The information (charge or number of electrons) stored in each pixel needs to be read out by circuitry at the edges of the CCD. To transfer a charge to a neighboring pixel, the voltage of the neighboring pixel needs to be increased, which allows the charge to be transferred to that pixel. These charges are continually transferred until all the charges are read out. The measured charge for each pixel is then converted into a voltage, which can then be converted into counts (also called "Analog-to-Digital Units" or ADUs) measured by each pixel. The number of counts measured for each pixel is directly related to the number of electrons that were stored by each pixel. The electronics keep track of the counts for each pixel.

CCDs are much more sensitive to light (quantum efficiencies of ~90%) than photographic plates and photomultiplier tubes. Due to CCDs having a much higher QE, the exposure times needed to acquire similar images are much shorter using a CCD. CCDs also acquire images in a digital format that can be easily analyzed using computers. CCDs can be used over and over while photographic plates can only be used once. However, CCDs can only detect light at wavelengths as long as ~1.1 µm.

1.6.5 Infrared Detectors

Infrared detectors primarily work through photoconductivity (Rogalski, 2002; Rieke, 2007). Photoconductivity is where the absorption of radiation increases the electrical conductivity of a material. Photons striking a semiconductor in an

Figure 1.5 Image of a charge-coupled device (CCD). Credit: NASA.

electric field free electrons that travel toward the electrodes and produce a current. Changes in the electrical current are then measured for each pixel of the array. The electrons are freed when the energy of the photons is equal to or larger than the binding energy of the electrons in the semiconductor crystal. Infrared detectors tend to be cooled to extremely low temperatures to reduce thermal electrical noise. InSb detectors are commonly used to detect infrared radiation between ~1 and ~5.6 μm. The semiconductors are made out of materials such as InSb and HgCdTe.

1.7 Observing Different Wavelengths of Light

To detect photons at specific wavelengths, filters (e.g., Bessell, 1990, 2005) or a spectrograph must be used. Filters transmit photons of light in a particular wavelength region (called a bandpass), while spectrographs disperse light into its component wavelengths. Filters tend to be used with photomultiplier tubes, CCDs, and infrared detectors, while spectrographs tend to be used with CCDs and infrared detectors.

1.7.1 Filters

Every filter has a bandwidth, which is the wavelength range in which the filter transmits light. Broadband filters have a bandwidth of less than 0.1 μm, intermediate band filters have a bandwidth between 0.007 and 0.04 μm, and narrow band filters have a bandwidth of less than 0.007 μm.

The UBV photometric system uses U (Ultraviolet) (mean wavelength of ~0.3663 μm), B (Blue) (~0.4361 μm), and V (Visual) (~0.5448 μm) filters. The UBV system is often called the Johnson or Johnson–Morgan system after the astronomers Harold Johnson (1921–1980) and William Morgan (1906–1994) who introduced this system (Johnson and Morgan, 1953). The UBVRI photometric system adds R (Red) (~0.6407 μm) and I (Infrared) (~0.7980 μm) filters. The UBVRI system is often called the Johnson–Cousins system after Alan Cousins (1903–2001) who introduced the R and I filters. The JHK system uses filters at J (~1.2 μm), H (~1.6 μm), and K (~2.2 μm). These filter systems were developed for observing stars but have been used extensively to study asteroids.

1.7.2 Spectrographs

Spectrographs disperse light onto a detector using a prism, a diffraction grating, or a grism (combination of a grating and a prism). Prisms tend to be triangularly shaped transparent material. A diffraction grating is a series of equally spaced slits on an opaque screen. However, prisms and gratings can introduce chromatic aberration. A grating and prism in concert (grism) will eliminate chromatic aberration since all light will be focused to the same focal point. The spectrograph allows the intensity of light at different wavelengths to be measured by the detector.

1.8 Observatories

Observatories house the telescope and detector and can be located on the ground, in space, or in the air (plane or balloon). The observatory also houses the computers and electronics that control the telescope. Observatories range from simple sheds to elaborate facilities.

Locating an observatory on a mountain enhances the seeing since there is less atmosphere that the light passes through, which resulted in less distorted images. Since infrared photons tend to be absorbed by the atmosphere before they reach the ground (Figure 1.2), infrared observatories also tend to be located on mountains. Mountains also tend to have very dark skies to observe, which allows for excellent seeing. For example, Mauna Kea in Hawaii, the home of a number of observatories, including the W. M. Keck Observatory and the NASA (National Aeronautics and Space Administration) Infrared Telescope Facility (IRTF), is at a height of ~4200 meters above sea level. Palomar Observatory on Palomar Mountain in California is located at a height of ~1700 meters above sea level.

1.8.1 Space Telescopes

Telescopes that study asteroids are located in space for a variety of reasons. Space-based telescopes are not affected by atmospheric turbulence. Space-based telescopes can observe asteroids in the ultraviolet and infrared since the Earth's atmosphere significantly absorbs ultraviolet and infrared light. X-ray and gamma rays are also absorbed by the Earth's atmosphere; however, asteroids do not have high enough fluxes of X-rays and gamma rays to be observable by space-based telescopes. Only orbiting spacecraft can detect significant fluxes of X-ray and gamma-ray photons from asteroids.

1.8.2 Hubble Space Telescope

The most famous space telescope is the Hubble Space Telescope (HST) (Figure 1.6) (e.g., Baker, 2015). Hubble has a 2.4-meter mirror and covers wavelengths from the ultraviolet to the near-infrared. Launched in 1990, HST has made significant minor planet discoveries (e.g., discovering four of Pluto's moons) due to its unprecedented resolution. Initially, HST had five instruments. The initial instruments were a Wide Field and Planetary Camera (WFPC), Goddard High Resolution Spectrograph (GHRS), High Speed Photometer (HSP), Faint Object Camera (FOC), and the Faint Object Spectrograph (FOS). When first launched, its images were slightly out of focus and blurry due to its primary mirror's curvature being slightly off. The primary mirror was too flat at its edges by 2.2 μm. In 1993, a servicing mission installed the Corrective Optics Space Telescope Axial Replacement (COSTAR) to correct for the primary mirror's aberration for the GHRS, FOC, and FOS. To include this instrument on HST, the HSP needed to be removed. During the same mission, the WFPC was also replaced with Wide Field and Planetary Camera 2 (WFPC2), which had its own optical corrective optics. The Near Infrared Camera and Multi-Object Spectrometer (NICMOS) replaced the GHRS in 1997. NICMOS is currently not working due to problems with the cooling system. The WFPC2 was then replaced with the Wide Field Camera 3 (WFC3) in 2009. The WFC3 is used for imaging. The WFC3 has two CCDs that detect radiation in the ultraviolet and visible and also a separate infrared detector.

1.8.3 Wide-field Infrared Survey Explorer

The Wide-field Infrared Survey Explorer (WISE) (Figure 1.7) was launched in December 2009. WISE initially observed the sky using four bands (3.4, 4.6, 12,

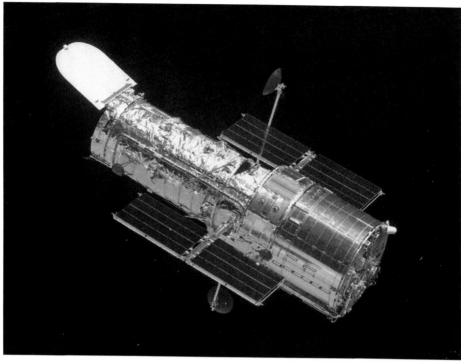

Figure 1.6 The Hubble Space Telescope as seen from the departing Space Shuttle Atlantis in 2009. The Hubble Space Telescope is 13.2 meters long. Credit: NASA.

22 μm) in the infrared. Each of the four infrared detectors contains one million pixels. The hydrogen coolant became depleted after 10 months, leaving only the 2.4 and 4.6 μm bands operable since the detector for the longer wavelengths could only obtain data at the original sensitivity at extremely low temperatures. After the hydrogen coolant was depleted, the mission was renamed NEOWISE for Near-Earth Object (NEO) WISE. NEOWISE is the asteroid detecting survey of the WISE telescope. NEOWISE was first extended for one month and then for an additional three months (October 2010 to February 2011). The WISE spacecraft transmitter was turned off, but the telescope was reactivated in September 2013 with NEOWISE observations restarting in December 2013.

One of the goals of the WISE/NEOWISE mission was to determine visual albedos and diameters of asteroids using infrared measurements (Sections 5.1.1–5.1.5). The WISE/NEOWISE mission has made ~2.3 million observations of ~159,000 objects, including ~34,500 discoveries. Over 200 near-Earth objects and over 20 comets were discovered.

Figure 1.7 The WISE spacecraft at Vandenberg Air Force Base. The WISE spacecraft is 2.85 meters long. Credit: NASA.

1.9 Magnitude

The visual brightnesses of asteroids (and stars and galaxies) are usually given in terms of magnitude. The magnitude system is based on the ancient Greek system developed by Hipparchus (*c.*190–*c.*120 BC), which divided stars into six magnitude groups. The brightest stars were called first magnitude stars. The second brightest stars were called second magnitude stars. Stars that could barely be seen by the ancient Greeks were called sixth magnitude stars. When Galileo first used the telescope, he was able to see stars that were fainter than sixth magnitude objects. He called the brightest of these objects seventh magnitude stars.

In the 1800s, photometric measurements of artificial "stars" that matched real stars in brightness showed that first magnitude stars were approximately 100 times brighter than a sixth magnitude star. These artificial "stars" were light that was projected into a telescope's field of view to match a real star in brightness. In 1856, Norman Pogson (1829–1891) defined that a 5 magnitude difference exactly equaled a 100 times difference in brightness. Therefore, a +1 magnitude star will be 2.512 times brighter than a +2 magnitude star since the fifth root of 100 is equal to 2.512. The formula for the ratio of the fluxes is then

$$\frac{F}{F_{ref}} = 2.512^{-(m-m_{ref})} \tag{1.12}$$

where F is the measured flux (or brightness) of the object, F_{ref} is the reference flux, m is the magnitude of your object, and m_{ref} is the reference magnitude, The reference flux is the flux from a star with a known magnitude, called a standard star. Flux is given in units of power (joules/second or watts) per unit area.

By taking the logarithm of both sides and rearranging the variables, the formula for calculating the magnitude (also called apparent magnitude) of an object becomes

$$m - m_{ref} = -2.5 \log_{10}\left(\frac{F}{F_{ref}}\right). \tag{1.13}$$

So by determining the ratio of the flux of an unknown object with the flux of a star with a known magnitude, the magnitude of any object can be determined. The relative flux can be written as

$$\frac{F}{F_{ref}} = 10^{-0.4(m-m_{ref})}. \tag{1.14}$$

The brightest objects have very negative magnitudes, while the faintest bodies have very positive magnitudes. The apparent magnitude of the Sun is −26.74 while the maximum apparent magnitude of Pluto is +13.65.

The reason that asteroids were first discovered with a telescope is that almost all asteroids are too faint to be seen with the naked eye, which only can see down to a magnitude of +6 in an extremely dark location. The most notable exception is (4) Vesta, which can be as bright as +5.1 at opposition; however, this apparent magnitude is still extremely difficult to see with the naked eye. Opposition is when the Earth and an object are approximately in a straight line as seen from the Sun. The object and the Sun are in "opposite" sides of the sky as seen from Earth. The body is roughly closest to the Earth during opposition and therefore should appear brightest. Near-Earth asteroid (99942) Apophis is predicted to have a magnitude of approximately +3 during its close approach in 2029 (Giorgini et al., 2008).

Because the amount of light reaching a body is an inverse function of the distance (*d*) squared and the light must travel to a minor planet and then be reflected back to Earth, the brightness of a distant minor planet from Earth will vary approximately as the inverse function of the distance of the body from the Sun to the fourth power (Ortiz et al., 2007; Jewitt, 2010).

Example 1.2

If Pluto (Body 1) has an apparent magnitude of 13.7 at a distance of ~30 AU from the Sun, what would be the estimate of the apparent magnitude of a Pluto-like body (Body 2) at ~120 AU?

Since the brightness of a distant minor planet varies approximately as the inverse function of the distance of the body from the Sun to the fourth power, the ratio of the brightnesses of the two distant minor planets $\left(\dfrac{F_2}{F_1} \right)$ will be

$$\frac{F_2}{F_1} \approx \left(\frac{d_1}{d_2} \right)^4 \tag{1.15}$$

where d_1 is the distance of Body 1 to the Sun and d_2 is the distance of Body 2 to the Sun. The formula for the relative brightnesses of Pluto at 30 AU and the Pluto-like body at 120 AU becomes

$$\frac{F_2}{F_1} \approx \left(\frac{30}{120} \right)^4 = \left(\frac{1}{4} \right)^4 = \frac{1}{256}. \tag{1.16}$$

So the Pluto-like body will be approximately 256 times fainter than Pluto. The formula for the magnitude difference (Equation 1.13) will become

$$m_2 - m_1 = -2.5 \log_{10} \left(\frac{F_2}{F_1} \right) \tag{1.17}$$

where $F_1 = F_{ref}$ and $m_1 = m_{ref}$. Substituting the values in the equation, the formula then becomes

$$m_2 - 13.7 = -2.5 \log_{10}\left(\frac{1}{256}\right), \tag{1.18}$$

which becomes

$$m_2 - 13.7 = 6. \tag{1.19}$$

The apparent magnitude of the Pluto-like body will then be +19.7.

1.9.1 Instrumental Magnitude

However when observing astronomical objects, what you actually measure is the counts from the object, not the actual flux. When you observe an object, what is first calculated is an instrumental magnitude (m_{inst}). The instrumental magnitude is not standardized to the magnitude of an astronomical body. The instrumental magnitude can then be calculated from the formula (Palmer and Davanhall, 2001)

$$m_{inst} = -2.5 \log\left(\frac{\left(\sum_{i=1}^{n} C_i\right) - nC_{sky}}{t} \times \frac{\text{seconds}}{\text{counts}}\right) + \text{offset} \tag{1.20}$$

where C_i is the counts in the ith pixel inside the aperture (a circular area), n is the number of pixels in the aperture, C_{sky} is the average count in a background sky pixel, t is the integration time in seconds, and *offset* is an arbitrary constant. This constant is usually a relatively large positive number, which will then produce a positive instrumental magnitude.

If a star has a known magnitude and is not variable, the *offset* can be chosen so that it produces an instrumental magnitude that is the same as the actual magnitude. All asteroids have magnitudes that will vary since they are not perfectly spherical and do not have the same albedo. If the image has a number of stars with known magnitudes and are not variable, the *offset* will be an average value, with an uncertainty, that best determines the actual magnitude of all stars. This offset will be different for every image.

1.9.2 Color Index

The magnitude of a minor planet can be measured at different wavelengths by using different filters (e.g., Wood and Kuiper, 1963). These observations are often presented as color indices by subtracting the magnitude measured using one filter from another. The magnitude at the longer wavelength is subtracted from the magnitude at the shorter wavelength to determine a color index (e.g., U–B, B–V). A positive color index indicates that the object is brighter at the longer wavelength, while a negative color index indicates that an object is fainter at the longer wavelength. Color indices for asteroids show a range values (e.g., Bowell and Lumme, 1979; Hollis, 1994), indicating that different asteroids can reflect light differently than other asteroids.

1.9.3 Photometry

Photometry is the study of how the brightness of a surface depends on the illumination and viewing geometry. The importance (Li et al., 2015) of photometric studies is that they give insight into the physical properties of the surface, allow corrections to be made so all observations are at the same viewing geometry, and allow for the prediction of asteroid magnitudes and reflectances at different phase angles. Different types of materials and different particles sizes will reflect light differently at different illumination and viewing geometries.

1.9.4 Magnitude Versus Phase Angle

The illumination of an asteroid at a phase angle of 0° is analogous to the illumination of the full Moon (Buchheim, 2010). The illumination of the asteroid at a phase angle of 90° would be analogous to the illumination of the first or third quarter phase of the Moon. However because asteroids orbit so far from the Sun, main-belt asteroids usually do not reach phase angles much greater than 20–30° as observed from Earth (Buchheim, 2010). However, near-Earth asteroids can be observed at much larger phase angles from Earth. Spacecrafts can also observe asteroids at larger phase angles than can be observed from Earth.

The magnitude (brightness) of an asteroid will be a function of phase angle. Asteroids are brightest at zero phase angle and becomes fainter at larger phase angles. Near zero phase angle, the asteroid becomes significantly brighter. This brightening at small phase angles for particulate (or rough) surfaces is called the opposition effect (or opposition surge). This name derives from the body being at opposition when this brightening occurs.

The opposition effect has been argued to be due to two effects (Hapke, 2002): shadow hiding and coherent backscatter. Shadow hiding is due to all shadows disappearing when an object is illuminated at zero phase angles. At all phase angles, shadows will be visible except at zero phase angle when each particle hides its own shadow. So as the phase angle gets close to zero, the brightness of an object will increase significantly. Coherent backscatter is due to multiply scattered light interfering constructively with each other as the light exits a medium near zero phase angle, causing a relative peak in brightness for observations near zero phase angle. The incident light must have a wavelength comparable to the particle size and be smaller than the distance between the scattering particles.

The reduced visual magnitude (V-magnitude) $[H(\alpha)]$ is the observed magnitude at a particular phase angle α if the body is at unit heliocentric (Sun–object) and geocentric (Earth–object) distances (Dymock, 2007). The reduced V-magnitude is the visual magnitude with the effect of distance removed. The reduced V-magnitude is only a function of phase angle. The reduced V-magnitude can be calculated using the formula

$$H(\alpha) = V_{obs}(\alpha) - 5\log\left[\frac{r\Delta}{(AU)^2}\right] \tag{1.21}$$

where $V_{obs}(\alpha)$ is the observed V-magnitude, r is the heliocentric distance in AU, and Δ is the geocentric distance of the asteroid in AU. How the reduced V-magnitude changes for a body will only be a function of phase angle since the changes in magnitude due to changing distances between the Earth and the Sun to the asteroid have been removed.

Example 1.3

An asteroid is observed to have a visual magnitude of +20.6 at a phase angle of 25.7° at a heliocentric distance of 1.989 AU and a geocentric distance of 1.319 AU. What would be its reduced V-magnitude at the same phase angle?

Substituting these values into *Equation 1.21* produces the formula

$$H(25.7°) = 20.6 - 5\log\left[\frac{(1.989\ AU)(1.319\ AU)}{(AU)^2}\right] \tag{1.22}$$

Solving the equation results in a reduced V-magnitude at a phase angle of 25.7° of +18.5 for the asteroid.

1.9.5 H-G Magnitude System

To predict the magnitude of an asteroid as a function of phase angle, the H-G magnitude system was developed and then adopted in 1985 by the IAU (Bowell et al., 1989). H is the absolute magnitude and G is the slope parameter.

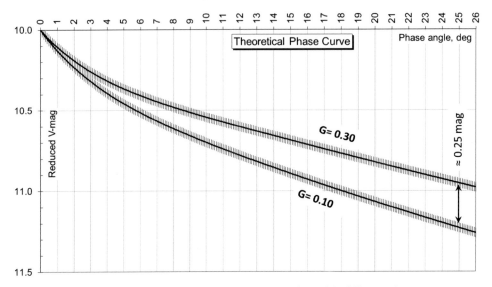

Figure 1.8 Theoretical phase curves for two asteroids with different slope parameters ($G = 0.10$ and $G = 0.30$) but absolute magnitudes of +10 (Buchheim, 2010). Note the steeper slope for the asteroid with a $G = 0.10$. Credit: Robert Buchheim, Altimira Observatory.

Absolute magnitude is the apparent magnitude of an object at a set distance. For stars, the set distance is 10 parsecs (32.6 light years). For asteroids, the absolute magnitude (H) is the apparent magnitude of the asteroid if the object is 1 AU from both the Sun and the Earth with a phase angle of 0° (so it is fully illuminated). Zero phase angle is where the Sun illuminates the body directly behind the observer. However, this configuration for determining the absolute magnitude is only theoretical since an asteroid cannot be both 1 AU from the Sun and the Earth and have a phase angle of 0°.

The slope parameter describes how strongly the measured magnitude (brightness) of an asteroid depends on the phase angle. As G decreases, the slope of the phase curve becomes steeper. For $G \approx 0$, the slope of the phase curve is very steep (changes dramatically with decreasing phase angle) while for $G \approx 1$, the slope is very shallow (changes gradually with decreasing phase angle). The theoretical phase curves for two asteroids with two different slope parameters are shown in Figure 1.8.

The usual default value for G is 0.15 but G is known to vary significantly for different asteroids (Lagerkvist and Magnusson, 1990). Higher-albedo bodies tend to have larger G values and, therefore, shallower phase curves. High-albedo bodies ($p_v = 0.46$) have an average G of 0.43 while low-albedo ($p_v = 0.06$) bodies have an average G of 0.12. However, less than 1% of known asteroids have measured

G values (Vereš et al., 2015). An accurate measurement of *G* requires a dense coverage of the phase curve over a large range in phase angle, which has not been done for most asteroids. Instead of *G*, it is often easier to measure the phase coefficient (β), which is the slope of the linear portion of the phase curve between $10°$ and $20°$ of phase angle (Dymock, 2007).

The absolute magnitude can be calculated from the formula

$$H = H(\alpha) + 2.5\log\left[(1-G)\Phi_1(\alpha) + G\Phi_2(\alpha)\right] \qquad (1.23)$$

where $\Phi_1(\alpha)$ and $\Phi_2(\alpha)$ are functions that describe the scattering off the surface (Dymock, 2007). These functions can be approximated as

$$\Phi_i(\alpha) = e^{-A_i\left(\tan\frac{\alpha}{2}\right)^{B_i}} ; i = 1,2 \qquad (1.24)$$

where $A_1 = 3.33$, $B_1 = 0.63$, $A_2 = 1.87$, and $B_2 = 1.22$. So by calculating the reduced V-magnitude from *Equation 1.21* from a measurement of the visual magnitude at a particular phase angle, the absolute magnitude can be determined using *Equation 1.23*.

If the absolute magnitude is already known (which is the case for most asteroids that have already been discovered), the effect of phase angle on the reduced visual magnitude can be determined by rearranging the components of *Equation 1.23*. The formula for calculating the reduced visual magnitude [$H(\alpha)$] at solar phase angle α for a body with a particular *H* and *G* will be

$$H(\alpha) = H - 2.5\log\left[(1-G)\Phi_1(\alpha) + G\Phi_2(\alpha)\right] \qquad (1.25)$$

This equation is valid for phase angles between $0°$ and $120°$ and for *G* values between 0 and 1. When calculating the values of the functions, the tangent is usually taken of the phase angle calculated in radians.

Example 1.4

For an asteroid with an *H* of +14, plot the phase curve from $0°$ to $30°$ for the reduced V-magnitude if the asteroid has a *G* of 0.15.

From *Equation 1.24*, the functions that describe the scattering off the surface are

$$\Phi_1(\alpha) = e^{-3.33\left(\tan\frac{\alpha}{2}\right)^{0.63}} \qquad (1.26)$$

and

$$\Phi_2(\alpha) = e^{-1.87\left(\tan\frac{\alpha}{2}\right)^{1.22}} \qquad (1.27)$$

Substituting all the values and formulas in *Equation 1.25*, the formula becomes

$$H(\alpha) = 14 - 2.5\log\left[(1-0.15)e^{-3.33\left(\tan\frac{\alpha}{2}\right)^{0.63}} + (0.15)e^{-1.87\left(\tan\frac{\alpha}{2}\right)^{1.22}}\right] \quad (1.28)$$

The phase angle in degrees can be converted to radians by the formula

$$\alpha(\text{radians}) = \frac{\pi}{180°}\alpha(°) \quad (1.29)$$

Substituting phase angles between 0° and 30° at 2° intervals results in the phase curve in Figure 1.9.

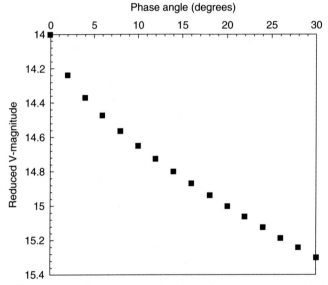

Figure 1.9 Phase curve for an asteroid with a slope parameter of 0.15 and an absolute magnitude of +14.

Even though it is called absolute, the absolute magnitude (and the slope parameter) for an asteroid can vary when observed at different oppositions (Dymock, 2007). Usually these values are averages of a number of observations at different oppositions. Since asteroids are not perfectly symmetrical and these objects are rotating, different aspects of the asteroid may be observed at different times in its orbit.

1.9.6 H-G_1-G_2 Magnitude System

A three parameter H-G_1-G_2 magnitude system (Muinonen et al., 2010) was adopted by the IAU in 2012. The absolute magnitude (*H*) is defined the same way in both this and the H-G system. The G_1 and G_2 parameters are new and are used with a number of functions to model the scattering off the surface and the opposition effect. This H-G_1-G_2 magnitude system better fits the phase curves of asteroids at low phase angles where backscattering occurs; however, this system is more complicated to implement. The H-G magnitude system is still used by many researchers (e.g., Reddy et al., 2015; Benishek and Pilcher, 2016) today due to its relative simplicity. For poorly sampled phase curves, a two parameter phase function (H-G_{12}) was also developed by Muinonen et al. (2010) since the G_1 and G_2 appear correlated for most asteroids.

1.10 Relationship between Albedos and Diameters

The observed magnitude of an asteroid will be a function of both the body's diameter and albedo. A larger body will be brighter than a smaller one with the same albedo since it will have more surface area, which will reflect more light. A higher albedo body will be brighter than a lower albedo object of the same size since it will also reflect more light.

The diameter (D) of an asteroid in km can be estimated if the absolute magnitude (H) and the visual geometric albedo (p_V) are known by using the relation (e.g., Fowler and Chillemi, 1992)

$$D = \frac{1329}{\sqrt{p_v}} 10^{-0.2H} \text{ km.} \qquad (1.30)$$

This formula is often used to roughly estimate a diameter by assuming an albedo since albedo information is not available for many asteroids.

Example 1.5

An asteroid has an absolute magnitude of +15 and a visual geometric albedo of 0.10. What is its estimated diameter?

Plugging in the values in *Equation 1.30* produces the formula

$$D = \frac{1329}{\sqrt{0.10}} 10^{-0.2(15)} \text{ km} = 4 \text{ km.} \qquad (1.31)$$

Solving the formula gives the asteroid's estimated diameter as approximately 4 km.

Questions

1) a) Compare the black body flux for a star with a surface temperature of 5800 K to the black body flux for an asteroid with an effective surface temperature of 250 K by drawing the relative shapes and intensities of both curves. Use wavelengths between ~0.1 and ~20 μm.

    b) What are the differences between the two curves?

2) A near-Earth asteroid has an effective surface temperature of ~250 K and was observed at a distance from the Sun of 1.10 AU. Assume the emissivity is 0.9. What is this object's Bond albedo?

3) Why are research observatories often located on mountains?

4) Give the advantages and disadvantages of photographic plates, photomultiplier tubes, and CCDs as detectors. Why is the CCD the primary type of detector that is used today to make visible astronomical observations?

5) If Pluto has an apparent magnitude of +13.7 at a distance of ~30 AU from the Sun, what would be the estimate of the apparent magnitude of a Pluto-like body at ~150 AU?

6) Show why a positive color index indicates that the object is brighter at the longer wavelength.

7) One asteroid has an absolute magnitude of +10. A second asteroid has an absolute magnitude of 17. Both have the same slope parameter (G). How many times brighter will the first asteroid be from the second asteroid at the same phase angle if they are both observed at the same distance from the Sun and the Earth?

8) An asteroid is observed to have a visual magnitude of +12.6 at a phase angle of 7.4° at a heliocentric distance of 2.638 AU and a geocentric distance of 3.538 AU. What would be its reduced V-magnitude at the same phase angle?

9) a) For an asteroid with an H of +14, plot the phase curve from 0° to 30° for the reduced V-magnitude if the body was a low-albedo object with a G of 0.12. Use the H-G magnitude system. When calculating the tangent, remember most computer programs want the phase angle inputted as radians and not as degrees.

    b) For an asteroid with an H of +14, plot the phase curve from 0° to 30° for the reduced V-magnitude if the body was a high-albedo object with a G of 0.43. Use the H-G magnitude system.

10) a) An asteroid has an H magnitude of +15 and an visual geometric albedo of 0.20. What is this asteroid's estimated diameter?

    b) An asteroid has an H magnitude of +15 and an visual geometric albedo of 0.05. What is this asteroid's estimated diameter?

2

Orbits and Discovering Minor Planets

2.1 History

All planets and minor planets orbit the Sun. But this concept has only been a prevailing idea for the last ~500 years. Since the beginning of civilization, all known celestial bodies were thought by almost everyone to revolve around the Earth since all the stars rise and set with fixed positions relative to each other. This model of the Universe is called the geocentric model. Notable people with a conflicting viewpoint included Philolaus (*c*.470–*c*.385 BC), a Greek philosopher who proposed that the Earth revolved around a central fire that was separate from the Sun, and Aristarchus (*c*.310 BC–230 BC), a Greek astronomer who proposed a heliocentric model where all the planets revolved around the Sun.

The Polish mathematician and astronomer Nicolaus Copernicus (1473–1543) published a book (*De revolutionibus orbium coelestium*, which is Latin for *On the Revolutions of the Celestial Spheres*) in 1543 where he proposed that all planets revolved around the Sun. However, Copernicus also proposed that all their orbits were perfect circles, which was not ultimately correct.

2.1.1 Kepler's Laws

The German astronomer Johannes Kepler, who also invented the Keplerian telescope, was able to determine the actual properties of planetary orbits. Kepler was the assistant of the greatest naked eye astronomer of his day, Danish astronomer Tycho Brahe (1546–1601). Kepler used Brahe's data on Mars, which has a very elliptical orbit, after Brahe's death to determine his three laws of planetary motion. Kepler's three laws of planetary motion are:

1) The orbit of each planet is an ellipse with the Sun at one focus and nothing at the other focus.
2) A line joining each planet and the Sun sweeps out equal areas in equal time.

3) The square of the period of each planet's orbit is proportional to the cube of its semi-major axis of its orbit. [The semi-major axis (a) is one-half of the length of the major axis of an object's elliptical orbit and is approximately the average distance from the Sun to the body and is usually given in terms of AU.]

Kepler determined that the orbit of each planet (or any other object in orbit around the Sun) was an ellipse and that its orbital velocity was largest when it was closest to the Sun and slowest when it was farthest from the Sun. When a body is closest to the Sun, the object is at perihelion (q). When a body is farthest from the Sun, the object is at aphelion (Q).

Kepler's third law is the basis for the formula

$$\frac{p^2}{(\text{yrs})^2} = \frac{a^3}{(\text{AU})^3} \tag{2.1}$$

where p is the period of the planet's orbit in years and a is the semi-major axis in AU. This formula is only valid for bodies in orbit around the Sun (or a star with the same mass as the Sun).

Example 2.1

How long will it take an asteroid with a semi-major axis of 4 AU to orbit the Sun once?

The semi-major axis (a) is 4 AU so Kepler's third law becomes

$$\frac{p^2}{(\text{yrs})^2} = \frac{(4 \text{ AU})^3}{(\text{AU})^3} \tag{2.2}$$

You must then solve for p. The formula then becomes

$$p = \sqrt{64} \text{ yrs.} \tag{2.3}$$

Therefore, the period (p) of the asteroid is 8 yrs.

2.1.2 Orbital Elements

Six orbital (also called Keplerian) elements (Figure 2.1) are needed to uniquely define an orbit of a planetary body. Six elements are needed since an orbit has six degrees of freedom. These six degrees of freedom are due to the orbit having three positions (x,y,z) in space and three velocities $\left(\dfrac{dx}{dt}, \dfrac{dy}{dt}, \dfrac{dz}{dt}\right)$. These orbital elements are the semi-major axis, eccentricity, inclination, argument of periapsis, longitude of

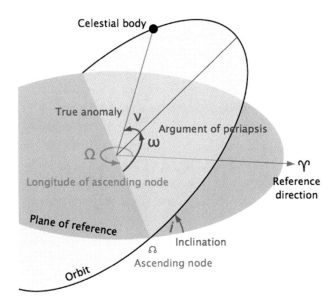

Figure 2.1 Five of the six orbital elements of a celestial body orbiting the Sun. The semi-major axis of the orbit, which is one-half of the major axis, is not shown. The orbital plane is in yellow. The ecliptic is in light gray. The reference direction points toward the vernal equinox (♈). Credit: Lasunncty. (A black and white version of this figure will appear in some formats. For the color version, please refer to the plate section.)

the ascending node, and the true anomaly. Calculating these orbital elements allows the position of the minor planet in the Solar System to be determined at any time.

The semi-major axis (a) is approximately the average distance of the body from the Sun. The eccentricity (e) defines the shape of the ellipse. The eccentricity of an ellipse varies between 0 and 1. A body having a nearly circular orbit has an eccentricity close to 0 while an object with nearly a parabolic orbit has an eccentricity close to 1. The semi-major axis and eccentricity give the two-dimensional structure of the orbit.

The inclination (i) is the angle to which the orbit is tilted relative to the ecliptic measured from where the object crosses the ecliptic plane going northward (ascending). An inclination between 0° and 90° indicates that the body is moving clockwise around the Sun (prograde) like all the major planets. An inclination between 90° and 180° indicates that the body is moving counter-clockwise (retrograde) around the Sun. Almost all minor planets (99.99%) have prograde orbits. The first numbered minor planet with a retrograde orbit was named (20461) Dioretsa (asteroid spelled backwards).

The argument of periapsis (ω) (also called argument of perihelion or perigee) is the angle between the object's closest approach to the Sun (perigee) and the object's ascending node. The inclination and the argument of periapsis give the three-dimensional structure of the orbit.

The longitude of the ascending node (Ω) is the angle measured counter-clockwise from the direction of the vernal equinox (Υ) to the point where the object crosses the ecliptic plane going northward. (Northward in our Solar System is defined as the same direction as north measured from the Earth's equator.) The vernal equinox point is the intersection of the ecliptic and the celestial equator where the Sun lies at on approximately March 20–21 of each year. The point of the vernal equinox is historically called the First Point of Aries but this point now is located in Pisces (MacRobert, 2006). This movement of the vernal equinox is due to the precession of the Earth's orbit. This element ties the position of the orbit to Earth.

The sixth element must specify the position of the body at a particular time (the epoch). A number of different elements have been used to characterize this position. One that is commonly used is the true anomaly (υ). True anomaly is the angle of the planet from the direction of periapsis and its current position. The true anomaly will be 0 radians (0°) at perihelion and increase over time over a range of 2π radians (360°). At aphelion, the true anomaly will be π radians (180°).

These orbital elements will change over time and are called osculating orbital elements. Gravitational interactions with other bodies (e.g., planets) will affect the orbital elements by since their orbits will be perturbed.

The inclination, argument of periapsis, and longitude of the ascending node are all measured relative to the ecliptic. The values of these orbital elements are dependent on the coordinate system used since the ecliptic plane is not fixed in position over time.

2.1.3 Forces

In 1687, English physicist Isaac Newton, who also built the first reflecting tele-scope, published the book *Philosophiæ Naturalis Principia Mathematica*, Latin for "Mathematical Principles of Natural Philosophy." This book introduced Newton's laws of motion and the law of universal gravitation, which he used to derive Kepler's laws of planetary motion. The law states that each point mass has an attractive force between another point mass that is directly related to the product of the masses of the bodies and inversely proportional to the square of the distance between the two bodies. The equation is usually written as

$$F_g = G\frac{m_1 m_2}{d^2} \tag{2.4}$$

where F_g is the attractive gravitational force between the bodies, m_1 and m_2 are the masses of the two bodies in kilograms, d is the distance between the bodies in meters, and G is the gravitational constant (6.67384×10^{-11} $m^3 kg^{-1} s^{-2}$), which was first measured by Henry Cavendish (1731–1810). For a body that has a spherical distribution of mass, the gravitational force for the body can be modeled as if the mass was all in the center of the body.

The strongest force on all the major and minor planets is due to the Sun, which is the most massive body in the Solar System. However, the reason that these bodies are not pulled into the Sun is that they formed out of a rotating disk of material, which produces an outward centripetal force. The centripetal force (F_c) on a body is

$$F_c = \frac{m_2 v^2}{r} \tag{2.5}$$

where v is the rotational speed in meters/second and r is the radius of the orbit in meters.

2.1.4 Proper Elements

Over time, the orbital elements of a body around the Sun will be affected by gravitational interactions with the planets and other minor planets. Proper orbital elements (e.g., Lemaitre, 1993) are orbital elements that change relatively slowly over time and are considered approximately constant. A number of different methods are used to compute these proper elements but all remove effect of the gravitational perturbations by the planets on the orbits of the asteroids (Knežević et al., 2002). Clusterings of asteroids in proper element space (semi-major axis, eccentricity, inclination) are called asteroid families, which are thought to be due to the breakup of the original parent bodies.

2.1.5 Tisserand Parameter

François Félix Tisserand (1845–1896) developed a parameter that would remain relatively unchanged during the gravitational interaction of a small body orbiting the Sun with a planet. The Tisserand parameter is derived from the Jacobi constant, which is the only known conserved quantity in a restricted three-body problem where one of the masses of the bodies is negligible compared to the other two bodies.

The Tisserand parameter (T_J) with respect to Jupiter is often used to roughly distinguish between asteroidal and cometary orbits. Asteroids tend to have T_J values greater than 3 and semi-major axes less than 5.2 AU (Jupiter's semi-major axis), while comets tend to have T_J values smaller than 3 (Jewitt et al., 2015). This parameter is also often used to identify whether a newly discovered comet is the same as a previously lost comet (Carusi et al., 1995). The Tisserand parameter with respect to Jupiter is

$$T_J = \frac{a_J}{a} + 2\left[(1-e^2)\frac{a}{a_J}\right]^{1/2} \cos(i) \qquad (2.6)$$

where a_J is the semi-major axis of Jupiter and a, e, and i are the osculating orbital elements of the body. The T_J for Jupiter is 3.

Example 2.2

A body is found to have a semi-major axis of 2.52 AU, eccentricity of 0.98, and an inclination of 12.6°. Is its orbit more consistent with that of an asteroid or of a comet?

Substituting these values in *Equation 2.6* produces the formula

$$T_J = \frac{5.2}{2.52} + 2\left[[1-(0.98)^2]\frac{2.52}{5.2}\right]^{1/2} \cos(12.6°) \qquad (2.7)$$

Remember that to compute the cosine of the inclination, the angle in degrees usually needs to be converted to radians using *Equation 1.29*. For this body, the value of T_J is 2.33, which is consistent with this body being a comet since its T_J is less than 3.

2.2 Minor Planet Terminology

Bodies in our Solar System are subdivided according to their orbits and sizes. Minor planets are objects that orbit the Sun and are not considered a planet due to their small sizes or a comet due to a lack of a coma. The first six planets from the Sun (Mercury, Venus, Earth, Mars, Jupiter, Saturn) have been known since historical times to move relative to the stars. Uranus (discovered in 1781) and Neptune (1846) were first discovered with telescopes. The inner four planets (Mercury, Venus Earth, Mars) are called the terrestrial ("Earth-like") planets, while the outer four (Jupiter, Saturn, Uranus, Neptune) are called the giant planets. The physical characteristics of the eight planets are given in Table 2.1.

Table 2.1 *Orbital and physical characteristics of the eight planets*

Planet	Semi-major axis (AU)	Diameter (km)	Mass (kg)	Brightest magnitude
Mercury	0.39	4879	3.30×10^{23}	−2.6
Venus	0.72	12 104	4.87×10^{24}	−4.9
Earth	1.00	12 742	5.97×10^{24}	–
Mars	1.52	6779	6.42×10^{23}	−3.0
Jupiter	5.20	139 822	1.90×10^{27}	−2.9
Saturn	9.56	116 464	5.68×10^{26}	−0.5
Uranus	19.22	50 724	8.68×10^{25}	+5
Neptune	30.11	49 224	1.02×10^{26}	+8

The brightest measured magnitudes of the planets are from Mallama (2011). The brightest measured magnitude for Saturn includes the contributions of its rings.

2.2.1 Minor Planets

Minor planets were also first discovered using telescopes since they are too small to be typically seen with the naked eye. Over 700 000 minor planets have been discovered to date. Minor planets with well-defined orbits are given a number and may be given a name [e.g., (5159) Burbine] to replace the provisional designation that was initially used to indicate when the body was discovered (e.g., 1977 RG). Moons around minor planets have also been discovered.

The difference between a minor planet and a comet is that comets have visible tails (comae) while asteroids do not. However, some minor planets have had tails appear later and some comets have had their tails disappear altogether, blurring any compositional distinction between these two types of bodies. Approximately 4000 comets have been discovered to date. There is also evidence for extrasolar minor planets around some white dwarf stars (e.g., Jura and Young, 2014) implying that minor planets are prevalent throughout the galaxy and, most likely, the Universe.

Minor planets are subdivided according to their orbits (Table 2.2), which are hoped to also subdivide them on their region of origin. Their orbits are usually defined by the body's semi-major axis.

2.2.2 Asteroids

The term "asteroids" usually refers to interplanetary bodies with semi-major axes that fall in the terrestrial planet region or out to the orbit of Jupiter. The main asteroid belt is the region between Mars and Jupiter where a significant number

Table 2.2 *Types of minor planets*

Type	Subtype	Definition
asteroids	near-Earth asteroids	$q < 1.3$ AU
	Mars-crossers	crosses the orbit of Mars, $q \geq 1.3$ AU
	main-Belt asteroids	$2 \text{AU} \leq a \leq 4.2$ AU
	Jupiter Trojans	$a \approx 5.2$ AU, 60° ahead and behind Jupiter
damocloids		$T_J \leq 2$, no coma or tail
dwarf planets		"round" but did not clear its neighborhood
Trojans	Uranus Trojans	$a \approx 19.2$ AU, 60° ahead and behind Uranus
	Neptune Trojans	$a \approx 30.1$ AU, 60° ahead and behind Neptune
centaurs		$5.52 \text{ AU} < a < 30.11$ AU
trans-Neptunian objects	Kuiper belt	$30 \text{ AU} \leq a \leq 50$ AU
	scattered disk object	$a > 50$ AU, $q > 30$ AU

The semi-major axis of an orbit is given by the symbol a and its perihelion is given by the symbol q.

of asteroids are located (Figure 2.2) and extends from ~2 to ~4.2 AU. The total mass of the asteroid belt is ~3 × 10²¹ kg (0.05% the mass of the Earth) (Kuchynka and Folkner, 2013). José Luis Galache has calculated (Chang, 2016) that there is approximately one asteroid greater than 100 meters in diameter for every ~450 trillion km³ of space in the asteroid belt, which is contary to the depiction of a dense asteroid belt in most movies (e.g., *The Empire Strikes Back*).

Near-Earth objects (NEOs) are objects (asteroids or comets) that pass relatively close to the Earth's orbit with perihelia less than 1.3 AU. Over 14 000 NEOs have been discovered to date. Near-Earth asteroids (NEAs) are the asteroids in this population. Mars-crossers are usually classified as asteroids that cross the orbit of Mars but have perihelia greater than or equal to 1.3 AU. Besides asteroids, minor planets include dwarf planets, trans-Neptunian objects, centaurs, damocloids, and Uranus and Neptune Trojans.

A plot of the distribution of asteroids versus semi-major axis shows a number of gaps (Figure 2.3). These gaps occur at resonances that arise when two periods or frequencies are in a simple numerical ratio. The most prominent ones are known as the Kirkwood gaps (Kirkwood, 1867), which are named after Daniel Kirkwood (1814–1895) who first noticed them. The most distinctive Kirkwood gaps are listed in Table 2.3. Kirkwood saw that these gaps are at semi-major axes where the ratio of the orbital periods of an asteroid and Jupiter is equal to the ratio of two small integers (4:1, 3:1, 5:2, 7:3, 2:1). These gaps occur at orbital (mean-motion) resonances of asteroids with Jupiter. Wisdom (1983) has

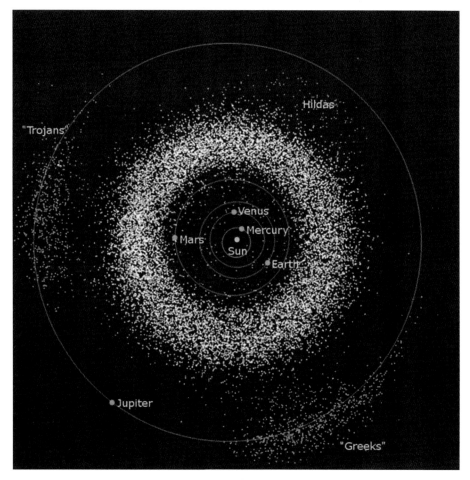

Figure 2.2 The orbits of the terrestrial planets and minor planets out to the orbit of Jupiter. The white points are asteroids in the main asteroid belt, the orange points are the Hilda asteroids, and the Trojan asteroids are in green. The orbits are for bodies that were discovered as of August 14, 2006. Credit: Mdf. (A black and white version of this figure will appear in some formats. For the color version, please refer to the plate section.)

shown that asteroids at the 3:1 mean-motion resonance have chaotic orbits that lead to increasing eccentricities that cause them to cross the orbits of Mars and Earth, which would either lead to impacts or gravitational scattering through close encounters. Mean-motion resonances can also result in clusterings of objects. For example, the 3:2 resonance (~3.97 AU) with Jupiter has a clustering of asteroids called the Hildas, which are named after (153) Hilda. In the outer Solar System, a number of trans-Neptunian objects have been found to be in mean-motion resonances with Neptune.

Table 2.3 *List of the most prominent Kirkwood gaps*

Resonance	Semi-major axis (AU)
4:1	2.06 AU
3:1	2.50 AU
5:2	2.82 AU
7:3	2.95 AU
2:1	3.27 AU

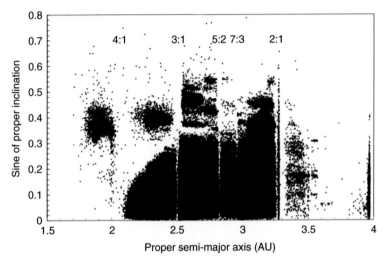

Figure 2.3 Plot of sine of proper inclination versus semi-major axis for ~500 000 main-belt asteroids. Mean-motion resonances and the v_6 secular resonance are labeled. The proper elements were calculated in 2014.

Example 2.3

Show that an asteroid located at the 3:1 Kirkwood gap with a semi-major axis of 2.50 AU is in a 3:1 resonance with Jupiter.

Using Kepler's third law, the period of the asteroid's orbit can be calculated with the formula

$$\frac{p^2}{(\text{yrs})^2} = \frac{(2.50 \text{ AU})^3}{(\text{AU})^3}. \tag{2.8}$$

The period for the asteroid will then be 3.95 years. For Jupiter with a semi-major axis of 5.20 AU (Table 2.1), the period of its orbit can be calculated with the formula

$$\frac{p^2}{(\text{yrs})^2} = \frac{(5.20 \text{ AU})^3}{(\text{AU})^3}.$$ (2.9)

The period of Jupiter is 11.86 years. The ratio of the orbits is then

$$\frac{11.86 \text{ years}}{3.95 \text{ years}} \approx \frac{3}{1}.$$ (2.10)

The locations of the four most prominent Kirkwood gaps can be used to used to break up the asteroid belt into the inner, middle, and outer belt (Figure 2.3). The inner part of the belt extends from 2.06 (4:1 resonance) to 2.50 AU (3:1 resonance), the middle part extends from 2.50 to 2.82 AU (5:2), and the outer part extends from 2.82 to 3.27 AU (2:1). However, there are asteroids in the main asteroid belt that extend past the 2:1 resonance out to ~4.2 AU.

In a plot of sine of proper inclination versus proper semi-major axis, the ν_6 secular resonance can be seen to cut through the asteroid belt (Figure 2.3). This resonance is due to the rate of precession of the asteroid's longitude of the periapsis equaling the rate of precession of Saturn's longitude of the periapsis. The longitude of the periapsis is the longitude (east-west direction relative to the Sun) measured from the vernal equinox at which periapsis would occur if the object's inclination was zero. The longitude of the periapsis is the sum of the longitude of the ascending node and the argument of periapsis, even though these two angles are in different planes. The value has been calculated by Froeschlé and Scholl (1987) as 28.06″/yr. The number "6" represents Saturn as the sixth planet from the Sun. The ν_6 resonance can change the eccentricity of an asteroid from 0 to almost 1.

2.2.3 Dwarf Planets

From its discovery in 1930 by Clyde Tombaugh (1906–1997) until 2006, Pluto was considered a planet. Pluto had always been considered anomalous compared to the major planets due to its small size and eccentric orbit, which crosses the orbit of Neptune. However, the discovery in 2005 of a body in the outer belt [later called (136199) Eris] with an estimated size larger than Pluto led to a heated discussion of whether Pluto should be considered a planet since Eris (and other bodies with similar sizes that have not yet been discovered) would then also be considered planets. Eris is now thought to be smaller than Pluto (Table 2.4).

Initially in 2006, a planet definition commission was assembled to decide whether Pluto was a planet (Gingerich, 2006). Their proposal was that a planet must have a nearly round shape due to having enough mass to be in hydrostatic

Table 2.4 *Orbital and physical characteristics of the dwarf planets*

Planet	Semi-major axis (AU)	Diameter (km)	Mass (kg)	Brighest magnitude
(1) Ceres	2.77	939	9.40×10^{20}	+6.6
(134340) Pluto	39.54	2374	1.30×10^{22}	+13.7
(136108) Haumea	43.22	1483	4.00×10^{21}	+17.3
(136199) Eris	67.78	2326	1.67×10^{22}	+18.7
(136472) Makemake	45.72	1428	–	+16.7

The diameter and mass of Ceres are from Russell et al. (2015). The diameter of Pluto is from Stern et al. (2015) and the mass of Pluto is from Brozović et al. (2015). The diameter of Haumea is from Lockwood et al. (2014) and its mass is from Ragozzine and Brown (2009). The diameter of Eris is from Sicardy et al. (2011) and the mass of Eris is from Brown and Schaller (2007). The diameter of Makemake is from Brown (2013).

equilibrium. This definition would have had Pluto, Ceres, the newly discovered 2003 UB_{313} (later named Eris), and Charon (Pluto's largest moon) as one of the 12 planets. Charon was considered a planet since Pluto and Charon rotate around a center of mass outside of both bodies due to Charon being relatively big compared to Pluto. Dynamicists then argued that a planet should be "the dominant object in its local population zone."

After a lot of discussion (Gingerich, 2006), the IAU finally decided by vote at its 2006 annual meeting in Prague that a planet "is in orbit around the Sun, has sufficient mass for its self-gravity to overcome rigid body forces so that it assumes a hydrostatic equilibrium (nearly round) shape, and has cleared the neighborhood around its orbit." Clearing the neighborhood means that similar-sized bodies are not found with similar orbits. Since Pluto and Eris are found in a region of the Solar System where there are a large number of bodies with similar orbits, these bodies are not considered to have cleared the neighborhood around their orbits and were not considered planets. Satellites were excluded as dwarf planets.

Pluto and Eris were then designated as dwarf planets by the IAU due to their nearly round shapes but not having cleared the neighborhoods around their orbits. Similarly, the largest body in the main asteroid belt is (1) Ceres, which is nearly round and is also considered a dwarf planet. Pluto [now listed as (134340) Pluto], (136199) Eris, (1) Ceres, (136108) Haumea, and (136472) Makemake are all now called dwarf planets (Table 2.4). The list of dwarf planets is expected to increase over time as more outer Solar System bodies are discovered and more of these bodies are found to be round.

Clearing the neighborhood around its orbit is a rather arbitrary term since a variety of bodies are known to cross almost every planet's orbit. Soter (2006) proposed a quantity that defines the degree to which a body dominates the other masses that

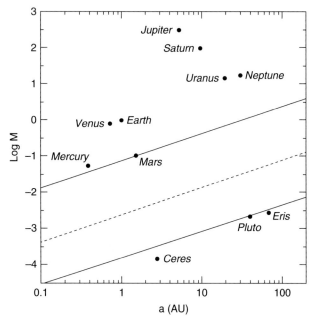

Figure 2.4 Mass in Earth masses versus semi-major axis for the eight planets, Ceres, Pluto, and Eris (Soter, 2006). The upper and lower lines represent values corresponding to those of Mars and Pluto, respectively. Any body above the dashed line will scatter a significant fraction of planetesimals out of its orbital zone within a Hubble time (~14 billion years), which is the approximate age of the Universe. Figure is reproduced by permission of the AAS. Credit: Steven Soter, American Museum of Natural History.

share its orbital zone. Soter (2006) proposed the planetary discriminant μ, which is defined by the formula

$$\mu = M\big/_m \tag{2.11}$$

where M is the mass of the primary body and m is the aggregate mass of all bodies that are within the orbital range of collisions with the primary body. Any value of μ greater than 100 is considered a planet by this definition (Figure 2.4). Margot (2015) used a more complicated formula for defining planets that could also be applied to extrasolar ones.

2.2.4 Trans-Neptunian Objects

Trans-Neptunian objects (TNOs) have semi-major axes that extend from Neptune's orbit (~30 AU) to the farthest reaches of the Solar System. Within this vast expanse of space is a region called the Kuiper belt. Kuiper belt objects (KBOs) (e.g., Pluto, Haumea, Makemake) have semi-major axes that extend from

~30 AU to about ~50 AU from the Sun. Scattered disk objects (e.g., Eris) have semi-major axes greater than ~50 AU and perihelia greater than 30 AU. These bodies are thought to have been gravitationally scattered by the giant planets. Except for Ceres in the main asteroid belt, all known dwarf planets are TNOs.

2.2.5 Damocloids

Damocloids are bodies without visible tails in comet-like orbits and are named after (5335) Damocles, the first asteroid discovered with such an orbit (Asher et al., 1994). A comet-like orbit is usually very eccentric (non-circular) and very inclined to the ecliptic compared to a typical asteroid orbit. Damocloids have a T_J less than or equal to 2.

2.2.6 Trojans

Trojans are bodies that share an orbit with another larger body but are found at either of two points of stability (called Lagrangian points) that are either ~60° ahead (L_4 point) or behind (L_5 point) the larger body's orbit. The best-known of these bodies are the Jupiter Trojans (Figure 2.2). Over ~6000 Jupiter Trojans are currently known; however, estimates of the total number of Jupiter Trojans with diameters greater than a kilometer is estimated to be around a million (Yoshida and Nakamura, 2005). However, Venus, Earth, Mars, Uranus, and Neptune also have Trojans. The Venus, Earth, Mars, and Jupiter Trojans are usually called asteroids. The Venus Trojan is only temporarily in a Trojan dynamical state (de la Fuente Marcos and de la Fuente Marcos, 2014). Only one Earth Trojan and a few Mars Trojans have been identified to date. Only one Uranus Trojan and over 10 Neptune Trojans are currently known, but millions of these bodies may exist (e.g., Horner and Lykawka, 2011).

2.2.7 Centaurs

Centaurs have semi-major axes that lie between Jupiter's and Neptune's orbits. Centaurs are named for the mythological beings that are half-human and half-horse. The name "centaur" is fitting since these objects have been found to have the properties of both asteroids and comets. Over 300 centaurs have currently been identified.

2.2.8 Other Types of Objects

Another term that is often used in discussing minor planets is meteoroid, which is an interplanetary object between ~10 µm and ~1 meter in space and is currently unobservable from Earth. Asteroids are bodies that are larger than meteoroids. If a meteoroid is actually observed in the sky, it will be called an asteroid. A meteorite is a natural fragment of any celestial body that has originated from outer space and landed on Earth. Micrometeorites are extremely small meteorites that are collected on the surface of the Earth. A meteor is the bright streak of light that is due to the heating of a meteoroid as it travels through the Earth's atmosphere.

2.3 Thermal Radiation Forces

For the longest time, the orbits of bodies in the Solar System were thought to be due just to gravity. However, the emission of thermal radiation from a body can also affect its orbit (e.g., Bottke et al., 2006).

2.3.1 Yarkovsky Effect

The Yarkovsky effect is the force acting on a rotating body in space due to the anisotropic emission of thermal photons (e.g., Hartmann et al., 1999). The Yarkovsky effect is named after Ivan Yarkovsky (1844–1902) who proposed this effect on small bodies at the beginning of the twentieth century (Beekman, 2006). He published his idea in a pamphlet, which was later read by Ernst Öpik (1893–1985).

The Yarkovsky effect occurs due to emitted photons having momenta (which is equal to E/c or hv/c). Any change in momentum results in a force applied over some time interval on the object. Any action force also has an equal reaction force in the opposite direction according to Newton's third law. The effect is analogous to the force due to the exhaust from a rocket engine propelling the rocket in the direction opposite of the exhaust. Differences in temperature on the surface will cause different parts of the surface to emit different fluxes of photons with different wavelength distributions. The area on the surface with the hottest surface temperature will emit the largest flux of photons (*Equation 1.3*) with the largest average momenta due to emitting more photons with higher frequencies (*Equation 1.5*). Heterogeneous emission due to these temperature differences will produce a net reaction force on the asteroid, which will primarily change the semi-major axis of an asteroid (Bottke et al., 2006). The reaction force will be in the opposite direction of the area with the highest surface temperature.

Analogous with the Earth, an asteroid will rotate once on its axis over one "asteroid day" and will also have "asteroid seasons" as different hemispheres receive different fluxes of sunlight in different parts of the asteroid's orbit around the Sun. The temperature over the asteroid's surface can also be thought of as analogous to the Earth's (Bottke et al., 2000). The diurnal (daily) temperature on a particular area of the surface will be hottest in the "afternoon" than at "noon." Also, the hottest temperatures on the asteroid's "northern hemisphere" will come after the "summer solstice" ("July–August") and the coldest temperatures will come after the "winter solstice" ("January–February"). These delays in heating or cooling of the surface are due to the body's thermal inertia, which is a measure of the resistance of a material to temperature changes. Thermal inertia is discussed further in Section 5.1.2. The higher the thermal inertia, the longer the delay in heating or cooling. The Yarkovsky effect has both a diurnal and seasonal component (e.g., Hartmann et al., 1999; Bottke et al., 2000, 2002, 2006).

The diurnal component is due to a net force on the asteroid from the higher temperature on the "afternoon" side, which results in the emission of photons with higher momenta from the "afternoon" side than other areas of the surface (Figure 2.5). This net force produces a reaction force in the opposite direction. The diurnal component causes the semi-major axis of the asteroid to either increase or decrease depending on the direction the body rotates. For a prograde rotator (rotating in the same direction as orbital motion), this reaction force is in the direction of orbital motion, which causes the body to speed up and increase its semi-major axis. For a retrograde rotator (rotating in the opposite direction to orbital motion), this reaction force is in the opposite direction of orbital motion, which causes the body to slow down and decrease its semi-major axis.

Example 2.4

Why does increasing the orbital velocity cause the semi-major axis to increase and decreasing the orbital velocity cause the semi-major axis to decrease?

The answer is analogous to changing the orbit of a satellite around the Earth (e.g., Hohmann transfer). For a circular orbit, the total orbital energy (E) of the system is equal to the kinetic energy plus the potential energy, which can be written by the equation

$$E = \frac{1}{2}m_2 v^2 - \frac{Gm_1 m_2}{a} \tag{2.12}$$

where m_2 is the mass of the asteroid, m_1 is the mass of the Sun, v is the orbital velocity, and a is the semi-major axis. The potential energy must be negative since a negative amount of work is done to bring a body closer to a larger body. If you equate the average centripetal (*Equation 2.5*) and gravitational (*Equation 2.4*) forces and solve for $m_2 v^2$, the equation

$$m_2 v^2 = G\frac{m_1 m_2}{a} \tag{2.13}$$

(a)

(b)

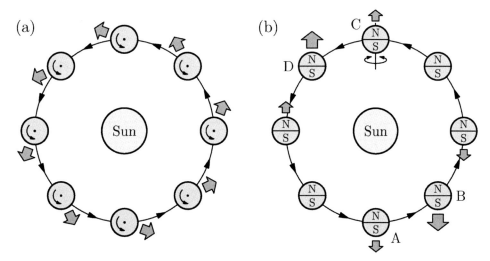

Figure 2.5 Plot of (a) the diurnal Yarkovsky effect and (b) the seasonal Yarkovsky effect. A circular orbit is assumed for both bodies. The obliquity is assumed to be 0° on the left side of the figure and 90° on the right side of the figure. For (a) the diurnal Yarkovsky effect, the body is heated by the Sun at "noon" (when the Sun is directly overhead). The body has prograde rotation. But due to thermal inertia, the maximum temperature will occur on the "afternoon" side, which "pushes" the object in the opposite direction. The body will accelerate in the direction of the gold arrow, which is in the same direction as orbital motion. This acceleration causes the semi-major axis of the body to increase. A body with retrograde motion will accelerate in the opposite direction to orbital motion and will have its semi-major axis decrease. For (b) the seasonal Yarkovsky effect during the "summer," the northern hemisphere will be hotter than the southern hemisphere. During "June" (A), the northern hemisphere will receive the most direct sunlight but due to thermal inertia, the body will be hottest in "July–September" (B). During "December" (A), the southern hemisphere will receive the most direct sunlight but due to thermal inertia, the body will be hottest in "January–February" (B). The body will accelerate in the opposite direction to orbital motion, which causes the semi-major axis of the body to decrease. Credit: David Vokrouhlický, Southwest Research Institute. (A black and white version of this figure will appear in some formats. For the color version, please refer to the plate section.)

is the result. Substituting for m_2v^2 in *Equation 2.12* results in

$$E = \frac{1}{2}\frac{Gm_1m_2}{a} - \frac{Gm_1m_2}{a} = -\frac{1}{2}\frac{Gm_1m_2}{a}. \tag{2.14}$$

If a force is applied to the system that increases the energy of the system by increasing the velocity, the change in energy will be equal to

$$\Delta E = E_2 - E_1 = -\frac{Gm_1m_2}{2}\left(\frac{1}{a_2} - \frac{1}{a_1}\right). \qquad (2.15)$$

Therefore for the energy change of the system to be positive, the semi-major axis of the body must increase. For the energy change of the system to be negative, the semi-major axis of the body must decrease.

The seasonal component is due to the delay in heating and cooling of the asteroid's hemispheres (due to thermal inertia) as it orbits around the Sun for an object that has "seasons" due to its axial tilt (obliquity) (Figure 2.5). This delay causes the strongest reaction force to occur during "July–August" and "January–February" of the object's orbit. The reaction force will weaken as the temperature difference between the hemispheres becomes smaller and strengthen as the temperature differences increases. Since the reaction force will always be in the opposite direction to orbital motion, the seasonal component of the Yarkovsky effect always causes the semi-major axis to decrease.

The obliquity, which is the tilt between the asteroid's equator and its orbital plane, will affect the strength of both the diurnal and the seasonal components of the Yarkovsky effect. The diurnal component is strongest when the rotation axis is perpendicular to the orbital plane (low obliquity), while the seasonal component is strongest when the rotation axis is in the orbital plane (high obliquity). The diurnal drift is almost always larger than the seasonal drift (Bottke et al., 2000; Vokrouhlický et al., 2000).

The diurnal and seasonal Yarkovsky effect can affect the orbits of asteroids ~30–40 km in size or smaller with significant orbital effects for objects in the size range of tens of meters to ~10 km (Bottke et al., 2006; Vokrouhlický et al., 2015). The diurnal drift is almost always larger than the seasonal drift (Bottke et al., 2000; Vokrouhlický et al., 2000). As expected, the Yarkovsky effect decreases with increasing distance from the Sun (Bottke et al., 2002). Typical drift rates for ~1-km diameter near-Earth asteroids are on the order of 10^{-3} AU Ma^{-1} (Nugent et al., 2012a, 2012b). Drift rates for ~5-km main-belt asteroids are on the order of 10^{-5} AU Ma^{-1} while ~20-km diameter main-belt asteroids are of the order of 10^{-6} AU Ma^{-1} (Bottke et al., 2006).

The first body whose orbit was determined to be affected by the Yarkovsky effect was the Earth-orbiting LAGEOS (Laser Geodynamics Satellite) (later called LAGEOS I) (Rubincam, 1987, 1990, 1993). LAGEOS is a brass sphere covered with retroreflectors and is a research satellite used for laser ranging. Infrared radiation heating from the Earth of the "south" part of the satellite causes a "north–south" temperature gradient, which creates a drag on the satellite due to seasonal Yarkovsky effect. This drag causes its orbit to decay, which was measured from

Earth. Radar ranging of NEA (6489) Golevka found a non-gravitational acceleration due to the Yarkovsky effect that affected its orbit (Chesley et al., 2003).

2.3.2 YORP Effect

The YORP (Yarkovsky–O'Keefe–Radzievskii–Paddack) effect is how the anisotropic emission of photons for a body with an asymmetric shape can change the rotation rate and pole orientation (obliquity) for a body due to the resulting torque. Rubincam (1998, 2000) named this effect in honor of Yarkovsky, John A. O'Keefe (1916–2000), Viktor Radzievskii (1911–2003), and Stephen Paddack, who all investigated this effect.

The emitted photons from any surface will be radiated normal to the surface. If the shape is asymmetric, the forces will not all cancel out, which results in a torque on the asteroid. The asteroid is assumed to have "windmill" asymmetry since the theoretical shape is often modeled with two wedges sticking out on the two opposite sides of the asteroid's equator. If the wedges are aligned in different directions relative to the Sun, the resulting reaction forces will not cancel each other out and a net torque will be applied to the asteroid. The YORP effect does not work on objects that are symmetrical in shape. A surface with large albedo variations on the surface could also have heterogeneous emission of photons that would produce a net torque (Vokrouhlický et al., 2015). The YORP effect causes a prograde-rotating asteroid to speed up and a retrograde-rotating asteroid to slow down. (e.g., Emery et al., 2015). The Binary YORP (BYORP) effect (Ćuk and Burns, 2005) can have a similar effect as YORP on a binary asteroid system.

Lowry et al. (2007) and Taylor et al. (2007) directly detected an increase in the rotation rate of an asteroid over a four-year span due to YORP. The observed asteroid was named (54509) YORP in honor of this discovery.

2.4 Celestial Coordinates

All stars, planets, and minor planets fall on the celestial sphere. The celestial sphere is the imaginary sphere surrounding the Earth on which all the stars, planets, and minor planets appear to reside. The ecliptic is the apparent path of the Sun on the celestial sphere. All the eight planets also fall on the ecliptic (within a belt 16° wide or 8° on each side).

2.4.1 Right Ascension and Declination

To determine an orbit, the right ascension (RA) (α) and declination (Dec) (δ) of the asteroid must be determined. The right ascension and declination are celestial

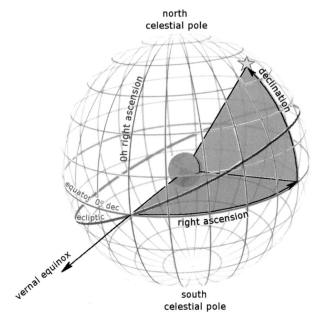

Figure 2.6 Right ascension and declination of a celestial object as seen from outside the celestial sphere. The celestial equator has a declination of 0°. The right ascension is 0 hours on the hour circle (great circle) that passes through the vernal equinox and the celestial poles. Credit: Tfr000. (A black and white version of this figure will appear in some formats. For the color version, please refer to the plate section.)

coordinates (also called equatorial coordinates) that can be used to locate any object on the celestial sphere (Figure 2.6). These celestial coordinates are geocentric with the center of the celestial sphere being the center of the Earth. Right ascension is the celestial equivalent of terrestrial longitude, while declination is the celestial equivalent of terrestrial latitude. The zero point for the right ascension is the vernal equinox, while the zero point for the declination is the celestial equator. The celestial equator is the extension of the Earth's equator onto the celestial sphere.

The units of right ascension are sidereal hours, minutes, and seconds, while the units of declination are degrees. Right ascension is given in units of time since stars were historically used to measure time. For right ascension, 24 hours is equivalent to 360° and one hour is equivalent to 15°. Finer units of declination are arcminutes [60 arcminutes (60′) in a degree] and arcseconds [60 arcseconds (60″) in an arcminute].

The right ascension is measured eastward along the celestial equator from the vernal equinox. The declination measures the angular distance of an object perpendicular to the celestial equator. For objects falling north on the celestial sphere relative to the celestial equator, the declination becomes positive, while the declination becomes negative for objects falling south on the celestial sphere.

Celestial coordinates will change over time due to the precession of the Earth's axis, which changes the position of the celestial equator and the vernal equinox. Right ascension and declination are given relative to a standard epoch. The currently used standard epoch is J2000.0. The previously used standard epoch was J1950.0. J refers to the Julian Date. Julian Day 0 is noon Greenwich Mean Time on January 1, 4713 BC. The Julian Date is the number of days since Julian Day 0. Joseph Scaliger (1540–1609) proposed the Julian Date system in 1583. Julian refers to Julius Caesar (100 BC–44 BC) who instituted the Julian calendar in 46 BC.

The shift in right ascension and declination values is small but significant over time (MacRobert, 2006). From 1950 to 2000, the shift along the ecliptic was only 0.7° and shifted less for other positions on the celestial sphere.

2.4.2 Azimuth and Altitude

The actual position of an object in the sky at a particular time is given by its azimuth and altitude. This is called the horizontal coordinate system. The azimuth is the angle between north and the star's perpendicular projection to the horizon from its position in the sky. The horizon is where the sky seems to meet the ground or sea. The ground is assumed to be perfectly flat with no mountains, valleys, or buildings in the way. The azimuth varies from 0° to 360°. An azimuth of 0° is directly north, 90° is directly east, 180° is directly south, and 270° is directly west.

Altitude is the angle between the object and the horizon. The altitude varies from 0° to 90°. The horizon has an altitude of 0°. At 90°, the object is at the zenith and directly overhead. Over an evening, the azimuth and altitude of a star or minor planet will change as the Earth rotates.

2.5 Discovery of Ceres

The search for a "missing planet" was motivated by predictions from the Titius–Bode Law, which was first formulated in the late 1700s. Not actually a law but a geometric progression that predicted the semi-major axes (a) of all the known planets. [The semi-major axis (*a*) is one-half of the major axis of an elliptical orbit and gives the average distance of a planet from the Sun.] In 1766, Johann Daniel Titius (1729–1796) first proposed this progression in an unattributed addition to a translation of Charles Bonnet's *Contemplation de la Nature* (French for "Contemplation of Nature") (Nieto, 1972). In the second edition of his astronomical compendium *Anleitung zur Kenntniss des gestirnten Himmels* (1772) (German for "Instruction for the Knowledge of the Starry Heavens"), Johann Bode (1747–1826) mentioned this progression in an unsourced footnote. In subsequent editions,

Table 2.5 *Semi-major axes of the eight planets and the*
predicted semi-major axes from the Titius–Bode law

Planet	Predicted semi-major axis (AU)	Actual semi-major axis (AU)
Mercury	0.4	0.39
Venus	0.7	0.72
Earth	1.0	1.00
Mars	1.6	1.52
	2.8	–
Jupiter	5.2	5.20
Saturn	10.0	9.56
Uranus	19.6	19.22
Neptune	38.8	30.11

his footnote later gave credit to Titius and subsequently the progression was known as the Titius–Bode Law.

Using the modern writing of the Titius–Bode Law, the semi-major axis of the planets follows the progression

$$a = \frac{(4+n)}{10} \text{ AU} \tag{2.16}$$

where $n = 0, 3, 6, 12, 24, 48, 96, 192$, etc. All the planets (Mercury, Venus, Earth, Mars, Jupiter, and Saturn) (Table 2.5) known at the time followed this progression rather well; however, there was no planet known between Mars and Jupiter at 2.8 AU. Until the discovery of Uranus by William Herschel (1738–1822) at the predicted Titius–Bode distance for a planet after Saturn, there was no concerted effort to try to discover if a planet actually existed between Mars and Jupiter. However after Uranus' discovery, Bode urged a search for the presumed "missing planet" between Mars and Jupiter.

One of the astronomers looking for this "missing planet" was Baron Franz Xavier von Zach. Since all known planets passed through the constellations of the zodiac, von Zach limited his telescopic observations to this region of the sky, called the ecliptic (Foderà Serio et al., 2002). The zodiac constellations (i.e., those that the Sun and planets pass through during the year) fall on the ecliptic. However, von Zach was unsuccessful in discovering this "missing" planet. To better cover the sky, he decided to organize in 1800 a group of 24 astronomers (which came to be known as the "Celestial Police") to systematically cover a 15° area of the zodiacal band.

One of the astronomers proposed to be one of the "Celestial Police" was Giuseppe Piazzi (1746–1826). However for an unknown reason, Piazzi was never formally invited. Piazzi would have been an obvious choice to be part of this group since he was the director of the Palermo Observatory, which was the most southerly European observatory. The observatory also had a telescope built by Jesse Ramsden (1730–1800) called a Ramsden Circle, which was thought to make the most precise measurements of the positions of stars at the time. The Ramsden Circle is a meridian circle. These telescopes are used for timing the passage of stars across the local meridian, which is the great circle that passes through the celestial poles on the celestial sphere. Stars could only be observed for two minutes a night (Cunningham, 2016).

As part of his work in updating a star catalog that was produced by Francis Wollaston (1731–1815), Piazzi discovered a faint (~8th magnitude) object in Taurus on January 1, 1801 that was not in the star catalog (Cunningham, 2016). Because there were a large number of inaccuracies in Wollaston's star catalog, the catalog had to be checked star by star. Piazzi then saw that that this object moved significantly relative to other stars when he observed it over the next few nights. He was initially cautious and called it a comet. But since it had no apparent coma and its movement in the sky was slow and uniform, he thought it could be a planet. Piazzi sent his observations to Bode who immediately thought the object was the missing planet. Piazzi observed this object for over a month before it was unobservable as it passed behind the Sun. However his ~20 observations of the object covered only 3° of the sky, which was not enough sky coverage to be able to determine a reliable orbit using the methods at the time (Foderà Serio et al., 2002).

Piazzi's observations were published in September of that year in *Montaliche Correspondez* (the primary astronomical journal at the time), which was edited by von Zach. Carl Friedrich Gauss (1777–1855) saw these observations and derived a mathematical method to predict Ceres' position from only a small number of observations. Gauss was able to predict Ceres' subsequent positions in the sky, which were published in *Montaliche Correspondez*. Von Zach rediscovered Ceres on December 7, 1801. Heinrich Olbers (1758–1840) confirmed this rediscovery on December 31, 1801.

Gauss developed a method for determining an orbit using only three observations (e.g., Marsden, 1977; Curtis, 2014). He first used observations taken by Piazzi on January 2, January 22, and February 11 and then used observations taken on January 1, January 21, and February 11 (Marsden, 1977). Gauss never published exactly how he computed Ceres' orbit in 1801 (Teets and Whitehead, 1999). Papers published later by Gauss included refinements to his original method. His method is based on the fact that the orbit of a body lies in a plane and that the body will sweep out equal areas in equal times.

2.6 Discovery and Naming of First Asteroids

Piazzi first proposed the name Ceres Ferdinandea for his discovery. In Italian, it is *Cerere Ferdinandea*. Ceres is the Roman God of Agriculture and patron goddess of Sicily, while Ferdinandea was in honor of King Ferdinand III (1751–1825) of Sicily. This king was also known as Ferdinand IV of Naples. The name was then shortened to just Ceres because Ferdinandea was not acceptable to other nations due the political nature of the name. It was known at the time that Ceres was much fainter than what would be expected for a planet at this distance from the Sun (Chambers and Mitton, 2013).

On March 28, 1802, Olbers observed a faint (7th magnitude) object in Virgo that was not previously there while attempting to observe Ceres (Foderà Serio et al., 2002). This body was discovered less than 0.5° from where Ceres was recovered in January of 1802 (Marsden, 1977). The body was observed for two hours and was found to move relative to the stars (Foderà Serio et al., 2002). After two days of observations, Olbers was sure it was a new planet. Gauss computed its orbit. It was subsequently named Pallas after Pallas Athena, which is another name for the Greek goddess Athena. Since Ceres and Pallas have similar semi-major axes, Olbers believed that these bodies were fragments of a larger planet that had broken up (Foderà Serio et al., 2002).

From their angular diameters, William Herschel estimated in 1802 the diameter of Ceres to be 259 km and of Pallas to be 238 km (Hughes, 1994), which was much smaller than the known planets. Due to their small sizes, Herschel argued that Ceres and Pallas were a new type of celestial object (Herschel, 1802) and was the first person to use the name asteroid (Cunningham and Orchiston, 2011). Herschel is credited as the discoverer of Uranus, which had been observed previously but had been identified as a star. However, Herschel did not think of the name himself. Herschel had asked his friend, Charley Burney Sr., about possible named for these bodies. Charles Burney Sr. then wrote his Greek-scholar son (Charles Burney Jr.) and asked him for suggestions. Charles Burney Jr. suggested the name "asteroids" from the Greek word *asteroeides* (star-like) due to their similarity in appearance to stars. Planets in the sky appear as disks while asteroids and stars appear as points of light. However, most astronomers believed Ceres and Pallas were planets (Hilton, 2006).

Karl Harding (1765–1834) then discovered a third object (Juno) in 1804. Juno is the Roman goddess who is the wife of Jupiter. Olbers then discovered a fourth object (Vesta) in 1807. Vesta is the Roman goddess of the hearth.

In 1811, Johann Hieronymus Schröter (1745–1816) estimated the diameter of Ceres as 2613 km, Pallas as 3380 km, and Juno as 2290 (Hughes, 1994).

These estimated diameters are approximately half the size of the planet Mercury (Table 2.1) and appeared to validate them as planets.

However, the fifth asteroid (Astraea) was not discovered until 1845. After 15 years of searching for asteroids, Karl Hencke (1793–1866) discovered Astraea (Chambers and Mitton, 2013). The sixth (Hebe), seventh (Iris), and eighth (Flora) were discovered two years later in 1847. The ninth (Metis) was discovered in 1848 and the tenth (Hygiea) in 1849. The eleventh (Parthenope), twelfth (Victoria), and thirteenth (Egeria) were discovered in 1850, and the fourteenth (Irene) and fifteenth (Eunomia) were discovered in 1851. All of these bodies received mythological names; however, there was a controversy concerning Victoria. Asteroid (12) Victoria was named after the Roman goddess of Victory, but the name appeared to also honor Queen Victoria (1819–1901) (Schmadel, 2003). This name was controversial since it appeared to honor a living person. The editor of the *Astronomical Journal*, Benjamin A. Gould (1824–1896), preferred the name Clio (Gould, 1852), which is the Greek muse of history. However, the first director of the Harvard College Observatory, William C. Bond (1789–1859), decided that the Victoria name was appropriate (Schmadel, 2003).

These bodies were thought of as planets and given astronomical symbols like the seven planets known at the time. The symbols for these bodies (Figure 2.7) became much more elaborate over time. However, the large number of objects being discovered led to the proposal of a much simpler system where each asteroid was listed with a circle with the number inside it that depicts the order of discovery (Gould, 1852). The use of these symbols and the realization that these bodies were much smaller (hundreds of kilometers in diameter) than the planets (Hughes, 1994) led to the labeling of these bodies as minor planets or asteroids (Hilton, 2006).

2.7 Discovering a Minor Planet

Discovering a minor planet today has come a long way from 200 years ago when these bodies were identified and celestial coordinates were determined with your eyes using a telescope like Piazzi and computing the orbit by hand like Gauss. Today minor planets are discovered using electronic detectors (CCDs) in conjunction with software that can identify these objects as they move relative to the stars and determine their celestial coordinates (e.g., Barrios et al., 2004; Warner, 2011; Puckett et al., 2016) and software that can compute their orbits (e.g., Granvik et al., 2009). This software can be purchased, found as open sourceware, or written specifically by the observing team.

Over 700 000 minor planets have been discovered so far with the numbers increasing daily with over 4000 objects discovered each month on average during

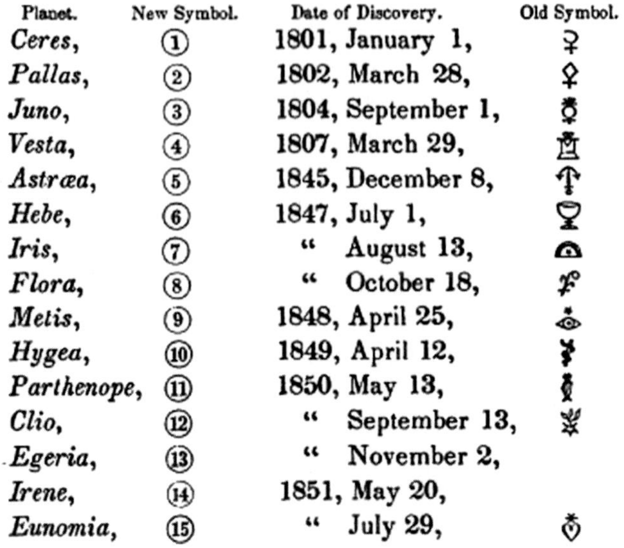

Planet.	New Symbol.	Date of Discovery.	Old Symbol.
Ceres,	①	1801, January 1,	♀
Pallas,	②	1802, March 28,	♀
Juno,	③	1804, September 1,	⚷
Vesta,	④	1807, March 29,	⚶
Astræa,	⑤	1845, December 8,	⚴
Hebe,	⑥	1847, July 1,	⚍
Iris,	⑦	" August 13,	☾
Flora,	⑧	" October 18,	♀
Metis,	⑨	1848, April 25,	⊙
Hygea,	⑩	1849, April 12,	⚕
Parthenope,	⑪	1850, May 13,	⚵
Clio,	⑫	" September 13,	⚥
Egeria,	⑬	" November 2,	
Irene,	⑭	1851, May 20,	
Eunomia,	⑮	" July 29,	⚷

Figure 2.7 Newly suggested symbols for minor planets from Gould (1852). The twelfth discovered asteroid is called Clio in Gould (1852); however, it was instead named Victoria. © AAS. Reproduced with permission of AAS.

the first half of 2016. Minor planets have been discovered by both professional astronomers with nationally supported telescopes and amateur astronomers with backyard telescopes (Di Cicco, 1996). However as more fainter and fainter bodies have been discovered, it has become harder and harder for amateurs to discover minor planets with their own telescopes. A citizen science project (Asteroid Zoo) is available on the web where anybody can look for minor planets by viewing multiple images of the sky using old sky survey data (Wall, 2014).

2.7.1 Minor Planet Center

The Minor Planet Center (MPC) is the clearinghouse for all asteroid astrometric observations. Astrometry is the precise measuring of the positions of objects in the sky. The MPC was founded in 1947 at the University of Cincinnati under the direction of Paul Herget (1908–1981) and then moved to the Smithsonian Astrophysical Observatory (SAO) at the Harvard–Smithsonian Center for Astrophysics (CFA) in Cambridge, Massachusetts in 1978 under the direction of Brian Marsden (1937–2010). MPC is currently funded by NASA. All astrometric observations of minor planets and comets are sent to the MPC so orbits (and, therefore, orbital elements) can be determined. Over 150 million observations have been processed by the Minor Planet Center to date.

The MPC publishes a variety of publications such as the Minor Planet Circulars (MPCs), Minor Planet Circulars Supplements (MPS), Minor Planet Circulars Orbit Supplements (MPOs), and Minor Planet Electronic Circulars (MPECs). Usually published once a month around the date of the full Moon, MPCs contain astrometric observations and orbits of minor planets (summarized by observatory code) and comets plus numbering of minor planets and naming citations. (All observatories making astrometric observations have a distinctive three character code assigned by the Minor Planet Center.) MPS contain the full minor planet observations, which are sorted by designation and date. MPOs contain orbital elements of newly numbered, newly linked (when objects observed at two separate times are found to be the same body), and newly identified minor planets plus one-line updated orbits for minor planets observed at one or multiple oppositions. MPECs announce discoveries and new orbits of unusual minor planets (e.g., near-Earth objects, trans-Neptunian objects) plus routine observations of comets.

2.7.2 Discovering a Minor Planet Today

To discover a minor planet, an image of the sky must first be taken (Figure 2.8). The asteroid will look just like a star in the image. Then another image of the exact same area of the sky must be taken. And then another image must be taken. By "blinking" the images, which is the displaying of the images in fast succession, asteroids can be found. Minor planets will be the objects that move relative to the stars, which are fixed in position relative to each other. This movement is due to minor planets being located so much closer to the Earth than the stars. The faster the object moves, the closer the body is to Earth. Di Cicco (1996) recommends taking three or more images at equal intervals between exposures over the course of an hour. To determine the six orbital elements, a minimum of three sets of observations must be done at three different times. If the six orbital elements are determined, the celestial coordinates of the minor planet can be predicted in the sky.

To "blink" photographic plates, the blink comparator was invented by Carl Pulfrich (1858–1927) in 1904. The blink comparator allows the user to rapidly switch viewing two different photographic plates. However, today the "blinking" of CCD images is done on a computer screen by rapidly displaying multiple images.

The right ascension (RA) (α) and declination (Dec) (δ) of the asteroid will be determined from the image. To do this, the centroid pixel position (to a fraction of a pixel) must be determined for the asteroid and the comparison stars. A comparison star is a star with a known right ascension and declination. The Minor Planet Center recommends at least three stars be used. Optimally the comparison stars should surround the minor planet in the image.

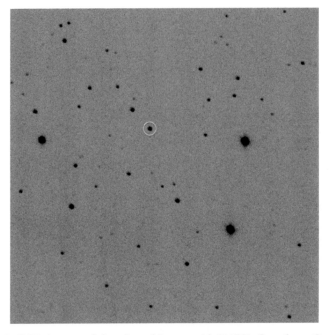

Figure 2.8 CCD image of the sky with asteroid (1492) Oppolzer. Oppolzer is circled in the image. Credit: Vishnu Reddy, University of Arizona.

A least squares plate reduction is done to determine the right ascension and declination of the minor planet. The spherical coordinate system of the celestial sphere must be projected onto a flat image plane that is linear. Plate constants are parameters in the transformation between pixel coordinates (x, y) and the right ascension and declination (α, δ) (Marsden, 1982). These transformations were originally used when using photographic plates, which is where the name of these constants is derived from.

The time of each observation is given as Coordinated Universal Time (UTC). UTC is defined using atomic clocks and replaced Greenwich Mean Time (GMT) as the primary time standard in the 1970s. UTC does not depart from GMT by more than one or two seconds (Guinot, 2011). GMT is the mean solar time at the Royal Greenwich Observatory in Greenwich, England through which the prime meridian passes. The prime meridian defines the great circle on the Earth where the longitude is $0°$.

The Minor Planet Center suggests that observations of potential new objects should be submitted nightly. Multiple positions (approximately three) taken over the course of an hour or so are submitted. The minor planet should then be observed over a second night, hopefully within a week of the first observation.

Table 2.6 *First letter of the provisional designation that indicates the half-month of an asteroid discovery*

A	January 1–15	N	July 1–15
B	January 16–31	O	July 16–31
C	February 1–15	P	August 1–15
D	February 16–28 or 29	Q	August 16–31
E	March 1–15	R	September 1–15
F	March 16–31	S	September 16–30
G	April 1–15	T	October 1–15
H	April 16–30	U	October 16–31
J	May 1–15	V	November 1–15
K	May 16–31	W	November 16–30
L	June 1–15	X	December 1–15
M	June 16–30	Y	December 16–31

2.7.3 Provisional Designation

When a minor planet is first discovered now, the object is given a provisional designation (e.g., 1973 EB, 1994 GT_9) by the MPC. The object must have been observed over two nights, but this does not have to be by the same observer. The first part of the provisional designation is the year the asteroid was discovered. Then there is a letter that indicates the half-month of the discovery (Table 2.6). The first part of each month starts on the first, while the second half of each month starts on the sixteenth. *I* is not used because it resembles a *1* and *Z* is not used because there are 24 half-months. The next letter (and often a subscript number) indicates the order of discovery (Table 2.7) within the half-month. For example, *A* indicates the first object discovered in the half-month, *B* is the second, *Z* is the twenty-fifth, A_1 is the twenty-sixth, Z_1 is the fiftieth, and A_2 is the fifty-first. To calculate the order of discovery from the second letter and subscript, multiply 25 times the subscript number and add the order of the alphabet of the second letter (without using the letter "I" in the calculation). For example, C_5 is the 128th object discovered in the half-month.

Example 2.5

What would be the provisional designation of the 235th minor planet discovered between August 1 and August 15, 1992?

The year of discovery is 1992. From Table 2.6, a minor planet discovered between August 1 and August 15 would be given the designation *P* to indicate the half-month. The designation that indicates the 235th discovered asteroid would be K_9 since *K* is the tenth letter in the alphabet if *I* is excluded and $25 \times 9 + 10 = 235$. Therefore, the provisional designation is 1992 PK_9.

Table 2.7 *Second letter of the provisional designation that*
indicates the order in the half-month of an asteroid discovery

A	1	O	14
B	2	P	15
C	3	Q	16
D	4	R	17
E	5	S	18
F	6	T	19
G	7	U	20
H	8	V	21
J	9	W	22
K	10	X	23
L	11	Y	24
M	12	Z	25
N	13		

Notable exceptions to this type of provisional designation are objects discovered during the Palomar–Leiden (P–L) Survey (PLS). The Palomar–Leiden Survey was a joint effort instituted in 1960 between the University of Arizona and Leiden Observatory. Observers at the University of Arizona made observations at Palomar Observatory, while researchers at Leiden Observatory analyzed the resulting photographs. Observations in 1971 (designated T-1), 1973 (T-2), and 1977 (T-3) were also made as part of the Palomar–Leiden Trojan Survey. The provisional designations used during these surveys are slightly different. The 4011th object discovered during the Palomar–Leiden Survey is given the designation 4011 P-L. The 3108th asteroid discovered during third Palomar–Leiden Trojan would be given the designation 3108 T-3.

2.8 Naming of Minor Planets

One of the coolest things about asteroids is that it is possible to get one named after you. Currently all comets are named after their discoverer [person(s), spacecraft, or observing program]. For a fee, many companies name stars after people, but these names are not officially recognized by any scientific organization. But asteroids can be named after a person (or place or thing) with this designation officially recognized by the IAU.

When the orbit of the body becomes well defined, the object is given a permanent designation (a sequentially issued number) by the Minor Planet Center. A body in a well-defined orbit usually has been observed over four oppositions. The name of the minor planet will either be its provisional designation or a name approved by the Committee on Small Body Nomenclature (CSBN) of the IAU.

The minor planet number and name (or provisional designation) will be written with the number in parentheses and then the name (or provisional designation) [e.g., (4) Vesta, (66391) 1999 KW$_4$].

Originally, minor planets were named after goddesses and female mythological people. The first minor planet not named after a female was (20) Massalia, which is the Greek name for the French city of Marseille. Massalia is also the first object in the Solar System given a non-mythological name. The asteroid (45) Eugenia was the first minor planet that was definitely named after a real person [Empress Eugenia di Montijo (1826–1920) who was the wife of Napoleon III] instead of a person of legend. Originally, all male asteroid names were feminized. For example, (51) Nemausa is a female version of Nemausus, a Celtic god of the French city Nîmes. Minor planet (54) Alexandra was the first asteroid named after a male, the German explorer Alexander von Humboldt (1769–1859).

The discoverer has the privilege for a ten-year period following the numbering of the object to name the object. Proposed names need to be 16 characters or less in length, pronounceable in some language, non-offensive, and not too similar to an existing name of a minor planet or natural planetary satellite. Previously, asteroids could have the same names as satellites. For example, (85) Io has the same name as the Galilean moon. It is also preferred that the name is one word; however, there are a number of asteroids with two-word names [e.g., (10958) Mont Blanc, which is named after the highest mountain in Europe]. The names of people or events principally known for political or military activities are unsuitable until 100 years after the death of the person or the occurrence of the event. Names of pet animals are discouraged. This rule was instituted after an asteroid was named after a cat that was named after Mr. Spock [(2309) Mr. Spock]. Names of a purely or principally commercial nature are not allowed. Names honoring persons, companies, or products for no more than success in business are also discouraged (Marsden, 2003).

Traditionally, asteroids are not named after their discoverer. An exception was made in the naming of (96747) Crespodasilva whose discoverer [Lucy Crespo da Silva (1978–2000)] tragically died a year after discovering the object.

There are a number of naming conventions for particular classes of minor planets (Table 2.8). For example, NEAs are usually given mythological names, but the name should not be associated with creation or underworld themes. One notable exception is NEA (4055) Magellan, which is named after the famed explorer Ferdinand Magellan (*c*.1480–1521). Bodies approaching or crossing the orbit of Neptune and in a stabilizing resonance (excluding Trojans) are given mythological names associated with the underworld.

Bodies sufficiently outside Neptune's orbit (trans-Neptunian objects) are typically given mythological names associated with creation gods. These objects include the dwarf planets Haumea and Makemake. Haumea is named after the

Table 2.8 *Naming conventions for minor planets*

Type	Convention
near-Earth asteroids	mythological names not associated with the Underworld or creation
Trojan asteroid (L4 point)	Greek heroes of Trojan War
Trojan asteroid (L5 point)	Trojan heroes of Trojan War
trans-Neptunian not in a 2:3 resonance	mythological names associated with the underworld
trans-Neptunian in a 2:3 resonance	mythological names associated with creation
centaurs	centaurs

Hawaiian goddess of fertility and childbirth. Makemake is named after the Easter Island god of fertility since it was discovered shortly after Easter.

The exception to this naming convention for trans-Neptunian objects are objects in a 2:3 mean-motion resonance with Neptune, which are given the name of mythological characters associated with the underworld. This naming convention was started since the first object discovered with this type of orbit was Pluto, the Roman god of the Underworld. For example, (28978) Ixion was a king in Greek mythology who was tied to a flaming spinning wheel and condemned to Tarturus (a dungeon of torment below Hades).

Jupiter Trojans are named after participants in the Trojan War in Greek mythology. Trojans at the Jupiter's L4 point are named after Greek characters and those at the L5 point are named after Trojan citizens. However, one Greek-named object[(617) Patroclus] was named among those at the L5 point and one Trojan-named object [(624) Hektor] was named among those at the L4 point.

Centaurs are named after the mythological centaurs. The object that started the centaur designation was (2060) Chiron, which was discovered in 1977 by Charles Kowal (1940–2011). Chiron is named for the famous centaur Chiron. Centaurs are Greek mythological creatures that have the body and legs of a horse but the head, arms, and torso of a human. The centaur designation for these objects is usually considered to be very appropriate since many of the objects have characteristics of both asteroids and comets. Kowal suggested that other objects with similar orbits be called centaurs. The second object to be given a centaur designation was (5145) Pholus, which was discovered in 1992. Pholus is the brother of Chiron. However, the first object to be discovered with this type of orbit was (944) Hidalgo, which was discovered in 1920. Hidalgo is named for Miguel Hidalgo y Castillo (1753–1811), a leader of the Mexican War of Independence, since the centaur naming convention had not yet been instituted.

The CSBN also approves the names for the satellites of minor planets (Marsden, 2003). Satellites should receive names closely related to the names of the primary body and be consistent with their relative sizes. For example, binary trans-Neptunian objects that have similar sizes should receive names of twins or siblings consistent with the current principle of using names of creation or underworld gods.

When an asteroid name is submitted to the CSBN for approval, a brief citation that supports the naming of the object must be included. There is a four-line limit [as printed in the Minor Planet Circular (MPC)] for the body of the citation. Examples of citations are shown in Table 2.9.

A number of minor planets have been named using a naming contest. The discoverer of the asteroid is usually involved with the contest. However, the CSBN still has to officially approve any name. Many of these contests have been to name spacecraft targets.

The target [(9969) 1992 KD] of the Deep Space 1 discovery mission was named Braille from a suggestion by Kerry Babcock during a Planetary Society sponsored contest. Braille is named after Louis Braille (1809–1852) who was the inventor of the system of reading and writing for the blind and visually impaired. Since Deep Space 1 was a technology demonstration mission, the theme of the contest was "inventors." The target [(101955) 1999 RQ_{36}] of the OSIRIS-REx sample return mission was named Bennu after a contest from a suggestion by nine-year-old Mike Puzio who argued that parts of OSIRIS-REx looked liked depictions of the Egyptian god Bennu. Bennu is often depicted as a gray heron. The Egyptian mythological name was appropriate since Osiris is the Egyptian god of the underworld.

The target [(162173) 1999 JU_3] of the Hayabusa2 mission was named Ryugu after a contest sponsored by JAXA (Japanese space agency) where 30 entries proposed Ryugu. The name originates from an ancient Japanese story "Urashima Tarō" where the main character (the fisherman Urashima Tarō) brought back a mysterious box from the Dragon's palace ("Ryugu-jo") at the bottom of the ocean. The name Ryugu symbolizes bringing "back a treasure" from a far-away place. Since samples of the asteroid are thought to contain water in hydrated silicates, the name Ryugu was picked since it also symbolizes water.

A number of special names were given minor planets with significant numbers. Minor planet (10000) Myriostos, which is the Greek word for "ten thousandth," was given its name to honor all astronomers who have discovered 10 000 minor planets with high-quality orbits. There was some discussion to give this minor planet number to Pluto; however at this time in 1999, Pluto was still considered a planet. The minor planet (15000) CCD was given its name to commemorate the detector that has revolutionized astronomy. The minor planet (100000) Astronautica was given its name to recognize the fiftieth anniversary of the start of the Space Age (defined

Table 2.9 *Examples of asteroid citations*

(4535) Adamcarolla = 1986 QV$_2$
Discovered 1986 Aug. 28 by H. Debehogne at the European Southern Observatory.
Adam Carolla (b. 1964) co-hosted Loveline, a syndicated radio program that
  has dispensed medical and relationship advice to millions of listeners for over
  20 years. Adam is a comedian who has also hosted numerous television programs
  and currently has his own morning radio show. The name was suggested by
  T. Burbine.

(4536) Drewpinsky = 1987 DA$_6$
Discovered 1987 Feb. 22 by H. Debehogne at the European Southern Observatory.
David Drew Pinsky ("Dr. Drew," b. 1958) cohosts Loveline, a syndicated radio
  program that dispenses medical and relationship advice to millions of listeners. He
  has also written a book detailing his experiences in treating patients at an addiction
  rehab clinic. The name was suggested by T. Burbine.

(4538) Vishyanand = 1988 TP
Discovered 1988 Oct. 10 by K. Suzuki at Toyota.
Viswanathan (Vishy) Anand (b. 1969) was India's first chess grandmaster. He went on
  to become the fifteenth undisputed world chess champion. In addition to his passion
  for chess, he is also an avid astrophotography enthusiast.

(5329) Decaro = 1989 YP
Discovered 1989 Dec. 21 by R. H. McNaught at Siding Spring.
Mario De Caro (b. 1963) is a philosopher who studies the philosophy of mind, the
  free-will controversy, and naturalism at the University of RomaTre. Mario has also
  worked on the mathematical philosophy of Galileo and argued that Galileo believed
  that reality is essentially mathematical. The name was suggested by T. Burbine.

(11092) Iwakisan = 1994 ED
Discovered 1994 Mar. 4 by T. Kobayashi at Oizumi.
Iwakisan is a composite volcano with a beautiful contour, dubbed the "Tsugaru Fuji."
  Located in the center of the Tsugaru plain, the mountain boasts a most exquisite
  landscape view from the peak. It is counted as one of the 100 most celebrated
  mountains in Japan.

(13957) NARIT = 1991 AG$_2$
Discovered 1991 Jan. 7 by R. H. McNaught at Siding Spring.
The National Astronomical Research Institute of Thailand (NARIT) was established
  in 2004 to commemorate the life and work of King Mongkut the "Father of Thai
  Science." NARIT operates a 2.4-m telescope and five regional telescopes in
  Thailand, and robotic telescopes in Chile, Australia, and China.

(19658) Sloop = 1999 RM$_{125}$
Discovered 1999 Sept. 9 by the Lincoln Laboratory Near-Earth Asteroid Research
  Team at Socorro.
Katie Michelle Sloop (b.1990) was a finalist in the 2003 Discovery Channel Young
  Scientist Challenge, a middle school science competition, for her botany and
  zoology project. She attended the St. Mary's School, Oak Ridge, Tennessee.

(52246) Donaldjohanson = 1981 EQ$_5$
Discovered 1981 Mar. 2 by S. J. Bus at Siding Spring.
American paleoanthropologist Donald Carl Johanson (b. 1943) discovered the fossil of
  a female hominin australopithecine known as "Lucy."

Table 2.9 (*cont.*)

(60000) Miminko = 1999TZ$_3$
Discovered 1999 Oct. 2 by L. Šarounová at Ondrejov.
"Miminko" is Czech word that expresses the unique stage of innocence at the beginning
of human life.

(115434) Kellyfast = 2003 TU$_2$
Discovered 2003 Oct. 5 by V. Reddy at Goodricke-Pigott.
Kelly E. Fast (b. 1968) is a program scientist for the MAVEN spacecraft at NASA.
She has measured planetary atmospheric compositions and winds, most notably the
ozone abundance on Mars, using the HIPWAC instrument at the NASA Infrared
Telescope Facility. She is a gifted singer, as featured in "Hotel Mauna Kea."

(236811) Natascharenate = 2007 RE$_{16}$
Discovered 2007 Sept. 12 by R. Gierlinger at Gaisberg.
Natascha Renate Gierlinger (b. 1997) is the daughter of the discoverer.

(284996) Rosaparks = 2010 LD$_{58}$
Discovered 2010 June 9 at WISE.
African-American Rosa Parks (1913–2005) refused to give up her seat to a white
passenger on a bus in Montgomery, Alabama in 1955. Her action forced the city to
end segregation of public buses and spurred efforts throughout the United States to
end segregation. She is known as the Mother of the Civil Rights Movement.

(289020) Ukmerge = 2004 TG$_{115}$
Discovered 2004 Oct. 12 by K. Cernis at Moletai.
Ukmerge is a city with 22 000 inhabitants in Vilnius County, Lithuania, located 78 km
northwest of Vilnius. Ukmerge was first mentioned in 1225. The city is located near
the confluence of the Vilkmerge River and the Sventoji River, and it obtained its
Magdeburg city rights in 1486.

(316201) Malala = 2010 ML$_{48}$
Discovered 2010 June 23 at WISE.
Malala Yousafzai (b. 1997) is a Pakistani human rights activist who advocates for the rights
of women and girls and worldwide access to education. She survived an assassination
attempt in 2012 and continued her activism. She won the Nobel Peace Prize in 2014.

(336694) Fey = 2010 AH$_{89}$
Discovered 2010 Jan. 8 at WISE.
Elizabeth Stamatina (Tina) Fey (b. 1970) is an American actor, writer, producer, and
comedian. She began her career with The Second City comedy troupe, then became
the head writer for Saturday Night Live, for which she has won numerous awards.

(432971) Loving = 2012 LJ$_{10}$
Discovered 2010 Feb. 11 at WISE.
Mildred (1939–2008) and Richard Loving (1933–1975) married in spite of anti-
miscegenation laws. They filed the lawsuit (Loving v. Virginia) that ultimately succeeded
in striking the laws down following a United States Supreme Court ruling in 1967.

(450931) Coculescu = 2008 EX$_{154}$
Discovered 2008 Mar. 11 by EURONEAR at the European Southern Observatory.
Nicolae Coculescu (1866–1952) was a Romanian astronomer who founded and
became the first director of the Bucharest Astronomical Observatory in 1908.
His interests and doctoral studies at Sorbona were in planetary perturbations and
celestial mechanics. He was also a professor at the University of Bucharest.

by the launching of the first artificial earth satellite on 1957 Oct. 4) since space is defined as beginning at an altitude of 100 000 meters above the Earth's surface.

When a number of sequential numbers are available (usually from discoveries by the same discover), these numbers are often given to individuals that have some type of relationship. There are minor planets (3350) Scobee, (3351) Smith, (3352) McAuliffe, (3353) Jarvis, (3354) McNair, (3355) Onizuka, and (3356) Resnik, which are named after the seven astronauts who died in the 1986 Challenger Space Shuttle disaster. There are minor planets (51823) Rickhusband, (51824) Mikeanderson, (51825) Davidbrown, (51826) Kalpanachawla, (51827) Laurelclark, (51828) Ilanramon, and (51829) Williemccool, which are named after the seven astronauts who died in the 2003 Columbia Space Shuttle disaster. There are minor planets (4147) Lennon, (4148) McCartney, (4149) Harrison, and (4150) Starr, which are named after the four members of the (8749) Beatles. There are minor planets (30439) Moe, (30440) Larry, and (30441) Curly plus (30444) Shemp, which are named after the members of the Three Stooges.

To honor science excellence among students, asteroids discovered by the LINEAR (Lincoln Near-Earth Asteroid Research) survey are named in honor of junior high school and high school students who have demonstrated excellence in select science competitions and their teachers (Evans et al., 2003). The exceptions are asteroids that have to follow a naming convention such as near-Earth asteroids. For example, (19658) Sloop in Table 2.9 is named after a finalist in a middle school science competition.

So what is the best way to get an asteroid named after you? Be successful and hopefully inspire an astronomer.

Questions

1) a) An asteroid has a semi-major axis of 3.2 AU? What is it orbital period?
   b) An asteroid has an orbital period of 5 years. What is its semi-major axis?
2) A body is found to have a semi-major axis of 3.08 AU, eccentricity of 0.31, and an inclination of 8.8°. Is its orbit more consistent with an asteroid or comet?
3) Show that an asteroid with a semi-major axis of 2.82 AU is in a 5:2 mean-motion resonance with Jupiter?
4) How are the diurnal and seasonal Yarkovsky effects different?
5) Why did the "Celestial Police" make observations on the ecliptic when they were looking for the missing planet?
6) a) What would be the provisional designation of the 105th minor planet discovered between May 16th and May 31st in 2016?
   b) What does provisional designation of 2016 TS$_{17}$ tell you about its discovery?

7) You are given the task to name a main-belt asteroid. Pick a name and write a citation. Follow all the minor planet naming rules. Make sure that the name has not been used previously and is not too close to a previously used name.

8) You are given the task to name a near-Earth asteroid. Pick a name and write a citation. Follow all the minor planet naming rules. Make sure that the name has not been used previously and is not too close to a previously used name.

9) You are given the task to name a trans-Neptunian object not in a 2:3 resonance with Neptune. Pick a name and write a citation. Follow all the minor planet naming rules. Make sure that the name has not been used previously and is not too close to a previously used name.

10) You are given the task to name a trans-Neptunian object in a 2:3 resonance with Neptune. Pick a name and write a citation. Follow all the minor planet naming rules. Make sure that the name has not been used previously and is not too close to a previously used name.

3

Meteorites, Minerals, and Isotopes

3.1 Meteorites

Relatively small fragments of asteroids exist on Earth as meteorites. Over 50 000 meteorites have been discovered to date. The importance of studying meteorites is that they allow the chemical and physical properties of asteroids to be studied in much more detail than these properties can be determined from Earth- or space-based observations. Meteorites show the geological diversity of the asteroid belt since they sample bodies that range from being fully differentiated to those that experienced minimal heating. Only meteorites allow absolute formation ages of asteroids to be determined.

3.1.1 Identifying a Meteorite

But how can a rock be determined to be a meteorite and not a terrestrial rock if you do not see the object fall from the sky? Classifying a rock as a meteorite is done by identifying surface features that are due to the effects of passing through the Earth's atmosphere and by finding features in the interior of the rock that could "only" form on another Solar System body besides Earth. Rocks that are thought to be meteorites but are not are often called "meteor-wrongs."

One identifying feature is the presence of fusion crust (Figure 3.1) on the outside of the meteorite. Fusion crust is due to the melting of the outer surface of the meteorite as it passes through the atmosphere. Also regmaglypts, which are oval depressions, are often visible on the surfaces of meteorites (Figure 3.2).

High abundances of metallic iron will also often cause a meteorite to be denser than a typical terrestrial rock. Another way to identify a meteorite is to test for the presence of nickel. Most meteorites contain metallic iron, which tends to be significantly enriched in nickel compared to terrestrial rocks. Most of the Earth's nickel is in the Earth's core since nickel is a siderophile ("iron-loving") element and tends

Figure 3.1 Image of ordinary (H4) chondrite Marilia. Note the distinctive black fusion crust. Credit: Gabisfunny.

Figure 3.2 Image of iron meteorite Murnpeowie. Note the regmaglypts on its surface. Credit: James St. John.

Figure 3.3 Slice of CV3 chondrite Allende. Note the chondrules present throughout the meteorite. Credit: Basilicofresco.

to bond with metallic iron, and most of the Earth's iron is in the core. Chondrules
(small round grains) are often present in the meteorite (Figure 3.3). Chondrules are
mm-sized grains that formed through rapid cooling of molten material.

3.1.2 Falls and Finds

Meteorites are identified as either falls of finds. Falls are meteorites that have been
collected after their passage through the atmosphere was observed. Finds are meteorites that have been found on or in the ground without their fall being observed.
Due to a shorter exposure to the Earth's atmosphere and to liquid water, falls are
expected to be more "pristine" than finds and have compositions more similar to
what would be found on their original parent body. Approximately 1300 meteorites
have been classified as falls to date. The largest known intact fall is the 60 000 kg
Hoba iron meteorite (Beech, 2013).

Objects have been known to fall from the sky since the beginning of man-
kind. Many cultures (e.g., Egyptian, Chinese, Greek, Roman) have descriptions
of meteors and meteorites dating back thousands of years (Dalrymple, 1991).
Meteorites were often worshiped. Meteorites have been found at Indian burial
grounds. Meteoritic metal has also been used as tools and weapons. A iron dagger
blade found at King Tutankhamun's (c. 1341 BC–1323 BC) tomb was found to be
meteoritic in origin due to its high nickel content (Comelli et al., 2016).

The first meteorite fall in the Western world with preserved fragments was the Ensisheim meteorite that fell in 1492 in Alsace, France (Marvin, 1992). A large explosion in the sky was heard as far as ~100 miles (~161 km) away. A ~280 pound (~127 kg) black stone was later recovered and stored in a local church. This fall was publicized with woodcuts. A poem was written by Sebastian Brant (1457–1521) that claimed the fall was a good omen for the "King of the Romans" (title used by the German king) Maximillian I (1459–1519).

However, most scholars during these times did not believe that rocks could fall from the sky. The person given credit for gaining scientific acceptance for meteorites being extraterrestrial objects was physicist Ernst Chladni (1756–1827). In 1794, Chladni published a book that proposed that meteorites originated in space (Marvin, 1996). The book was titled *Über den Ursprung der von Pallas gefundenen und anderer ihr ähnlicher Eisenmassen und über einige damit in Verbindung stehende Naturerscheinungen* (German for "On the Origin of the Pallas Iron and Others Similar to it, and on Some Associated Natural Phenomena"). Chladni argued that masses of stone and iron can fall from the sky as seen by observed fireballs. His hypothesis was validated by a number of witnessed falls that occurred between 1794 and 1798. Chemist Edward Charles Howard (1774–1816) argued that the high Ni contents found in the metal of his analyzed meteorites were evidence that they were unlike terrestrial rocks. What cemented that rocks could fall from space was the L'Aigle meteorite shower in Normandy, France in 1803 where over 3000 fragments fell from the sky. Jean-Baptiste Biot (1774–1862) investigated and concluded that the fragments were extraterrestrial.

3.1.3 Discovering Meteorites

Meteorites fall randomly all over the world. Iron meteorites tend to be the easiest to find since they tend to appear anomalous compared to terrestrial rocks and tend to be much more resistant to terrestrial weathering than the silicate-rich meteorites.

However, the best places to find meteorites on the Earth are deserts, since meteorites tend to appear different from the local rocks and tend to weather more slowly there due to the lack of liquid water. Antarctica, which is considered a desert due to its lack of precipitation, is considered the premier place to find meteorites (Figure 3.4) since the meteorites are well preserved in the ice and also the glaciers move and collect the meteorites in specific locations. These locations are called blue ice fields. These areas appear blue because air bubbles in the ice have been squeezed out, which causes the ice crystals to enlarge. These large ice crystals absorb all visible wavelengths of light (except blue) more efficiently so the ice will appear blue from the reflected light. The glaciers tend to move until they reach an obstacle, such as a mountain. This movement tends to collect the

Figure 3.4 A meteorite being collected in Antarctica by the ANSMET team. A picture is being taken of the meteorite in the ice. Credit: US Antarctic Search for Meteorites program/Katie Joy. (A black and white version of this figure will appear in some formats. For the color version, please refer to the plate section.)

meteorites in one place. Katabatic (downslope) winds erode the ice, exposing the meteorites. Meteorites (which tend to be gray in color) are also easier to spot in the ice. Evatt et al. (2016) argues that the underrepresentation of collected iron and stony-iron meteorites in Antarctica compared to falls is due to heating of these meteorites due to solar radiation, which would cause them to sink below the ice surface.

Since 1969, Japanese expeditions to Antarctica have discovered over 15 000 meteorites. Japanese meteorites are stored and curated at the National Institute of Polar Research (NIPR) (Kojima, 2006). The 1969 discovery of nine meteorites was accidental and found four different types of meteorites. This diversity indicated something special about Antarctica as a place to collect meteorites. A 1974 Japanese expedition found 663 meteorites over only two weeks. These 663 meteorites also included a number of diverse types, cementing Antarctica's place as the premier place to find meteorites. Approximately 35 000 meteorites have been recovered from Antarctica, which represents 66% of the world's collected specimens (Evatt et al., 2016).

ANSMET (Antarctic Search for Meteorites) is the annual American expedition (Cassidy, 2003; Marvin et al., 2015) which has recovered over 20 000 meteorites in Antarctica since 1976 (Righter, 2016). It is hard to determine how many of these meteorites are from distinct falls since many of the meteorites are paired, meaning that they originated from the same body before it broke up before or after impact. An area that contains meteorites from a single fall is called a strewn field. Other deserts, such as the Sahara, are also prime places to find meteorites, since the lack of precipitation allows the meteorites to survive longer before they disintegrate due to weathering and the slower surface changes allows meteorites to stay visible longer. The meteorites that are collected by ANSMET are first transferred to the Antarctic Meteorite Curation Facility at NASA Johnson Space Center (JSC) in Houston, Texas.

Meteorites are named for the geographic locality where they are found such as a nearby town or geographic feature (such as a river or mountain). For example, the New Orleans meteorite fell in New Orleans, Louisiana while the Whetstone Mountains meteorite was found in the Whetstone mountain region. However, special designations are used for localities where significant numbers of meteorites are recovered. Meteorites from Antarctica are given a name that indicates a geographical locality (such as Allan Hills) then an identification number (e.g., 84001), which would be written as Allan Hills 84001 or ALH 84001. The identification number includes a two digit number indicating the start of the field season where the meteorite was collected (e.g., 84 for 1984) and then a three digit number (including zeros) to indicate the order of when it was collected (e.g., 001 for the first meteorite collected during the field season). The Martian meteorite Allan Hills 84001 (or abbreviated as ALH 84001) was the first meteorite collected from Allan Hills during the 1984–1985 Antarctic field season. Japanese Antarctic meteorites have a similar naming system (e.g., Yamato 980003 or Y-980003) but use a dash between the letter and the number.

At JSC, the Antarctic meteorites are first removed from the Teflon bags that they are collected in. Pictures of the meteorites are taken (Figure 3.5). Thin sections are made. A thin section is a ~30 μm thick slice of a meteorite that is glued to a glass slide. This thin width for the slice allows light to pass through the meteorite so

Figure 3.5 Image of Antarctic CM2 chondrite LAP 03718. Credit: NASA.

it can be studied using a petrographic microscope. Petrographic microscopes are different than typical microscopes since they have a setting that just allows polarized light to pass through the thin section. Many different minerals appear differently when viewed using polarized and polarized light (Nesse, 2012). Most of the ANSMET meteorites are ultimately transferred to the Smithsonian Institution in Washington, D.C.

Originally, meteorites from Morocco and adjacent countries were named Northwest Africa (NWA) and then given a number (e.g., NWA 796) unless reasonable proof of the find location was known. Starting in 2015 with NWA 10001, any meteorite thought to be originating from Morocco, Western Sahara, Algeria, Tunisia, Mauritania, or Mali may now be given the NWA acronym. Table 3.1 lists commonly used special designations for meteorites found in dense collection areas.

The Meteorite Nomenclature Committee of the Meteoritical Society is responsible for the naming of all meteorites. The Meteoritical Society is a non-profit organization founded in 1933 to promote the study of extraterrestrial materials. For a meteorite to receive an official name, part of the meteorite's mass (20% of the

Table 3.1 *Commonly used designations for the names of meteorites from dense collection areas*

Name	Abbreviation	Country or region
Acfer		Algeria
Allan Hills	ALH	Antarctica
Asuka	A	Antarctica
Catalina		Chile
Dar al Gani	DaG	Libya
Dhofar	Dho	Oman
Dominion Range	DOM	Antarctica
El Médano		Chile
Elephant Moraine	EET	Antarctica
Frontier Mountain	FRO	Antarctica
Graves Nunataks	GRA	Antarctica
Grosvenor Mountains	GRO	Antarctica
Grove Mountains	GRV	Antarctica
Hammadah al Hamra	HaH	Libya
Jiddat al Harasis	JaH	Oman
LaPaz Icefield	LAP	Antarctica
Larkman Nunatak	LAR	Antarctica
Lewis Cliff	LEW	Antarctica
Lucerne Valley	LV	California, USA
MacAlpine Hills	MAC	Antarctica
Meteorite Hills	MET	Antarctica
Miller Range	MIL	Antarctica
Northwest Africa	NWA	Northwest Africa
Patuxent Range	PAT	Antarctica
Pecora Escarpment	PCA	Antarctica
Queen Alexandra Range	QUE	Antarctica
Ramlat as Sahmah	RaS	Oman
Reckling Peak	RKP	Antarctica
Roach Dry Lake	RhDL	Nevada, USA
Roberts Massif	RBT	Antarctica
Roosevelt County	RC	New Mexico, USA
Sahara		Sahara
Sayh al Uhaymir	SaU	Oman
Yamato	Y	Antarctica

total mass or 20 g, whichever is the smaller amount) must be deposited in a well-curated meteorite collection.

The Meteoritical Bulletin Database is maintained by the Meteorite Nomenclature Committee and contains records of all known meteorites. These records include classifications, place and year of discovery, and whether it was observed to fall.

The law regarding who owns a meteorite after its discovery depends on the country (Schmitt, 2002). In the United States, the landowner owns the meteorite find. A find on federal government property is owned by the federal government and can be acquired by the Smithsonian Institution. Meteorites are considered part of the 1906 Antiquities Act, which was passed to protect archaeological sites on federal land, since they are "objects of scientific interest." For example, two prospectors searching for a legendary lost Spanish gold mine found the Old Woman iron meteorite (~2753 kg) on federal land (Plotkin et al., 2012). However because of this Antiquities Act, a lawsuit from the prospectors for ownership of the meteorite was dismissed in court. The main mass of the meteorite was later loaned to the Bureau of Land Management's Desert Discovery Center in Barstow, California, while a large slice was kept for study and curation at the Smithsonian.

3.2 Minerals

All known natural stable elements are found in meteorites. These elements tend to be found within minerals. A mineral is a naturally occurring, homogeneous inorganic solid having a definite chemical composition and characteristic crystalline structure. This crystalline structure has a repeating pattern, which is called a crystal lattice.

True minerals cannot be made through biologic activity. For example, ice found naturally would be a mineral, while ice produced in a freezer would not be a mineral. The Commission on New Minerals and Mineral Names of the International Mineralogical Association is in charge of approving all new mineral names (De Fourestier, 2002). The choice of name for a new mineral is primarily the responsibility of the discoverer. New minerals can be named after people (living or dead). For example, joegoldsteinite ($MnCr_2S_4$)] is named after iron meteorite researcher Joseph Goldstein (1939–2015) (Isa et al., 2015).

Minerals can be broken up into a variety of classes such as silicates, native elements, sulfides, halides, and oxides. The names of these classes reflect the element that is part of the crystal structure of each member of the class. Except for the native elements, these minerals contain cations (positively charged ions) and anions (negatively charged ions).

Mineral formulas are written to give the relative atomic proportions of different elements in a mineral. The relative atomic proportions are usually given as integers. Cations precede anions in the formula. Polymorphs are minerals with the same chemical formula but different crystal structures.

3.2.1 Mineral Properties

Because different minerals have different chemical compositions and different crystal structures, they will also have different physical properties (e.g., hardness, cleavage). Over 300 minerals (Rubin, 1997a, 1997b) have been identified in meteorites out of the approximately ~5000 known minerals that are found on Earth. Examples of extraterrestrial minerals found in meteorites are listed in Table 3.2. Minerals common in terrestrial rocks but relatively rare in meteorites include quartz (SiO_2), calcite ($CaCO_3$), and biotite group members [$K(Mg,Fe)_3(AlSi_3O_{10})(F,OH)$].

3.2.2 Silicates

Silicates are the most abundant type of mineral in the Earth's crust and [with iron–nickel (FeNi) alloys] one of the most abundant minerals in meteorites. Silicates contain both silicon and oxygen. The basic unit of all silicates is the $(SiO_4)^{4-}$ tetrahedron where the four oxygen ions surround the silicon atom. The silicates have a wide variety of structures due to the possible sharing of different numbers of oxygens of the tetrahedron. Mafic silicates are rich in magnesium (*ma*) and/or iron (*fic* from ferric). Felsic silicates are less rich in magnesium and/or iron and are named after feldspar (*fel*) and silica (*sic*).

3.2.2.1 Olivine

Olivine group minerals, which have an isolated $(SiO_4)^{4-}$ tetrahedron, are one of the commonest types of silicates in meteorites. Large olivine grains usually have an olive-green color, which is the reason for its name. The primary olivine minerals found in meteorites are fayalite (Fe_2SiO_4) and forsterite (Mg_2SiO_4). Fayalite is often abbreviated as Fa, while forsterite is often abbreviated as Fo. The $(SiO_4)^{4-}$ tetrahedrons are not directly linked to each other (do not share electrons) and instead are linked by the cations (e.g., Fe^{2+}, Mg), which are found in octahedral sites. Ca-rich olivines include monticellite ($CaMgSiO_4$) and kirschsteinite ($CaFeSiO_4$).

Olivine is usually found as a solid solution between fayalite and forsterite. Minerals that are part of a solid solution series have elements that can substitute relatively freely for each other due to similar ionic radii and ionic charges. For fayalite and forsterite, Fe^{2+} and Mg^{2+} can substitute freely for each other due to their similar ionic radii and ionic charges (+2).

Olivine compositions are usually written in terms of fayalite (or forsterite) mole percentage (mol%). The mass of one mole of a substance is equal to the mean molecular mass of the substance in grams. The mole percentage is the same quantity as the atomic percentage, which is the percentage of one atom relative to the total number of atoms. For example, an olivine grain that is Fa_{10} (or Fo_{90}) contains 10 mol% Fe and 90 mol% Mg.

3.2.2.2 Pyroxene

Pyroxenes are another common type of silicate mineral. Pyroxenes generally have the formula $XYSi_2O_6$ where X and Y are the two cation sites. X is called

Table 3.2 *Examples of extraterrestrial minerals found in meteorites*

Mineral	Formula	Group	Class
alabandite	MnS	galena	sulfide
albite	$NaAlSi_3O_8$	feldspar	silicate
anorthite	$CaAl_2Si_2O_8$	feldspar	silicate
antigorite	$Mg_3Si_2O_5(OH)_4$	serpentine	silicate
augite	$(Ca,Na)(Mg,Fe,Al,Ti)$ $(Si,Al)_2O_6$	clinopyroxene	silicate
calcite	$CaCO_3$	calcite	carbonate
chamosite	$(Fe^{2+},Mg,Fe^{3+})_5Al(Si_3Al)$ $O_{10}(OH,O)_8$	chlorite	silicate
chromite	$FeCr_2O_4$	spinel	oxide
chrysotile	$Mg_3Si_2O_5(OH)_4$	serpentine	silicate
clinochlore	$(Mg,Fe)_5Al(Si_3Al)$ $O_{10}(OH)_8$	chlorite	silicate
corundum	Al_2O_3	corundum-hematite	oxide
cronstedtite	$Fe^{2+}_2Fe^{3+}_2SiO_5(OH)_4$	serpentine	silicate
diamond	C	carbon polymorph	native element
diopside	$CaMgSi_2O_6$	clinopyroxene	silicate
dolomite	$CaMg(CO_3)_2$	dolomite	carbonate
enstatite	$Mg_2Si_2O_6$	orthopyroxene	silicate
fayalite	Fe_2SiO_4	olivine	silicate
ferrosilite	$Fe_2Si_2O_6$	orthopyroxene	silicate
forsterite	Mg_2SiO_4	olivine	silicate
graphite	C	carbon polymorph	native element
halite	$NaCl$	halite	halide
hedenbergite	$CaFeSi_2O_6$	clinopyroxene	silicate
hypersthene	$(Mg,Fe)_2Si_2O_6$	orthopyroxene	silicate
ilmenite	$FeTiO_3$	ilmenite	oxide
kamacite	$Fe_{0.9}Ni_{0.1}$	iron–nickel	native element
kaolinite	$Al_2Si_2O_5(OH)_4$	clay	silicate
kirschsteinite	$CaFe(SiO_4)$	olivine	silicate
lizardite	$Mg_3Si_2O_5(OH)_4$	serpentine	silicate
magnetite	$Fe^{3+}_2Fe^{2+}O_4$	spinel	oxides
melilite	$(Ca,Na)_2(Al,Mg,Fe)$ $(Si,Al)_2O_7$	melilite	silicate
niningerite	$(Mg,Fe,Mn)S$	galena	sulfide
oldhamite	$(Ca,Mg)S$	galena	sulfide
orthoclase	$KAlSi_3O_8$	feldspar	silicate
pentlandite	$(Fe,Ni)_9S_8$	pentlandite	sulfide
perovskite	$CaTiO_3$	perovskite	oxide
pigeonite	$(Ca,Mg,Fe)(Mg,Fe)Si_2O_6$	clinopyroxene	silicate
pyrrhotite	$Fe_{(1-x)}S$, $x = 0 - 0.2$	pyrrhotite	sulfide
ringwoodite	$Mg_2(SiO_4)$	olivine	silicate
saponite	$Ca_{0.1}Na_{0.1}Mg_{2.25}Fe_{0.75}Si_3Al$ $O_{10}(OH)_2•4(H_2O)$	clay	silicate
spinel	$MgAl_2O_4$	spinel	oxide

Mineral	Formula	Group	Class
sylvite	KCl	halite	halide
taenite	$Fe_{0.8}Ni_{0.2}$	iron–nickel	native element
tetrataenite	$Fe_{0.5}Ni_{0.5}$	iron–nickel	native element
tochilinite	$6Fe_{0.9}S \bullet 5(Mg,Fe)(OH)_2$	hydrated sulfide	sulfide
troilite	FeS	troilite	sulfide

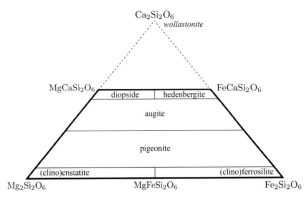

Figure 3.6 Image of the pyroxene quadrilateral. The three apexes of the triangle are the $Mg_2Si_2O_6$ (enstatite), $Fe_2Si_2O_6$ (ferrosilite), and wollastonite ($Ca_2Si_2O_6$) compositions. The pyroxene compositions in the upper part of the triangle do not exist in nature. Clinoenstatite and clinoferrosilite are polymorphs of enstatite and ferrosilite, respectively. Credit: Angrense.

the M2 site and can contain ions such as Fe^{2+}, Mg^{2+}, and Ca^{2+}, while Y is called the M1 site and can contain ions such as Fe^{2+} and Mg^{2+}. The M2 site is slightly larger than the M1 site. Al^{3+} can also substitute for Si^{4+}, but another substitution must occur to compensate for the difference in charge. Pyroxenes can be broken up into the orthopyroxene and clinopyroxene groups. Orthopyroxenes contain low contents of Ca and have an orthorhombic crystal structure, while clinopyroxenes contain high contents of Ca and have a monoclinic crystal structure. Orthopyroxenes include enstatite ($Mg_2Si_2O_6$), ferrosilite ($Fe_2Si_2O_6$), and hypersthene [$(Mg,Fe)_2Si_2O_6$]. Clinopyroxenes include pigeonite [$(Ca,Mg,Fe)(Mg,Fe)Si_2O_6$], diopside ($MgCaSi_2O_6$), augite [$(Ca,Na)(Mg,Fe,Al,Ti)(Si,Al)_2O_6$], and hedenbergite ($CaFeSi_2O_6$).

The variation in pyroxene compositions can be seen in the pyroxene quadrilateral (Figure 3.6). The pyroxene quadrilateral is the region where pyroxenes are stable with specific compositions in a pyroxene ternary diagram.

3.2.2.3 Feldspar

Another group of silicate minerals found in meteorites are feldspars. Feldspar minerals include anorthite ($CaAl_2Si_2O_8$), albite ($NaAlSi_3O_8$), and orthoclase ($KAlSi_3O_8$). Solid solutions between albite and anorthite are called plagioclase feldspar, while solid solutions between orthoclase and albite are called alkali feldspar.

3.2.2.4 Phyllosilicates

Phyllosilicates (also called sheet silicates) have the basic structural unit of Si_2O_5 and contain hydroxyl (OH) or water (H_2O) in their crystal structure. Depending on the crystal structure, phyllosilicates are subdivided into a variety of groups (serpentine, clay mineral, mica, chlorite).

A wide variety of serpentine group minerals have been found in meteorites. Antigorite, chrysotile, and lizardite are polymorphs of each other and have the formula $Mg_3Si_2O_5(OH)_4$. Cronstedtite contains Fe^{3+} and has the formula $Fe^{2+}_2Fe^{3+}_2Si$ $O_5(OH)_4$. Clay minerals found in meteorites include kaolinite [$Al_2Si_2O_5(OH)_4$] and saponite[$Ca_{0.1}Na_{0.1}Mg_{2.25}Fe_{0.75}Si_3AlO_{10}(OH)_2\bullet4(H_2O)$]. Chlorite minerals found in meteorites include chamosite [$(Fe^{2+},Mg,Fe^{3+})_5Al(Si_3Al)O_{10}(OH,O)_8$] and clinochlore [$(Mg,Fe^{2+})_5Al(Si_3Al)O_{10}(OH)_8$].

3.2.3 Native Elements

3.2.3.1 FeNi

Native elements include minerals that contain only one element and alloys, which are minerals composed of two or more metals. The most abundant alloys in meteorites are kamacite (FeNi) and taenite (FeNi). Kamacite contains 5–10 wt% Ni, while taenite contains more than 20 wt% Ni. These two alloys also have different crystal structures, with kamacite having a body-centered cubic structure, while taenite has a face-centered cubic structure. A body-centered cubic (bcc) structure has atoms at each corner of the cube with also an atom at the center, while the face-centered cubic (fcc) structure has atoms at each corner and at the center of each face of the cube. A fine-grained mixture of kamacite and taenite is called plessite.

In many iron meteorites, the kamacite and taenite are found together in what is called a Widmanstätten pattern (or sometimes called the Thomson structure) (Figure 3.7). Guglielmo Thomson (1760–1806) in 1804 and Count Alois von Beckh Widmanstätten (1753–1849) in 1808 independently discovered the Widmanstätten pattern. The Widmanstätten pattern is the interweaving of kamacite and taenite bands that is apparent in some irons when the meteorite is sliced and washed with a weak acid solution (such as nitric acid). Ni is slightly more resistant to the acid

Figure 3.7 Example of Widmanstätten pattern in the iron meteorite Harriman (Of). The light bands are kamacite, while the darker bands are taenite. A large troilite inclusion is apparent. Credit: Chuck Sutherland.

than Fe, so the taenite (which is richer in Ni) is more resistant to the etching the kamacite (which is poorer in Ni). On the etched meteorite, the light bands are kamacite, while the darker bands are taenite.

The Widmanstätten pattern forms by the nucleation and growth of kamacite from taenite during slow cooling (Goldstein et al., 2009). As the metallic iron decreases in temperature, more kamacite starts growing into the surrounding taenite, which forms the Widmanstätten pattern. To determine cooling rates of iron meteorites, the growth of the Widmanstätten pattern and the distribution of Ni in the kamacite and taenite must be modeled. Typical cooling rates for fractionally crystallized irons are ~100–10000 °C/Ma for time periods of 10 Ma or less at temperatures of 500–700 °C (Goldstein et al., 2009).

3.2.3.2 Native Elements

The commonest native element found in meteorites is graphite (C). Graphite is very dark (opaque). Diamonds have also been identified in meteorites such as ureilites (Kunz, 1888; Urey, 1956; Nagashima et al., 2012), carbonaceous chondrites (Kebukawa et al., 2014), and iron meteorites (Foote, 1891). The diamonds tend to have sizes that range from nanometers to microns (Karczemska, 2010).

3.2.4 Sulfides

Sulfides contain S^{2-} as the anion. Common sulfides found in meteorites include troilite (FeS), pyrrhotite $(Fe_{1-x}S)$ ($x = 0 - 0.17$), and pentlandite $[(Fe,Ni)_9S_8]$. All the minerals tend to be very opaque (dark).

3.2.5 Halides

Halides contain halogen ions (e.g., F^-, Cl^-) as the anion. Halides include halite (NaCl) and sylvite (KCl) but these minerals are relatively rare in meteorites but could be commoner on asteroid surfaces. The presence of these minerals in meteorites was originally very controversial because they were always found as microscopic grains that could have been terrestrial contamination. However, relatively large grains (up to 3 mm in diameter) of purple halite were discovered in the H5 chondrite fall Monahans (1998) (Zolensky et al., 1999), which was broken apart with a hammer. [Monahans (1998) was the second meteorite to fall in Monahans, Texas and that is why it is given the (1998) designation.] Backscattered electron imaging discovered sylvite grains within the halite. Halides may be very common in chondritic material in space but are possibly being destroyed through exposure to water after fragments of these bodies land on Earth.

3.2.6 Oxides

Oxides contain O^{2-} as the anion. Common oxides in meteorites include those in the spinel group and melilite group. Spinel group minerals have the general formula AB_2O_4 with B being Al for aluminum spinels, Fe for iron spinels, and Cr for chromium spinels. The spine group mineral spinel has the formula $MgAl_2O_4$, magnetite has the formula $Fe^{3+}_2Fe^{2+}O_4$, and chromite has the formula $FeCr_2O_4$. Melilite group minerals have the general formula $X_2Y(Z_2O_7)$. The mineral melilite has the formula $(Ca,Na)_2(Al,Mg,Fe)(Si,Al)_2O_7$.

3.2.7 Organic Compounds

Organic compounds are molecules that contain carbon (and almost always hydrogen) but are not considered minerals. A wide variety of organic molecules have been identified in meteorites (e.g., Sephton, 2002; Martins, 2011). These organic compounds can also contain oxygen, nitrogen, and sulfur (Pizzarello et al., 2006; Pizzarello and Shock, 2010). This complex and insoluble macromolecular material has compositions similar to $C_{100}H_{70}N_3O_{12}S_2$ (Hayatsu et al., 1980) and $C_{100}H_{46}N_{10}O_{15}S_{4.5}$ (Pizzarello et al., 2001). Carboxylic acids and (Martins, 2011) and polycyclic aromatic hydrocarbons (PAHs) have been identified in meteorites (e.g., Elisa et al., 2005). Carboxylic acids contain a carboyxl group [C(O)OH], while PAHs contain only carbon and hydrogen and are composed of multiple rings.

Amino acids (Kvcnvolden et al., 1970) and nucleobases (Hayatsu et al., 1975; Martins et al., 2008; Chan et al., 2016) have also been identified in meteorites. Amino acids are the building blocks of proteins, while nucleobases are the building blocks of DNA (deoxyribonucleic acid) and RNA (ribonucleic acid). Besides carbon and hydrogen, amino acids contain $-NH_2$ and $-COOH$ functional groups. Nucleobases are one-ring or two-ring organic compounds containing nitrogen atoms. Amino acids have been identified in a large number of different types of meteorites (Burton et al., 2011, 2013, 2015). These organic-bearing meteorites may have supplied the Earth with the ingredients for life.

3.3 Forming Minerals

Minerals initially form through ions coming together in a variety of environments. What first occurs is nucleation where a small number of the ions form the crystal lattice pattern that is characteristic of the resulting mineral's crystal structure. More ions will then start coming together, which causes the crystal to grow larger.

3.3.1 Condensation

One way a mineral forms is through condensation in the solar nebula. The Solar System is generally thought to have formed from the gravitational collapse of a fragment of a giant molecular cloud ~4.6 billion years ago. These giant molecular clouds get their names from their vast size and low temperatures, which allows for the formation of molecular hydrogen (H_2). The cloud is primarily composed of hydrogen with a much lower abundance of helium. Minerals can sublimate from a gas.

Grossman (1972) did the first detailed calculations of the condensation sequence for different minerals for a cooling gas with a solar composition at a fixed low pressure. This condensation sequence is shown in Table 3.3. Later calculations with different starting compositions, at different pressures, and using more elaborate models have the roughly the same condensation sequence (Ebel, 2006); however, the temperatures where different minerals condense (and also disappear) will differ.

As the temperature decreases, the minerals that form tend to be more oxidized (less reduced). A more oxidized mineral has cations with higher oxidation states. The oxidation state is the total number of electrons that an atom will either gain or lose to form a chemical bond to another atom. Magnetite, which contains Fe^{3+}, is more oxidized than ferrous olivines and pyroxenes, which contain Fe^{2+}. These minerals are more oxidized than FeNi, which contain Fe^0.

Table 3.3 *Condensation sequence for different minerals from Grossman (1972)*
for a gas of solar composition at 10^{-3} atm. The temperatures in parentheses are
the temperatures where the condensate disappears.

Temperature (K)	Mineral
1758 (1513)	corundum (Al_2O_3)
1647 (1393)	perovskite ($CaTiO_3$)
1625 (1450)	melilite group ($Ca_2Al_2SiO_7$–$Ca_2MgSi_2O_7$)
1513 (1362)	spinel ($MgAl_2O_4$)
1471	FeNi
1450	diopside ($CaMgSi_2O_6$)
1444	forsterite (Mg_2SiO_4)
1362	anorthite ($CaAl_2Si2O_8$)
1349	enstatite (Mg_2Si2O_6)
<1000	alkali-bearing feldspar [$(Na,K)AlSi_3O_8$–$CaAl_2Si_2O_8$]
<1000	ferrous olivines [$(Mg,Fe)_2SiO_4$], ferrous pyroxenes [$(Mg,Fe)_2Si_2O_6$]
700	troilite (FeS)
405	magnetite (Fe_3O_4)

3.3.2 CAIs

The high-temperature condensates (corundum, perovskite, melilite, spinel, diop-side, anorthite) are major components of calcium–aluminum-rich inclusions (CAIs) (Figure 3.2), which are refractory inclusions found in almost all chondritic meteorites (e.g., MacPherson et al., 2005). Refractory inclusions primarily contain minerals with high vaporization temperatures (Connolly, 2005). CAIs have diam-eters on the order of a mm to a cm in size. CAIs tend to have the oldest calculated ages of any analyzed material from our Solar System.

MacPherson et al. (2005) notes that CAIs must have been isolated from the nebular gas or they would have reequilibrated with the cooling gas and potentially formed new mineral phases. MacPherson et al. (2005) also notes that most CAIs (and potentially all) have been reprocessed from their original composition by pro-cesses such as melting, which altered the original precursor material.

CAIs are typically broken up into type A, B, and C (Scott and Krot, 2005). Type A are melilite-spinel rich. Type A CAIs are also subdivided into fluffy and compact. Fluffy Type A CAIs have much more complex structures and experi-enced more heating (Kornacki and Cohen, 1983). Type B CAIs are composed of pyroxene, melilite, spinel, and sometimes plagioclase feldspar. There are also forsterite-bearing type B. Type C CAIs are melilite-poor and pyroxene–plagioclase rich.

3.3.3 AOAs

Amoeboid olivine aggregates (AOAs) are another type of refractory inclusion (e.g., Komatsu et al., 2001). AOAs are fine-grained irregularly shaped objects that have sizes comparable to chondrules and CAIs. They have mineralogies of forsterite and FeNi plus a refractory component (e.g., diopside, spinel, anorthite). The mineralogy of AOAs is consistent with being a high-temperature condensate.

3.3.4 Metamorphism

Minerals on the parent body can also be produced through metamorphism of existing minerals (Krot et al., 2006; Huss et al., 2006). Metamorphism can change a mineral into another mineral through alteration due to heat, fluids, and/or pressure (shock through impact). Almost all chondritic meteorites have experienced varying degrees of thermal and/or aqueous alteration that have altered the original minerals that condensed out of the solar nebula. Aqueous alteration of meteorites produces minerals such as phyllosilicates, magnetite, pyrrhotite, pentlandite, and carbonates (Bullock et al., 2005; Krot et al., 2006).

The composition of a mineral can also be changed through metamorphism (Huss et al., 2006). As the temperature increases, the rate of diffusion of different cations increases between minerals, which tends to equilibrate the compositions of the minerals.

Impacts can also form new minerals (Sharp and DeCarli, 2006). Stöffler et al. (1991) did a shock classification of ordinary chondrites using the effects on olivine and plagioclase. At shock pressures between ~30–35 and ~45–55 GPa, plagioclase is converted to maskelynite (a glass). Glass is amorphous (non-crystalline) and cooled extremely quickly, while maskelynite is a glass formed out of plagioclase feldspar (Chen and El Goresy, 2000). At shock pressures between ~45–55 and ~75–90 GPa, plagioclase is converted to normal glass and olivine is converted to ringwoodite. Whole-rock melting occurs at pressures greater than ~75–90 GPa

3.3.5 Melting and Crystallization

Another way to form minerals is through the cooling of a melt, which allows minerals to crystallize. Cooling from a melt could occur after flash melting in the solar nebula or after melting on the original parent body. Olivine, pyroxene, and feldspar can all be formed through cooling of a melt. The faster the melt cools, the smaller the crystal size of the resulting mineral.

On their original parent bodies, almost all chondrites experienced heating, while achondrites, primitive achondrites, irons, and stony-irons were heated to the point of melting. Achondrites and most irons formed on bodies that differentiated. Differentiation is when denser material (e.g., metallic iron) sinks to the core, while lighter material (e.g., plagioclase feldspar) rises to the surface.

Melting occurs when the temperature rises high enough for the bonds holding a mineral together start to break. Exactly when a mineral melts will be a function of the temperature and also the pressure it experiences with each mineral melting at different temperatures and pressures. The melting point of a mineral is also the same temperature where minerals will crystallize during cooling. However, rocks melt at lower temperatures than the melting temperatures of their constituent minerals. After melting, the material will then cool and minerals will start to crystallize.

Partial melting is the dominant process during the differentiation of planetary bodies. Differentiation is the result of gravity where denser molten material migrates toward the core and lighter molten material rises toward the surface. On Earth, differentiation forms a basaltic or andesitic crust, peridotite mantle, and a metallic iron core. Basalt, andesite, and peridotite are all igneous rock types. Basalt and andesite are rich in feldspar, while peridotite is rich in olivine.

Partial melting occurs because different minerals have different melting temperatures. Fractional crystallization is the segregation of minerals that crystalize out of a melt. Cumulates are igneous rocks that form through the accumulation of these crystals.

3.3.6 *Chondrules*

Chondrules are produced by the quenching (rapid cooling) of material that has flash melted. Most chondrites are primarily composed of chondrules, matrix, and CAIs with chondrules usually making up 15–80% of a chondrite by volume (Friedrich et al., 2015). Chondrules are often broken up into type I and II (Hewins, 1997). Type I are FeO-poor and volatile-poor, while type II are FeO-rich and approximately solar in composition.

Numerous scenarios have been invoked to produce chondrules and there is no consensus on their formation mechanism. Ciesla (2005) notes that any chondrule formation theory must explain a number of chondrule properties. These properties include chondrules generally being 0.1–1 mm in diameter, they make up a large volume of chondritic meteorites, the chondrule precursor material was at rather low temperatures, they were rapidly heated to ~1700–2100 K, they cooled at rates in the range 10–1000 K/hr, compound chondrules (two chondrules fused together) formed in some meteorites, and some chondrules accreted dusty rims where the dust is thought to be from the solar nebula.

The two basic scenarios (Sears, 2004) for forming chondrules are is through rapid heating in the solar nebula or during impacts by planetesimals (e.g., Johnson et al., 2015) and then cooling. It may be possible that both scenarios formed chondrules.

The X-wind model (Shu et al., 1996, 1997) proposes that chondrules and CAIs formed close to the Sun and were then transported away from the Sun by an X-wind. X-winds are bipolar jets from young stellar objects (YSOs) (also known as protostars). The X-wind derives its name from the high X-ray emission from YSOs compared to young stars. These X-winds are produced through the interaction of the magnetosphere of the YSO with the unmagnetized protoplanetary disk.

3.3.7 Precipitation

Another way to form minerals is through precipitation from water, which could produce evaporite minerals such as halite and sylvite. Precipitation from a liquid would only occur on the parent body and is assumed to be a relatively rare process on asteroids. This process occurs as water on a chondritic body leaches elements out of the minerals and chondrules glasses, enriching the water in these elements (e.g., Na, K, Cl, Br, Al, Ca, Mg, Fe) (Rubin et al., 2002). As the liquids evaporate, the fluids become supersaturated with these elements, allowing these halide minerals to precipitate at low temperatures (< 373 K). Evaporite minerals are relatively rare in meteorites.

3.4 Terrestrial Weathering

When a meteorite passes through the atmosphere and lands on Earth, it will experience some terrestrial weathering, which alters minerals in the meteorite and produces new minerals. Some of the common minerals produced by terrestrial weathering in meteorites (Rubin, 1997a; Bland et al., 2006) include goethite [$FeO(OH)$], hematite (Fe_2O_3), and calcite ($CaCO_3$). The formation of minerals such as goethite and hematite will give terrestrially weathered meteorites a reddish tinge.

Two systems are currently used to note how much terrestrial weathering a meteorite has experienced. For Antarctic hand samples, a system from A (minor rustiness) to B (moderate rustiness) to C (severe rustiness) is used. The letter E is used to note where evaporate minerals are visible to the naked eye.

For thin sections, Wlotzka (1993) proposed a system (Table 3.4) from W0 to W6 where W0 are the least altered and W6 are the most altered. Limonite staining is the production of the reddish-colored iron oxides.

Table 3.4 *Wlotzka (1993) scale for weathering in meteorite thin sections*

Weathering grade	Alteration
W0	No visible oxidation of metal or sulfide. A limonitic staining may already be noticeable in transmitted light.
W1	Minor oxide rims around metal and troilite, minor oxide veins.
W2	Moderate oxidation of metal, about 20–60% being affected.
W3	Heavy oxidation of metal and troilite, 60–95% being replaced.
W4	Complete (>95%) oxidation of metal and troilite, but no alteration of silicates.
W5	Beginning alteration of mafic silicates, mainly along cracks.
W6	Massive replacement of silicates by clay minerals and oxides.

3.5 Meteorite Classification

Meteorites are broken up into a number of types based on mineralogy. The fall statistics and basic mineralogies for meteorite types are given in Table 3.5.

Meteorites can be broadly classified by how much metallic iron they contain. Stony meteorites primarily contain silicates, iron meteorites are predominately iron, and stony-iron meteorites contain approximately equal amounts of silicates and metal and have also undergone differentiation.

Stony meteorites can be broken up into chondrites and achondrites. Chondrites are meteorites that have not undergone significant degrees of melting on their parent bodies. The term "chondrite" comes from the chondrules that are found in almost all chondrites (the exception being the CI chondrites).

Achondrites do not contain chondrules since they have undergone significant amounts of melting on their parent body, which would have destroyed any chondrules that were originally present. Primitive achondrites are achondrites that only have partially melted and elemental compositions intermediate between chondrites and achondrites.

Some chondritic meteorites do contain significant amounts of metal but are still considered stony since they have not experienced considerable amounts of melting. These types of meteorites include the H, enstatite, CB, and CR chondrites.

Meteorite groups contain five or more members. Members of a meteorite group have closely similar mineralogies, whole-rock chemistries, and oxygen isotopic characteristics. The hope is that members of the same meteorite group originate on the same parent body. The term subgroup is often used to break up a group when some members have slightly different mineralogical or isotopic characteristics. Grouplets are groups that have fewer than five members. Clans include two or more groups of meteorites that have a petrologic relationship. For example, the howardite, eucrite, and diogenite (HED) meteorites are a clan of

Table 3.5 *Fall percentages and primary mineralogy (Grady et al., 2014) for each meteorite type. Fall statistics are calculated using data for classified meteorites from the Meteoritical Bulletin Database.*

Type	Fall percentage (%)	Mineralogy
L	37.2	olivine, orthopyroxene, clinopyroxene, metallic iron, troilite
H	33.8	olivine, orthopyroxene, clinopyroxene, metallic iron, troilite
LL	8.2	olivine, orthopyroxene, clinopyroxene, metallic iron, troilite
HED	5.8	pyroxene, plagioclase feldspar
iron	4.6	metallic iron, troilite, graphite
CM	1.5	phyllosilicates, tochilinite
L/LL	1.0	olivine, orthopyroxene, clinopyroxene, metallic iron, troilite
aubrite	0.8	enstatite, forsterite, plagioclase feldspar
EH	0.8	enstatite, metallic iron, troilite
EL	0.7	enstatite, metallic iron, troilite
CV	0.7	olivine, orthopyroxene, pentlandite, troilite, magnetite, spinel
mesosiderite	0.7	metallic iron, orthopyroxene, pigeonite, plagioclase
CO	0.6	olivine, orthopyroxene, troilite
ureilite	0.6	olivine, pigeonite, graphite
CI	0.5	phyllosilicates, magnetite, troilite
Martian	0.5	olivine, orthopyroxene, clinopyroxene
pallasite	0.4	metallic iron, olivine
C2-ungrouped	0.3	phyllosilicates, olivine
CR	0.3	phyllosilicates, metallic iron
H/L	0.3	olivine, orthopyroxene, clinopyroxene, metallic iron, troilite
acapulcoite/ lodranite	0.2	olivine–orthopyroxene, clinopyroxene, plagioclase feldspar
CK	0.2	olivine, orthopyroxene, clinopyroxene, magnetite
angrite	0.1	diopside-hedenbergite, olivine, anorthite
C3-ungrouped	0.1	olivine, orthopyroxene
CB	0.1	metallic iron, olivine, orthopyroxene
K	0.1	olivine, orthopyroxene, clinopyroxene, plagioclase feldspar
R	0.1	olivine, orthopyroxene, plagioclase, pyrrhotite, pentlandite
winonaite	0.1	orthopyroxene, olivine, plagioclase feldspar, metallic iron
brachinite	all finds	olivine, clinopyroxene
CH	all finds	olivine, metallic iron, orthopyroxene
lunar	all finds	orthopyroxene, clinopyroxene, olivine, ilmenite

achondrite meteorites that appear to have originated on the same parent body (e.g., Mittlefehldt, 2015).

A number of terms are used to describe meteorites. Matrix is the dark, fine-grained material in meteorites. Meteorite breccias contain fragments of minerals and rocks embedded together in a fine-grained matrix and are produced during impacts. Regolith breccias contain evidence (e.g., solar wind gases, solar-flare tracks) that they are from the surface of an asteroid. Monomict meteorites sample only a single lithology, while polymict meteorites contain two or more lithologies. Impact melts, as the name suggests, are melts produced during impacts that cooled quickly to form glass.

3.5.1 Classifying Meteorites

Meteoriticists can usually identify and roughly classify a meteorite by looking at a hand sample. However, meteorites are classified in much more detail by analyzing a thin section of the sample using an electron microprobe (Jansen and Slaughter, 1982) or scanning electron microscope (SEM), which allow different minerals to be identified, mineral compositions to be determined, and abundances of different minerals to be calculated. These instruments bombard a sample with electrons, which can collide with inner-shell electrons of different atoms and cause them to scatter. These electron vacancies can be filled with outer-shell electrons, which releases X-ray photons as the electrons go to a lower energy state. Since different atoms emit X-rays with different characteristic energies during this process, measuring the energies of the emitted X-rays can be used to determine what elements are presents. The strength of the measured characteristic X-rays is a function of the abundance of that particular element in the material.

3.5.2 Oxygen Isotopes

Besides composition, oxygen isotopes are also used to classify meteorites. Abundances of isotopes are measured using a mass spectrometer. A mass spectrometer consists of an ion source, a mass analyzer, and a detector. The ion source converts the material to ions. Using electric and magnetic fields, the analyzer is able to exert forces on the ions as they travel though the fields. The electric field accelerates the ions, while the magnetic field deflects the ions according to their mass to charge ratio. The magnetic field is varied, so ions with different mass to charge ratios can be measured by a detector. Isotopic ratios are determined because of the difficulty in determining bulk concentrations. An isotopic standard is also measured with known isotopic ratios is also measured.

The stable isotopes of oxygen are ^{16}O, ^{17}O, and ^{18}O. Oxygen isotopic studies of meteorites was pioneered by Robert Clayton and Toshiko Mayeda (1923–2004) who found that different meteorite groups tend to have distinct ratios of oxygen isotopes ($^{17}O\,/^{16}O$ and $^{18}O\,/^{16}O$). These ratios (Franchi et al., 1999) are usually given as

$$\delta^{17}O = \left[\frac{\left(\frac{^{17}O}{^{16}O}\right)_{SAM} - \left(\frac{^{17}O}{^{16}O}\right)_{VSMOW}}{\left(\frac{^{17}O}{^{16}O}\right)_{VSMOW}} \right] \times 1000 \qquad (3.1)$$

and

$$\delta^{18}O = \left[\frac{\left(\frac{^{18}O}{^{16}O}\right)_{SAM} - \left(\frac{^{18}O}{^{16}O}\right)_{VSMOW}}{\left(\frac{^{18}O}{^{16}O}\right)_{VSMOW}} \right] \times 1000 \qquad (3.2)$$

where *SAM* refers to the sample and *VSMOW* refers to "Vienna Standard Mean Ocean Water." *VSMOW* is an international water standard that defines the isotopic composition of fresh water. The term "Ocean Water" refers to the evaporation of ocean waters in the hydrologic cycle. Originally, oxygen isotopic values were calculated versus *SMOW* ("Standard Mean Ocean Water"), which was an earlier used water standard. The symbol δ refers to the deviation in parts per thousand. In plots of $\delta^{18}O$ versus $\delta^{17}O$, all rocks from Earth fall on a line (called the terrestrial fractionation line), which has a slope of approximately 0.52. Rocks from the Moon also plot on this line, implying that the Earth and Moon are genetically related. This slope is due to mass-dependent fractionation where the relative abundances of the isotopes are a function of the masses of the isotopes. Material with low $\delta^{17}O$ and $\delta^{18}O$ values are enriched in ^{16}O, while material with high $\delta^{17}O$ and $\delta^{18}O$ values are depleted in ^{16}O. The offset from the terrestrial fractionation line ($\Delta^{17}O$) is given as

$$\Delta^{17}O = \delta^{17}O - 0.52 \times \delta^{18}O. \qquad (3.3)$$

Chondritic meteorites plot above, below, and on the terrestrial fractionation line (Figure 3.8). The H, L, and LL ordinary chondrites (OCs) and the R chondrites plot above the terrestrial fractionation line (TFL) in distinct regions. Enstatite chondrites (ECs) plot on the TFL. Carbonaceous chondrites (CCs) plot below the TFL except for the CI chondrites, which plot slightly above.

The oxygen isotopic values for components of carbonaceous chondrites (CAIs, AOAs, chondrules, fine-grained matrix) plot a line with a slope of 0.94 (Figure 3.8)

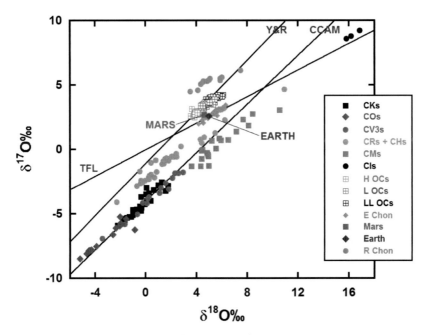

Figure 3.8 Oxygen isotope plot ($\delta^{18}O$ versus $\delta^{17}O$) for chondrites. The average oxygen isotopic values for the Earth and Mars are also plotted. TFL is the terrestrial fractionation line, CCAM is the carbonaceous chondrite anhydrous mineral line, and Y&R is the Young and Russell (1998) slope 1 line. Credit: Richard Greenwood, Open University. (A black and white version of this figure will appear in some formats. For the color version, please refer to the plate section.)

and is called the carbonaceous chondrite anhydrous mineral (CCAM) line (Clayton et al., 1977; Krot et al., 2010). CM, CV, CO, and CK chondrites fall on the CCAM line, while CR and CH chondrites tend to fall above. This line is argued to be a mixing line between ^{16}O-rich solids and ^{16}O-poor gas. The Young and Russell (1998) (Y&R) line is based on a line fit through measurements of a CAI from the CV3 chondrite Allende that passes through OCs. The line is thought to represent either mixing or fractionation between ^{16}O-rich reservoirs and ^{16}O-poor reservoirs and links the CAIs with the ordinary chondrites.

Achondritic meteorites such as HEDs, acapulcoites/lodranites, winonaites, and main-group (M-G) pallasites, fall in regions below the TFL (Figure 3.9). Aubrites plot on the TFL and have indistinguishable oxygen isotopic values from ECs. Ureilites have a much wider spread of oxygen isotopic values than other achondritic groups (Clayton and Mayeda, 1988) and would plot on the CCAM line.

3.5.3 Presolar Grains

Presolar grains are material that formed before our Sun formed and is durable enough to have survived being incorporated into chondritic planetesimals (e.g.,

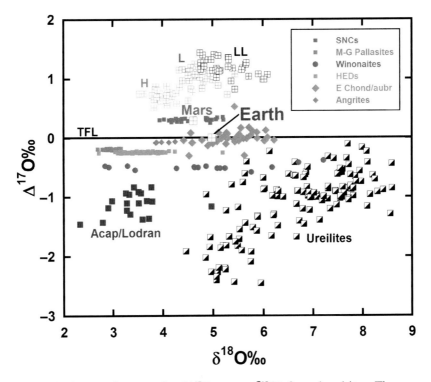

Figure 3.9 Oxygen isotope plot ($\Delta^{17}O$ versus $\delta^{18}O$) for achondrites. The average oxygen isotopic values for the Earth, Mars (from Martian meteorites), and ordinary chondrites (H, L, LL) are also plotted. SNCs (Shergottites, Nakhlites, Chassignites) are Martian meteorites. Enstatite chondrites (E Chond) and aubrites (aubr) have indistinguishable oxygen isotopic compositions and are plotted using the same symbols. M-G pallasites are main-group pallasites. HEDs are howardites, eucrites, and diogenites. Credit: Richard Greenwood, Open University. (A black and white version of this figure will appear in some formats. For the color version, please refer to the plate section.)

Davis, 2011; Nguyen and Messenger, 2011). These grains tend to be smaller than a micron in size and are found as diamond, silicon carbide, graphite, silicon nitride, oxides (aluminum oxide, hibonite), and silicates. These grains tend to have anomalous isotopic compositions compared to meteorites. Isotopic ratios that have been used to identify presolar grains include $^{12}C/^{13}C$, $^{14}N/^{15}N$, $^{17}O/^{16}O$, $^{18}O/^{16}O$, $^{29}Si/^{28}Si$, and $^{30}Si/^{28}Si$. Analyses of presolar grains give insight on the stars that contributed dust to the cloud fragment that was the precursor to our Solar System and what processes formed our Solar System.

3.5.4 Chondrites

Most chondrites can be primarily classified into three broad classes (ordinary, enstatite, and carbonaceous) on the basis of similar bulk mineralogical

Table 3.6 *Mineralogical properties that are often used to classify chondritic meteorites from Krot et al. (2005). Fayalite contents for the ordinary chondrites are from Nakamura et al. (2011). Chondrule mean diameters are from Friedrich et al. (2015).*

Type	CAIs + AOAs (vol%)	chondrules (vol%)	matrix (vol%)	metal (vol%)	Chondrule mean diameter (mm)	Fa (mol%)
CI	≪1	≪1	>99	0	–	highly variable
CM	5	20	70	0.1	0.27	highly variable
CO	13	48	34	1–5	0.15	highly variable
CR	0.5	50–60	30–50	5–8	0.70	1–3
CH	0.1	~70	5	20	0.02	<1–36
CB	<0.1	30–40	<5	60–70	0.20	2–3
CV	10	45	40	0–5	0.90	highly variable
CK	4	15	75	<0.01	0.90	<1–47
H	<0.1	60–80	10–15	8	0.45	16–20
L	<0.1	60–80	10–15	4	0.50	22–26
LL	<0.1	60–80	10–15	2	0.55	26–32
EH	<0.1	60–80	<0.1	10.1	0.23	0.4
EL	<0.1	60–80	<0.1	10.2	0.50	0.4
K	<0.1	27	60	7.4	0.5–1.1?	2.2
R	<0.1	>40	30	<0.1	0.40	38

characteristics. A much smaller percentage of chondrites are found in the R group and K grouplet. Approximate mineralogies of the different chondrite groups are given in Table 3.5 and the mineralogical properties that are often used to classify the meteorites are given in Table 3.6.

Ordinary chondrites (Figures 3.1 and 3.10) are the commonest type of meteorite to fall to Earth, hence, they are given the name "ordinary." Ordinary chondrites are broken up into three classes: H, L, and LL. H chondrites are named due to their relatively "High total iron" concentrations. L chondrites have relatively "Low total iron" concentrations, while LL have even lower iron concentrations with "Low total iron" and "Low metal." Total iron concentration refers to the concentration of all iron (metallic iron and oxidized iron in the silicates) in the meteorite. The current meteorite classification system is based on the work of Prior (1920). In this system, H chondrites were originally called bronzite chondrites, L chondrites were called hypersthene chondrites, and LL chondrites were called amphoterites.

Enstatite chondrites are predominately composed of FeO-free enstatite and metallic iron. Metallic iron abundances are significantly higher than H chondrites. Enstatite chondrites are broken up into EH (High iron) and EL (Low iron) chondrites.

Carbonaceous chondrites (Figures 3.2 and 3.5) are relatively dark (low-albedo) meteorites that have not undergone significant amounts of heating. The term "carbonaceous" (carbon-rich) is misleading since not all carbonaceous chondrites are rich in carbonaceous material. The different carbonaceous chondrite groups are named using a "C" for carbonaceous and then a capital letter usually taken from the first letter of a type specimen (a prominent member that is often the first discovered) of that group.

Chondrites are subdivided (Van Schmus and Wood, 1967; Tait et al., 2014) according to how much they have been aqueously altered or thermally metamorphosed using a number between 1 and 7. Increasing numbers (4 through 7) indicate chondrites that have experienced more thermal alteration due to being exposed to higher temperatures. Decreasing numbers (2 through 1) indicate chondrites that have experienced more aqueous alteration with type 1 having higher water contents than type 2. Chondrites designated with a 3 have not been altered. Types 1, 2, and 3 are called unequilibrated chondrites since they have not experienced significant thermal metamorphism to equilibrate the silicate compositions. Types 4 through 7 are called equilibrated chondrites since thermal metamorphism causes the olivine and pyroxene compositions to become relatively homogeneous. The olivines and pyroxenes tend to become more homogeneous with increasing type. The chondrules become less well defined as the type increases from 2 to 7. The amount of matrix decreases as the type increases from 1 to 7.

There is a subdivision (3.00 to 3.9 by increments of 0.1) for the least thermally metamorphosed chondrites. A few chondrites have also been classified as 3.05 and 3.15. Chondrites classified as 3.00 are the least altered and are hoped to be the best representative of the solid material in the solar nebula (Hubbard and Ebel, 2015). The classification of chondrites as 3.00, 3.05, or 3.15 is done by determining the average and standard deviation of the chromium content of olivine grains (Grossman and Brearley, 2005), which will be sensitive to small degrees of metamorphism. The most studied of these meteorites is LL3.00 Semarkona, which is the first of these relatively rare type of meteorites to be identified (e.g., Grossman and Wasson, 1981). There is a subdivision (2.0 to 2.9) for the least aqueously altered carbonaceous chondrites with the more aqueously altered chondrites having lower numbers.

3.5.4.1 Ordinary Chondrites

Ordinary chondrites are primarily composed of olivine, pyroxene, and metallic iron (Brearley and Jones, 1998). The ordinary chondrites can be differentiated according to their different mineral chemistries and different metallic iron abundances. H chondrites have the lowest iron contents (Fa_{16-20}, $Fs_{14.5-18}$) in their silicates (olivine and orthopyroxene), L chondrites have slightly higher iron contents

Figure 3.10 Polished slice of the ordinary chondrite (H5) Faucett. The highly reflective grains are metallic iron. Credit: Randy Korotev, Washington University.

(Fa_{22-26}, Fs_{19-22}) in their silicates, and LL chondrites have the highest iron contents (Fa_{26-32}, Fs_{22-26}) in their silicates. Also, H chondrites have the highest abundance of metallic iron, L chondrites have lower abundances, and LL chondrites have the lowest abundance. The metallic iron grains in ordinary chondrites are usually a millimeter in diameter or smaller. These metallic iron grains are very apparent in slices of ordinary chondrites (Figure 3.10) due to their highly reflective nature. Chondrules are abundant but CAIs are small and rare (Table 3.6). Besides olivine, orthopyroxene, and metallic iron, ordinary chondrites also contain minerals such as clinopyroxene, plagioclase feldspar, and troilite.

Each ordinary chondrite group also plots in a specific region above the terrestrial fractionation line of the oxygen isotope diagram. Either each member of each ordinary chondrite group originates on the same parent body or parent bodies with similar oxygen isotopic compositions. The average chondrule sizes for ordinary chondrites are all approximately 0.5 mm (Friedrich et al., 2015).

The large number of ordinary chondrite samples available for study has led to considerable research on these meteorites. Studying meteorites such as the unshocked unequilibrated LL3.00 Semarkona give the best insight on the "original" characteristics of the solar nebula.

3.5.4.2 EH and EL Chondrites

EH and EL chondrites are composed primarily of FeO-free enstatite and metallic iron plus troilite (Keil, 1989). Enstatite chondrites have higher metallic iron abundances than H chondrites. EH chondrites have higher metallic iron contents than EL chondrites. The metal in EH chondrites is also enriched in Si relative to EL chondrites. EH chondrites also have smaller average chondrule sizes (0.23 mm) compared to EL chondrites (0.5 mm) (Friedrich et al., 2015). Both EH and EL chondrites fall on the terrestrial fractionation line.

3.5.4.3 R Chondrites

R chondrites are predominately composed of olivine and are named after the type specimen Rumuruti (Brearley and Jones, 1998). R chondrites plot above the ordinary chondrites on the oxygen isotopic diagram. One unusual R chondrite (LAP 04840) contains a number of relatively rare OH-bearing silicates such as magnesiohornblende $[Ca_2Mg_4Al_{0.75}Fe^{3+}_{0.25}(Si_7AlO_{22})(OH)_2]$ and phlogopite $[KMg_3(AlSi_3O_{10})(OH)_2]$ (McCanta et al., 2008).

3.5.4.4 K Chondrites

K chondrites have high abundances of matrix, metal abundances similar to H chondrites, and low FeO-contents of the silicates (Weisberg et al., 1996). K chondrites are named after the type specimen Kakangari. K chondrites also plot in a specific region of the oxygen isotopic diagram that is below the terrestrial fractionation line.

3.5.4.5 CI Chondrites

CI chondrites are named after the type specimen Ivuna. CI chondrites are extremely dark and also extremely friable (Figure 3.11). CI chondrites are predominately composed of fine-grained phyllosilicates (Brearley and Jones, 1998). CI chondrites do not contain intact chondrules or CAIs. Of all known meteorite groups, the CI chondrites have elemental compositions that most closely match the Sun's photosphere (Figure 3.12) and are believed to be representative of the composition of the solar nebula (e.g., Anders and Grevesse, 1989; Lodders, 2003). Exceptions include very volatile elements and noble gases, which tend to be lost from meteorites, which are enriched in the Sun's photosphere. Lithium tends to be destroyed in the Sun and is enriched in meteorites relative to the Sun.

3.5.4.6 CM Chondrites

CM chondrites are named after the type specimen Mighei. CM chondrites contain abundant phyllosilicates (Brearley and Jones, 1998). Slices of CM chondrites show

Figure 3.11 Image of CI1 chondrite Ivuna. Note the extremely dark appearance of the meteorite. © The Trustees of the Natural History Museum, London.

a dark matrix containing both chondrules and CAIs (Figure 3.13). CM chondrites contain abundant intergrowths of phyllosilicates and tochilinite, which is often called PCP for poorly characterized phase (Mackinnon and Zolensky, 1984).

The best-studied CM chondrite is Murchison (Sephton, 2002; Sephton et al., 2004; Schmitt-Kopplin et al., 2010) due to the over 100 kg of material that was recovered. Murchison is a fragment of a large fireball that exploded near Murchison, Australia on September 28, 1969. Fortuitously, this fireball occurred at a time when a number of laboratories had already been built to study samples of the Moon that were returned by the Apollo missions. Apollo 11 had been launched on July 16, 1969 and had returned samples back to Earth on July 24. Murchison was found to contain extraterrestrial amino acids (Kvenvolden et al., 1970) and nucleobases (Hayatsu et al., 1975; Martins et al., 2008).

3.5.4.7 CO Chondrites

CO chondrites are named after the type specimen Ornans. CO chondrites are primarily composed of olivine, pyroxene, magnetite, and sulfides (Howard et al., 2014). CO chondrites have average chondrule sizes of ~0.15 mm (Friedrich et al., 2015).

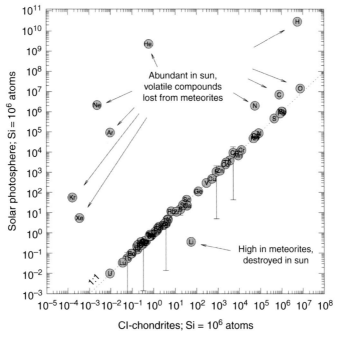

Figure 3.12 Elemental abundances in the solar photosphere and CI1 chondrites. Note the 1:1 ratio for the solar photospheric and CI1 abundances except for very volatile elements lost in meteorites and Li, which is destroyed in the Sun. The image is from Lodders and Fegley (2010). Credit: Katharina Lodders and Bruce Fegley Jr, Washington University.

Figure 3.13 Slice of CM2 chondrite Murchison. Note the abundant chondrules and CAIs. Credit: Randy Korotev, Washington University.

3.5.4.8 CV Chondrites

CV chondrites are named after the type specimen Vigarano. Slices of CV chondrites show a gray matrix with abundant chondrules and CAIs (Figure 3.2). CV chondrites are composed primarily of olivine (Brearley and Jones, 1998; Bland et al., 2004). CV chondrites tend to have larger chondrules (~0.9 mm) and CAIs than CO chondrites and also have a larger proportion of matrix (Hutchison, 2004).

CV chondrites are subdivided into three subgroups ($CV3_{red}$, $CV3_{oxA}$, $CV3_{oxB}$) based on the degree of oxidation that the meteorite experienced (Weisberg et al., 1997). Matrix/chondrule ratios increase going from $CV3_{red}$ to $CV3_{oxA}$ to $CV3_{oxB}$. The symbol *red* is for reduced, *ox* is for oxidized, *A* is for Allende, and *B* is for Bali. Metal/magnetite ratios tend to decrease going from $CV3_{red}$ to $CV3_{oxA}$ to $CV3_{oxB}$. $CV3_{oxB}$ chondrites also contain abundant phyllosilicates. Since Bali-like and Allende-like clasts are found in reduced CV chondrites, Krot et al. (2000) argues that all three types of CV chondrites originated on the same heterogeneous parent body.

The best-known CV chondrite is Allende since over 100 kg of material was recovered and it contained abundant CAIs (Mason and Taylor, 1982) that could be compositionally studied and dated to give insight on the early history of the Solar System. The Allende meteorites are fragments of a large fireball that exploded over Mexico on February 8, 1969. This fireball occurred before the Murchison fireball and at a time when a number of laboratories were being readied to study samples of the Moon. Allende is considered the best and most studied meteorite in history.

3.5.4.9 CR Chondrites

CR chondrites are named after the type specimen Renazzo. CR chondrites contain abundant metal (~5–8 vol%) (Scott and Krot, 2014), which is often on the surfaces of chondrules, and also abundant phyllosilicates. One of the most unusual CR chondrites is Kaidun, which contains a wide spectrum of material that ranges including carbonaceous chondritic material, enstatite chondritic material, and melted clasts. CR chondrites have average chondrules sizes of 0.7 mm (Friedrich et al., 2015).

3.5.4.10 CK Chondrites

CK chondrites are named after the type spectrum Karoonda. CK chondrites are the most highly oxidized "anhydrous" carbonaceous chondrite group (Huber et al., 2006). These meteorites are relatively free of metallic iron, their fayalite contents are ~Fa_{30}, and they contain abundant magnetite and pentlandite. CK chondrites are the only type of carbonaceous chondrites that span type 3 to 6 (CK3 to CK6) (Kallemeyn et al., 1991).

CK chondrites have a number of mineralogical and isotopic similarities with CV chondrites. CK and oxidized CV3 chondrites have the same range on the CCAM line (Greenwood et al., 2010). CK and CV3 chondrites also have similar opaque mineralogies and major and trace element geochemistries. Greenwood et al. (2010) argue that both chondrites originated on a single stratified parent body with regions of $CV3_{red}$, $CV3_{oxA}$, and $CV3_{oxB}$ on the surface and deeper layers of CK3, CK4, CK5, and CK6 going toward the core.

3.5.4.11 CB Chondrites

CB chondrites are named after the type spectrum Bencubbin (Weisberg et al., 2001). CB chondrites contain more than ~60 vol% metallic iron (Figure 3.14). CB chondrites are divided into the Cb_a and CB_b subgroups based on a number of different compositional and isotopic characteristics (Weisberg et al., 2001). CB_a contain 60 vol% metal, cm-sized silicate chondrules, 5–8% Ni in the metal, no refractory inclusions, and extreme enrichment in ^{15}N. CB_b chondrites contain greater than 70 vol% metal, millimeter-sized chondrules, a larger range of Ni in the metal, refractory inclusions are present, and less enrichment in ^{15}N. These meteorites appear to have formed by impacts due to melt regions being found between the metal and chondrules (Weisberg et al., 2001).

3.5.4.12 CH Chondrites

CH chondrites are the one type of carbonaceous chondrite that is not named after a type specimen. They are instead named after "High metal" due to their high proportion of metal (~20 vol%) (Scott and Krot, 2014). The metal grains typically range in size from mm- to cm-sized. CH chondrites have extremely small average chondrule sizes of 0.02 mm (Friedrich et al., 2015).

The CR, CB (CB_a and CB_b), and CH chondrites are part of the CR clan of meteorites (Krot et al., 2002). This clan was proposed since these meteorites have a lot of mineralogical and isotopic similarities, including being very metal-rich, having a solar Co/Ni ratio, being depleted in moderately volatile lithophile (oxygen-loving) elements, having bulk oxygen isotopic compositions that fall on a mixing line, having large ^{15}N enrichments, having Fe-poor anhydrous mafic silicates, and containing a very hydrated matrix.

3.5.4.13 Ungrouped Carbonaceous Chondrites

A number of carbonaceous chondrites do not fall within a group and are called ungrouped C chondrites or C ungrouped (e.g., Choe et al., 2010). One of the most studied is C2 chondrite Tagish Lake (Brown et al., 2000). The Tagish Lake meteorite samples were fragments of a fireball that exploded over British Columbia, Canada and landed on and near Tagish Lake. The total mass collected was between 5 kg and 10 kg. Slices of Tagish Lake show an extremely dark matrix with small light-colored inclusions. Tagish Lake has an extremely low amino acid content compared to CI and CM chondrites (Kminek et al., 2002).

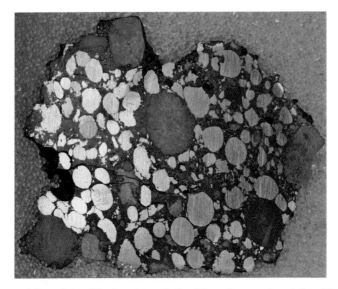

Figure 3.14 Slice of the CB chondrite Gujba. Note the metal nodules. The slice is 4.6 × 3.8 cm in size. Credit: James St. John.

3.5.4.14 Prior's Rules

The mineralogies of chondrites are broadly consistent with Prior's Rules. Prior's Rules (Prior, 1916) stated that that the smaller the percentage of FeNi metal, the higher the nickel concentration in the metal and the higher the iron-content in the olivine and pyroxenes. This is due to Ni being more siderophile ("iron-loving") than the Fe, so Ni tends to partition into the metal (Russell and Grady, 2006). Also, the more oxidized a body is, the more FeO-rich are the silicates and the smaller concentration of metallic iron. This trend can be seen mineralogically. The enstatite chondrites have the most FeO-poor silicates and the highest metallic iron concentration. Among the ordinary chondrites, the H chondrites have the most FeO-poor silicates and the highest metallic iron concentrations. L chondrites have more FeO-rich silicates and lower concentrations of metallic iron. LL chondrites have more FeO-rich silicates than L chondrites and lower concentrations of metallic iron. R chondrites contain more FeO-rich silicates than ordinary chondrites and virtually no metallic iron.

3.5.5 Achondrites

Achondrites are from bodies that have experienced differentiation. Achondrites do not contain chondrules, which would have been erased during melting and differentiation.

Figure 3.15 Image of a fragment of the Bishopville aubrite. Note the white color of the FeO-free enstatite. The fragment is 2.2 cm across. Credit: James St. John.

3.5.5.1 Angrites

Angrites are named after the type specimen Angra dos Reis. Most angrites are composed predominantly of Al-Ti diopside-hedenbergite (formally called fassaite), anorthite, and Ca-rich olivine (including sub-calcic kirschsteinite) (Mittlefehldt et al., 2002). The exception is Angra dos Reis, which is composed almost entirely of just Al-Ti diopside-hedenbergite (e.g., Prinz et al., 1977).

Angrites are thought to originate from the crust of a differentiated body that formed under oxidizing conditions. Jurewicz et al. (1991, 1993) found that melting experiments of carbonaceous chondritic material under low oxygen fugacities (reducing conditions) resemble eucrites, while those under high oxygen fugacities (oxidizing conditions) resemble angrites. Oxygen fugacity is equivalent to the partial pressure of oxygen in a particular environment (Albarède, 2011).

3.5.5.2 Aubrites

Aubrites are predominately composed of virtually FeO-free enstatite (Watters and Prinz, 1979) (Figure 3.15). This enstatite is much more FeO-poor than enstatite found on Earth and has an extremely white color due to the absence of transition metal cations. Aubrites are named after the type specimen Aubres. Other accessory minerals in aubrites include plagioclase, diopside, forsterite, kamacite, and troilite. Minor constituents in aubrites include a variety of sulfides such as troilite, oldhamite, daubréelite, and ferromagnesian alabandite and the phosphide schreibersite. Aubrites are thought to originate from the mantles of very reduced

Figure 3.16 Slice of the NWA 3359 eucrite. The slice is 3.6 cm across at its widest.
Credit: James St. John.

differentiated bodies. Aubrites are thought to be formed from the melting of an
enstatite chondrite-like parent body that had a number of relatively small compositional differences (e.g., higher troilite to metallic iron ratio, higher diopside
content) from known enstatite chondrites (Keil, 1989).

3.5.5.3 HEDs

HEDs (howardites, eucrites, diogenites) are a clan of achondritic meteorites that
appear to come predominately from the same parent body. Howardites are named
after meteorite researcher Edward Howard (1774–1816). The name eucrites originates from the Greek word *eukritos*, which means "easily distinguished." *Eukritos*
was chosen due to the large grain size of the silicate minerals. Diogenites are
named after Diogenes of Apollonia (fifth century BC) who is given credit for first
suggesting that meteorites could originate from space.

Mineralogically, eucrites contain primarily anorthic plagioclase and low-Ca
pyroxene (Figure 3.16) with augite exsolution lamellae, while diogenites are
predominately magnesian orthopyroxene (Mittlefehldt et al., 1998). Monomict
eucrites and diogenites sample only a single lithology, while polymict eucrites and
diogenites contain two or more lithologies. Howardites are also polymict achondrites. The arbitrary dividing line (Delaney et al., 1983) is that polymict eucrites
contain less than 10% diogenitic material, while howardites contain more than

10% diogenitic material. Howardites are mixtures of eucritic and diogenitic material and are the conclusive evidence that all three types of meteorites come from the same parent body. Eucrites originate from the crust of a differentiated body, diogenites originate from deeper depths, and howardites are impact-produced mixtures.

Eucrites can be broken up into a number of subtypes (Mittlefehldt et al., 1998). Cumulate eucrites appear to be the result of slow cooling and crystal settling with preferred orientations and relatively large grain sizes. Noncumulate (basaltic) eucrites are consistent with cooling much faster with much smaller grain sizes.

A number of eucrites (Yamaguchi et al., 2002; Bland et al., 2009; Gounelle et al., 2009; Scott et al., 2009) have been found to have oxygen isotopic compositions distinct from "typical" HEDs, implying a number of differentiated parent bodies with eucritic crusts. The identification of these "anomalous" eucrites is consistent with the large number of iron meteorite groups, grouplets, and ungrouped irons that are found in our meteorite collections, which Wasson (1995) argues is evidence for ~70 disrupted differentiated parent bodies in the asteroid belt.

3.5.6 Primitive Achondrites

Primitive achondrites have experienced melting and some melt migration but did not undergo full-scale differentiation (Mittlefehldt et al., 1998), which results in these meteorites having near-chondritic bulk compositions.

3.5.6.1 Acapulcoites/Lodranites

Acapulcoites and lodranites are a clan of primitive achondritic meteorites. Acapulcoites are named after the type specimen Acapulco, while lodranites are named after the type specimen Lodran. Both groups of meteorites are primarily composed of metallic iron, olivine, and orthopyroxene (Mittlefehldt et al., 1998). Lodranites are coarser grained than acapulcoites. McCoy (1994) and McCoy et al. (1996, 1997) studied these two groups of meteorites and argue that they both originated on the same parent body. McCoy (1994) found that acapulcoites formed by low degrees of partial melting without silicate partial melting, while lodranites experienced higher degrees of partial melting with silicate partial melting. Acapulcoites were heated to lower temperatures than lodranites, which resulted them experiencing lower degrees of partial melting. The higher degrees of partial melting for the lodranites resulted in more melt migration for these meteorites that depleted them in plagioclase and troilite. Several acapulcoites have relict chondrules. Mafic silicate compositions of acapulcoites and lodranites are intermediate between enstatite and H chondrites.

3.5.6.2 Brachinites

Brachinites are named after the type specimen Brachina. Brachinites are pre-dominately composed of olivine (Fa_{27-36}) with some clinopyroxene (Keil, 2014). Brachinites are commonly thought to be ultramafic residues from various low degrees of partial melting.

3.5.6.3 Ureilites

Ureilites are named after the type specimen Novo-Urei. Ureilites are primarily composed of olivine and pyroxene (pigeonite, orthopyroxene, and/or augite) with a significant amount of carbon phases (Mittlefehldt et al., 1998). Interestingly, ureilites can contain more carbon than carbonaceous chondrites. The carbon can be found as amorphous carbon, graphite, and diamond. Ureilites are subdivided based on the dominant pyroxene (olivine-pigeonite, olivine–orthopyroxene, augite-bearing).

3.5.6.4 Winonaites

Winonaites are named after the type specimen Winona. They have oxygen isotopic compositions similar to silicate inclusions in IAB iron meteorites (Mittlefehldt et al., 1998). Like acapulcoites and lodranites, mafic silicate compositions of winonaites are intermediate between enstatite and H chondrites.

3.5.7 Irons

Iron meteorites are composed primarily of metallic iron, consisting of kamacite and taenite (Mittlefehldt et al., 1998). Irons were originally classified according to the presence or absence of the Widmanstätten pattern (Figure 3.7).

Hexahedrites have low-Ni concentrations and do not have Widmanstätten patterns, while ataxites have high Ni concentrations and also do not have Widmanstätten patterns. Hexahedrites are named after the hexagonal structure of the kamacite. The name ataxite is derived from the Greek word for structureless. Octahedrites have an intermediate Ni concentration and do have a Widmanstätten pattern. The name octahedrites is derived from the octahedral structure of the crystals.

Presently, irons are classified according to groupings in logarithmic plots of Ni, Ga, Ge, and Ir. There are currently 14 of these chemical groups. The groups are IAB, IC, IIAB, IIC, IID, IIE, IIF, IIG, IIIAB, IIICD, IIIE, IIIF, IVA, and IVB (Goldstein et al., 2009). The Roman numerals I–IV (Goldberg et al., 1951; Lovering et al., 1957) were used to discriminate iron meteorites in order of decreasing contents of gallium and germanium. Letters were then used to subdivide these groups. The IIIAB group has approximately 220 members, while a few of the groups (IIC, IIF, IIG, IIIF) have fewer than ten members (Goldstein et al., 2009).

Magmatic iron meteorites (IIAB, IID, IIIAB, IVA, IVB) have similar elemental fractionation trends that are consistent with forming through fractional crystallization of a large body that melted (Chabot and Haack, 2006). These iron meteorites have cooling rates consistent with forming in the core of a differentiated body (Goldstein et al., 2009). These differentiated bodies must have been disrupted to create the iron meteorites that we find on Earth today.

Silicate-bearing irons (IAB, IIICD, IIE) [often called non-magmatic even though the metallic iron must have been at least partially molten (Ruzicka, 2014)] do not have large fractionations in Ir and more variable Ni, Ga, and Ge concentrations. These silicate-bearing irons contain silicate abundances that can range from minor to ~50% (Ruzicka, 2014). The silicates in these irons often have chondritic compositions. Silicate-bearing irons are thought to have formed through partial melting or impact melting.

H chondrites have been linked with the IIE irons, unusual irons that contain abundant silicate inclusions. Similarities between oxygen isotopic compositions (Clayton and Mayeda, 1978; McDermott et al., 2011), silicate mineralogies (Casanova et al., 1995), and the nonmagmatic (did not form in a core) origin of the metal in IIE irons (Wasson and Wang, 1986) argue for a possible relationship.

3.5.8 Stony-Irons

Stony-irons are meteorites that contain approximately 50–50 mixtures of silicates and metallic iron (Mittlefehldt et al., 1998).

3.5.8.1 Mesosiderites

Mesosiderites are stony-iron meteorites that are breccias of silicates and FeNi metal (Mittlefehldt et al., 1998). The name "mesosiderites" originates from the Greek word *mesos* for "half" and *sideros* for "iron." The silicates are similar in mineralogy to those found in HEDs.

The mineralogy of mesosiderites is consistent with the mixture of crustal (basalt) and core (iron) material from a differentiated body (or bodies) but a relative absence of mantle (olivine) material. To form mesosiderites, Scott et al. (2001) proposes that a ~200–400 km asteroid with a molten core was disrupted by a ~50–150 km projectile.

3.5.8.2 Pallasites

Pallasites are stony-iron meteorites that are named after the German biologist Peter Pallas (1741–1811) who in 1772 studied a Russian meteorite (later called Krasnojarsk) that was used by Chladni as one of his examples of rocks that fell from space. Pallasites are composed of relatively large olivine grains that are embedded

Figure 3.17 Slice of the Esquel pallasite. Note the greenish olivine grains.
Credit: Doug Bowman. (A black and white version of this figure will appear in
some formats. For the color version, please refer to the plate section.)

in FeNi metal (Figure 3.17) (Mittlefehldt et al., 1998). The metal often shows a
Widmanstätten pattern when etched. Pallasites contain approximately two-thirds
by volume olivine and one-third FeNi metal. Pallasites contain ~5% accessory
minerals such as chromite and troilite (Buseck, 1977).

Pallasites are broken up into the main group (M-G) and the Eagle Station subgroup after the Eagle Station pallasite. The Eagle Station subgroup contains more
Fe-rich olivines with different oxygen isotopic compositions than main-group pallasites (Scott, 1977; Clayton and Mayeda, 1978).

Pallasites have long been argued to originate at the core–mantle boundary of
a differentiated body (e.g., Boesenberg et al., 2012). However, Yang et al. (2010)
determined that metallographic cooling rates below 975 K for a number of main-
group pallasites ranged from ~2.5 to ~18 K/Ma. This cooling rate range does not
appear to be consistent with an origin from the core–mantle boundary since samples from this boundary should have indistinguishable cooling rates.

3.5.9 Cometary Meteorites

Meteorites from comets also appear possible. Campins and Swindle (1998) predicted that cometary meteorites would be dominated by highly unequilibrated
anhydrous silicates and high concentrations of C and N or be similar (or be) to CI1
chondrites. Gounelle et al. (2008) argue that some type 1 (and possibly some type
2) carbonaceous chondrites could possibly have originated on a comet.

Table 3.7 *Major meteor showers from the American Meteor Society (2016)*

Name	Time of year	Associated object	Type
Quadrantids	January 1–10	(196256) 2003 EH_1	asteroid
Lyrids	April 16–25	C/1861 G1 (Thatcher)	comet
Eta Aquarids	April 19–26	1P/Halley	comet
Delta Aquarids	July 21–August 23	96P/Machholz	comet
Alpha Capricornids	July 11–August 10	169P/NEAT	comet
Perseids	July 13–August 26	109P/Swift–Tuttle	comet
Orionids	October 4–November 14	1P/Halley	comet
Southern Taurids	September 7–November 19	2P/Encke	comet
Northern Taurids	October 19–December 10	2P/Encke	comet
Leonids	November 5–30	55P/Tempel–Tuttle	comet
Geminids	December 4–16	(3200) Phaethon	asteroid
Ursids	December 17–23	8P/Tuttle	comet

3.6 Meteors

Meteors are the bright streaks of light in the sky due to the disintegration of meteoroids in the atmosphere. Meteor showers are when a large number of meteors appear to radiate from one point of the night sky. Meteor showers are named (Table 3.7) for the nearest constellation that the meteors appear to radiate from (Jenniskens, 2006). If more than one shower appears to originate from the constellation, the meteor shower is usually named after the brightest star with a Greek letter that the shower appears to originate from (e.g., Eta Aquarids and Delta Aquarids). Northern and Southern (e.g., Northern Taurids and Southern Taurids) refers to branches of a shower above and below the ecliptic.

Meteor showers are related to specific comets or asteroids. These showers occur at approximately the same time each year when the Earth passes through a comet's or asteroid's dust trail. The asteroids linked with meteors tend to be thought of as extinct comets (e.g., Jenniskens, 2004).

The composition of meteors can be determined through meteor spectroscopy. Emission lines are produced as electrons in the atoms are excited to higher energy levels and then drop back down to the lower energy levels, emitting photons at specific energies. The intensity and wavelengths of the lines can be used to determine the abundances of different elements in the meteors (e.g., Trigo-Rodriguez et al., 2003). Trigo-Rodriguez et al. (2003) analyzed the spectra of 13 meteors and found their elemental compositions were most similar to CI and CM meteorites and interplanetary dust particles (IDPs).

3.6.1 Fireballs

Fireballs are brighter than usual meteors (magnitudes of −4 or greater). One of the most famous fireballs occurred on October 9, 1992 with a peak visual magnitude of −13. Because it occurred on a Friday night over the east coast of the United States, a large number of videos of fireballs were taken during football games. A meteorite (called the Peekskill meteorite) was found the next day after it had crashed into a red 1980 Chevrolet Malibu. The meteorite was later sold for $69,000 and the car was sold for $10,000.

Bolides are extremely bright fireballs (visual magnitudes of −14 or greater) that explode in the atmosphere. Bolides are also called airbursts. From 1994 to 2013, the U.S. government recorded over 500 bolides (Dean, 2015). Superbolides have visual magnitudes greater than −17.

One recent superbolide occurred on April 22, 2012. The fragments that were found near Coloma, California. were called Sutter's Mill and were classified as CM chondrites (Fries et al., 2014; Zolensky et al., 2014). Approximately 1 kg of material was recovered.

A superbolide exploded in the atmosphere over Chelyabinsk, Russia on February 15, 2013 (Popova et al., 2013). This was one of the best-characterized fireballs due to the video surveillance in the Chelyabinsk area (Borovička et al., 2016). The body was estimated to have a diameter of ~20 meters and moving at ~19 km/s when first detected (Popova et al., 2013). A shock wave first developed at ~90 km from the ground and fragmentation started at ~83 km. The equivalent energy of the object was estimated to be ~500 kT (Brown et al., 2013; Pilger et al., 2015) and was painful to look at since it was so bright, with a magnitude of −27.3 ± 0.5 at peak brightness. This peak brightness occurred at ~30 km from the ground. People experienced sunburns and retinal burns, and a few experienced concussions, cuts, and bruises from the explosion. Over 100 kg of meteoritic material was recovered. These fragments were classified as LL5 chondrites and called Chelyabinsk.

In an interesting coincidence, the Chelyabinsk superbolide occurred 16 hours before a predicted close encounter (~27,700 km) of NEA (367943) Duende (2012 DA_{14}) with the Earth. However, these bodies had very different orbits and could not be dynamically related with each other (Chodas and Chesley, 2014).

3.7 Meteorite Densities

The density (ρ) of any object is its mass divided by its volume. The mass (M) of the meteorite is determined using a scale. The volume is calculated using the method of Archimedes (*c.*287 BC–*c.*212 BC), where the volume of an object can be determined by measuring the volume of the material that is displaced when an object is totally immersed in the material. To determine the bulk volume (V_b), the meteorite is usually immersed in a container of glass beads of a known mass.

The grain volume (V_g) is determined using the ideal gas law (Consolmagno et al., 2008) with a gas such as helium. Helium is ideal to use since it is small enough to quickly penetrate any cracks in the sample and will not react with the meteorite since it is an inert gas. The meteorite volume is determined using two chambers of known volume that are connected using a valve. With the connecting valve closed, the meteorite is then placed in one of the chambers and then He is introduced into one of the chambers.

The bulk density (ρ_b) of the meteorite will be $\dfrac{M}{V_b}$ and the grain density (ρ_g) will be $\dfrac{M}{V_g}$. The bulk density will always be lower than the grain density since the calculated bulk density includes pore spaces, while the calculated grain density should not.

Porosity is the void fraction of a material. The porosity (P) of the meteorite will then be

$$P = \left(1 - \frac{\rho_b}{\rho_g}\right) \times 100\%. \tag{3.4}$$

Average grain and bulk densities (g/cm³) and porosities for meteorites are given in Table 3.8. The density of meteoritic iron is ~7.6–7.8 g/cm³ (Consolmagno et al., 2008).

3.8 Magnetic Field

Many meteorites show evidence of being exposed to a magnetic field. A magnetic field is the region around a moving charge or magnetic material that can exert a force on a charged particle. The strength of a magnetic field is usually given in terms of teslas. A particle with a charge of 1 coulomb (the charge that is due to ~6.241×10^{18} protons) that passes through a magnetic field of 1 tesla with a speed of 1 m/s will experience a force of 1 newton. The magnetic field strength on the Earth's magnetic equator is ~31 μT.

The magnetic field for a planet or star is generated through the motion of a highly conductive fluid (the dynamo). The Earth's magnetic field is generated by its convecting molten iron core, while the Sun's magnetic field is generated by the motion of its conducting plasma. Iron (e.g., tetrataenite) and iron oxide minerals (e.g., magnetite, hematite, pyrrhotite) can preserve the strength and direction of a magnetic field. Paleomagnetism is the study of these remnant magnetic fields.

Table 3.8 *Average bulk and grain densities (g/cm³) and porosities for different chondritic and achondritic groups (Macke, 2010). The average grain density and porosity for ordinary, enstatite, and CO chondrites are for falls.*

Meteorite group	Average grain density (kg/m³)	Average bulk density (kg/m³)	Average porosity (%)
H	3350	3710	9.5
L	3300	3580	8.0
LL	3180	3520	9.5
CB	5250	5650	3.9
CI	1570	2420	34.9
CK	2900	3580	17.7
CM	2270	2960	22.2
CO	3100	3360	7.6
CV (all)	2970	3610	17.7
CV_o	2870	3640	21.4
CV_r	3360	3480	3.1
EH	3480	3610	3.7
EL	3580	3660	2.1
howardite	2850	3260	12.5
eucrite	2840	3190	10.9
diogenite	3100	3430	9.2
aubrite	2900	3210	8.7
angrite	3210	3420	6.2
ureilite	3220	3360	4.0
acapulcoite	3460	3690	6.2
lodranite	3530	3740	5.4
brachinite	3480	3550	1.5
winonaite	3240	3600	7.9

A meteorite can be exposed to a magnetic field on its parent body or in the solar nebula. Chondrites, achondrites, stony-irons, and irons all show evidence of being exposed to a magnetic field. The CV chondrite Allende was found to have been exposed for a few million years to a 20 μT or greater magnetic field for 9–10 Ma after the formation of the solar nebula (Carporzen et al., 2011). Due to the directional stability of the field and the time of occurrence, Allende is argued to have formed on a body with a convecting iron core. CM chondrites were exposed to a field of ~2 μT from the solar nebula or the parent body during parent body aqueous alteration (Cournede et al., 2015). Chondrules in the LL3.0 Semarkona meteorite were found to be magnetized in a nebular magnetic field of 54 ± 21 μT (Fu et al., 2014).

A eucrite was found to have been exposed on its parent body [usually assumed to be (4) Vesta] to a surface magnetic field of greater than 2 μT when it cooled

3.69 Ga (billion years) ago (Fu et al., 2012). This magnetism is argued to be due to a remnant magnetic field that was the result of a previously conducting metallic iron core. The Esquel and Imilac pallasites were exposed to magnetic fields that could have lasted hundreds of millions of years (Bryson et al., 2015). Evidence for a dynamo on the IVA iron meteorite parent body has been found from the scatter in the determined paleomagnetic field directions for the measured samples (Weiss et al., 2016).

3.9 Radioactive Dating

Why are almost all meteorites thought to be fragments of asteroids? One reason is that almost all meteorites have crystallization ages between 4.56 Ga and 4.57 Ga. A crystallization age measures the time when a mineral formed and is usually thought of as the age of the meteorite.

Dating of meteorites is done by determining the relative abundances of radioactive isotopes (the parent isotope) and the stable isotopes that are produced by the decay (the daughter isotope). These radioactive isotopes are not stable and thus decay over time. Radiogenic isotopes are those that produced during radioactive decay, while nonradiogenic isotopes are not produced by radioactive decay. Radioactive dating can be used to determine the crystallization age, the time (impact age) when the meteorite was liberated from its original parent body, the time the meteorite has been within a few meters of the surface in space (cosmic-ray exposure age), and the time (terrestrial age) the meteorite has spent on Earth.

Dating is done on particular minerals. To determine an accurate age, it must be assumed that neither the parent nor the daughter isotope has been added or removed from the minerals. Gaining parent isotopes or losing daughter isotopes lowers the calculated age. Losing parent isotopes or gaining daughter isotopes raises the calculated age. The closure temperature is the temperature where a mineral has cooled down so there is no further diffusion of parent or daughter isotopes out of the mineral. The formation age of a meteorite is calculated when the measured mineral has reached this closure temperature.

The half-life $(t_{1/2})$ is the time it takes for half of a quantity of an isotope (parent) to decay into another nuclide (daughter). The daughter isotopes can also be radioactive. Half-lives range from 10^{-24} seconds to much longer than the age of the Universe. Isotopes must have relatively long half-lives to be able to be used in determining the ages of meteorites. The formula for the amount of the parent isotope during radioactive decay is usually given as

$$N(t) = N_0 e^{-\lambda_d t} \tag{3.5}$$

where $N(t)$ is the number of parent atoms at time t, λ_d is the decay constant where $\lambda_d = \ln(2)/t_{1/2}$, and N_0 is the original amount of parent atoms. The amount of the daughter $[D(t)]$ will then be

$$D(t) = N_0 - N(t) = (e^{\lambda_d t} - 1)N_0. \tag{3.6}$$

Radioactive decay is exponential and is a function of the time elapsed. If no daughter was present in the mineral originally, the ratio of the parent to the daughter will be directly related to the formation age of the mineral. The mineral will have a parent–daughter ratio of 100:0 when it first forms if no daughter was originally present. After one half-life, the mineral will have a parent–daughter ratio of 50:50. After two half-lives, the mineral will have a parent–daughter ratio of 25:75. After three half-lives, the parent–daughter ratio will be 12.5:87.5.

The amount of the daughter will then be

$$D(t) = (e^{\lambda_d t} - 1)N(t) + D_0 \tag{3.7}$$

where $D(t)$ is the number of daughter atoms at time t and D_0 is the original number of daughter atoms in the sample. However, isotopic ratios can be determined much more precisely than absolute abundances. So by dividing both sides of the equation by the number of atoms of a nonradiogenic stable isotope of the daughter $[D_i]$, the equations become

$$\frac{D(t)}{D_i} = (e^{\lambda_d t} - 1)\frac{N(t)}{D_i} + \frac{D_0}{D_i}. \tag{3.8}$$

This is the equation of a straight line ($y = mx + b$) where the slope (m) is ($e^{\lambda_d t} - 1$) and the y-intercept (b) is D_0/D_i. This straight line is called the isochron and the formula is called the isochron equation. So if you plot $D(t)/D_i$ (y on the plot) versus $N(t)/D_i$ (x) for a number of measurements of different minerals with different initial abundances of the radioactive parent, the slope ($e^{\lambda_d t} - 1$) of the line will allow you to calculate the age (t) of the sample. At *time* = 0 (when the rock forms), the slope of the line will be 0. As the time increases, the slope of the line will also increase. This intercept is called the initial ratio.

A list of isotopic systems used for dating are given in Table 3.9. Half-lives and decay constants are calculated by measuring the count rate of the isotope at a number of different times. Isotopes with relatively long half-lives are best used to determine absolute ages, while isotopes with relatively short half-lives are best used to determine relative ages (e.g., Kleine and Rudge, 2011). Stable isotopes used for dating are produced during exposure in galactic cosmic rays and can be used to determine cosmic-ray exposure ages. Isotopes produced during cosmic-ray exposure

Table 3.9 *Commonly used isotopes for dating meteorites*

Parent	Daughter	Half-life (yrs)	Decay constant (λ_d) (yrs^{-1})	Type
^{3}He	stable	–		cosmic-ray exposure
^{10}Be	^{10}B	1.39×10^6	4.99×10^{-7}	terrestrial
^{14}C	^{14}N	5730	1.210×10^{-4}	terrestrial
^{21}Ne	stable	–		cosmic-ray exposure
^{26}Al	^{26}Mg	7.17×10^5	9.67×10^{-7}	relative, terrestrial
^{36}Cl	^{36}Ar, ^{36}S	3.01×10^5	2.30×10^{-6}	terrestrial
^{36}Ar	stable	–		cosmic-ray exposure
^{40}K	^{40}Ca, ^{40}Ar	1.251×10^9	5.541×10^{-10}	absolute, impact
^{41}Ca	^{41}K	1.02×10^5	6.80×10^{-6}	terrestrial
^{53}Mn	^{53}Cr	3.7×10^6	1.9×10^{-7}	relative
^{81}Kr	^{81}Br	2.29×10^5	3.03×10^{-6}	terrestrial
^{87}Rb	^{87}Sr	4.88×10^{10}	1.42×10^{-11}	relative
^{107}Pd	^{107}Ag	6.5×10^6	1.1×10^{-7}	relative
^{129}I	^{129}Xe	1.57×10^7	4.41×10^{-8}	relative
^{147}Sm	^{143}Nd	1.06×10^{11}	6.54×10^{-12}	absolute
^{176}Lu	^{176}Hf	3.85×10^{10}	1.80×10^{-11}	absolute
^{182}Hf	^{182}W	8.90×10^6	7.79×10^{-8}	relative
^{187}Re	^{187}Os	4.16×10^{10}	1.67×10^{-11}	absolute
^{232}Th	^{208}Pb	1.405×10^{10}	4.933×10^{-11}	relative
^{235}U	^{207}Pb	7.038×10^8	9.849×10^{-10}	absolute
^{238}U	^{206}Pb	4.468×10^9	1.551×10^{-10}	absolute

with short half-lives can used to determine terrestrial ages. The impact age is the age when an impact occurs. Shock heating causes the loss of isotopes from the crystal lattice or isotopic exchange between different minerals (Bogard, 1995).

3.9.1 Rubidium–Strontium Dating

Rubidium–strontium (Ribs) dating is based on the decay of ^{87}Rb to ^{87}Sr with a half-life of 48.8 Ga by beta decay where a neutron transforms into a proton and releases an electron and an antineutrino. The decay can be written as

$$^{87}\mathrm{Sr} = {}^{87}\mathrm{Rb}\left(e^{\lambda_{87}t} - 1\right) + {}^{87}\mathrm{Sr}_0 \tag{3.9}$$

where ^{87}Sr$_0$ is the initial value of ^{87}Sr, ^{87}Sr is the value at time t, and ^{87}Rb is the value at time t. The decay constant for ^{87}Rb is given as λ_{87}. The most used half-life is 4.88×10^{10} yrs with a corresponding decay constant of 1.42×10^{-11} yrs^{-1}. The nonradiogenic stable isotope of strontium that is measured is ^{86}Sr. The isochron equation for this system becomes

Meteorites, Minerals, and Isotopes

$$\frac{^{87}\text{Sr}}{^{86}\text{Sr}} = \left(e^{\lambda_{87}t} - 1\right)\left(\frac{^{87}\text{Rb}}{^{86}\text{Sr}}\right) + \left(\frac{^{87}\text{Sr}}{^{86}\text{Sr}}\right)_0 \tag{3.10}$$

where $^{87}\text{Rb}/^{86}\text{Sr}$ and $^{87}\text{Sr}/^{86}\text{Sr}$ are the measured isotopic ratios and $(^{87}\text{Sr}/^{86}\text{Sr})_0$ is the initial isotopic ratio.

Example 3.1

Measured values for $^{87}\text{Rb}/^{86}\text{Sr}$ and $^{87}\text{Sr}/^{86}\text{Sr}$ for the Binda howardite are given in Table 3.10. What is the age of this meteorite?

Table 3.10 *The $^{87}Rb/^{86}Sr$ and $^{87}Sr/^{86}Sr$ values measured for different samples of the Binda howardite (Birck and Allegre, 1979)*

Sample	$\dfrac{^{87}\text{Rb}}{^{86}\text{Sr}}$	$\dfrac{^{87}\text{Sr}}{^{86}\text{Sr}}$
total rock 1	0.00643	0.69936
total rock 2	0.00647	0.69932
pyroxene (H1)	0.0249	0.70040
plagioclase (H)	0.00192	0.69907
pyroxene (H2)	0.0194	0.70019
pyroxene (HL)	0.01159	0.69954
separate 1	0.0168	0.69985
separate 2	0.0734	0.70271

The best-fit line through the points gives an equation of $\dfrac{^{87}\text{Sr}}{^{86}\text{Sr}} = 0.0508\left(\dfrac{^{87}\text{Rb}}{^{87}\text{Sr}}\right) + 0.699$ (equivalent to $y = 0.0508x + 0.699$) (Figure 3.18). Therefore,

$$e^{\lambda_{87}t} - 1 = 0.0508 \tag{3.11}$$

and

$$e^{\lambda_{87}t} = 1.0508. \tag{3.12}$$

The age (t) will therefore be

$$t = \ln(1.0508) / \lambda_{87}. \tag{3.13}$$

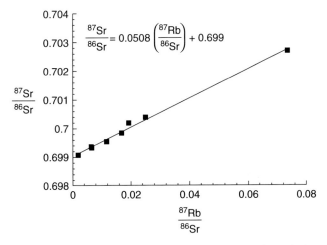

Figure 3.18 Plot of the Rb–Sr isochron for the Binda howardite (Birck and Allegre, 1979).

For a decay constant of 1.42×10^{-11} yrs^{-1}, the age will then be 3.49×10^9 yrs. This value is slightly different than the published age from Birck and Allegre (1979) due to fitting without using the published uncertainties. Since dating of other howardites yields ages of ~4.5×10^9 yrs, this age must have been reset by an impact, which would have remelted the rock.

This dating method is often used to date feldspars since rubidium is found as a trace element in these minerals since it can replace K or Na in the feldspar. Rubidium is also found as a trace element in olivines and pyroxenes. Since the Rb–Sr system can be reset by shock heating due to impacts, caution must be used in interpreting Rb–Sr ages.

3.9.2 Lead–Lead Dating

Lead–lead (Pb–Pb) dating uses isotopes of lead to date meteorites. Two different uranium isotopes decaying to two different stable Pb isotopes (Schoene, 2014). The decay chain of ^{238}U ultimately produces the stable ^{206}Pb with a half-life of 4.47 Ga. The decay chain of ^{235}U ultimately produces the stable ^{207}Pb with a half-life of 0.704 Ga. The decay equations can be written as

$$^{206}\text{Pb} = {}^{238}\text{U}\left(e^{\lambda_{238}t} - 1\right) + {}^{206}\text{Pb}_0 \tag{3.14}$$

and

$$^{207}\text{Pb} = {}^{235}\text{U}\left(e^{\lambda_{235}t} - 1\right) + {}^{207}\text{Pb}_0. \tag{3.15}$$

The decay constant for ^{238}U is given as λ_{238} and the decay constant for ^{235}U is given as λ_{235}.

The only nonradiogenic isotope of lead that is measured is ^{204}Pb. The isochron equations are

$$\frac{^{206}\text{Pb}}{^{204}\text{Pb}} = \left(e^{\lambda_{238}t} - 1\right)\left(\frac{^{238}\text{U}}{^{204}\text{Pb}}\right) + \left(\frac{^{206}\text{Pb}}{^{204}\text{Pb}}\right)_0 \tag{3.16}$$

and

$$\frac{^{207}\text{Pb}}{^{204}\text{Pb}} = \left(e^{\lambda_{235}t} - 1\right)\left(\frac{^{235}\text{U}}{^{204}\text{Pb}}\right) + \left(\frac{^{207}\text{Pb}}{^{204}\text{Pb}}\right)_0. \tag{3.17}$$

If the ages determined by two or more isotopic systems are the "same," the ages are considered concordant.

This dating method is often used to date zircons on Earth since they contain trace amounts of uranium and thorium. Zircons are also very resistant to mechanical and chemical weathering and metamorphism, which keeps the concentrations of radiogenic isotopes from being disturbed. Zircons also contain high relatively high concentrations of U and small concentration of Pb. However, zircons are relatively rare in chondrites but much more abundant in achondrites, Martian meteorites, and terrestrial rocks.

If you divide the two isochron equations (*Equations 3.16 and 3.17*) (Schoene, 2014), the resulting formula is

$$\frac{\dfrac{^{207}\text{Pb}}{^{204}\text{Pb}} - \left(\dfrac{^{207}\text{Pb}}{^{204}\text{Pb}}\right)_0}{\dfrac{^{206}\text{Pb}}{^{204}\text{Pb}} - \left(\dfrac{^{206}\text{Pb}}{^{204}\text{Pb}}\right)_0} = \left(\frac{^{235}\text{U}}{^{238}\text{U}}\right)\frac{\left(e^{\lambda_{235}t} - 1\right)}{\left(e^{\lambda_{238}t} - 1\right)}. \tag{3.18}$$

The above equation can then be rearranged to form the Pb–Pb isochron equation

$$\frac{^{207}\text{Pb}}{^{204}\text{Pb}} = a\left(\frac{^{206}\text{Pb}}{^{204}\text{Pb}}\right) + b \tag{3.19}$$

where

$$a = \left(\frac{^{235}\text{U}}{^{238}\text{U}}\right)\frac{\left(e^{\lambda_{235}t} - 1\right)}{\left(e^{\lambda_{238}t} - 1\right)} \tag{3.20}$$

and

$$b = \frac{^{207}\text{Pb}}{^{204}\text{Pb}} - \left(\frac{^{235}\text{U}}{^{238}\text{U}}\right)\frac{\left(e^{\lambda_{235}t} - 1\right)}{\left(e^{\lambda_{238}t} - 1\right)}\left(\frac{^{206}\text{Pb}}{^{204}\text{Pb}}\right)_0.$$ (3.21)

This Pb–Pb isochron equation is often used for dating terrestrial rocks.

For dating meteoritic material, this equation is usually rewritten as the inverse Pb–Pb isochron

$$\frac{^{207}\text{Pb}}{^{206}\text{Pb}} = b\left(\frac{^{204}\text{Pb}}{^{206}\text{Pb}}\right) + a$$ (3.22)

where a and b are the same variables as before. The reason for rewriting the equation is that ^{204}Pb abundances may be extremely low for meteoritic material, which will lead to large measurement errors for the isotopic ratios that will lead to large uncertainties in the calculated age. For the inverse Pb–Pb equation, the ^{207}Pb/^{206}Pb values are plotted on the y-axis and ^{204}Pb/^{206}Pb values are plotted on the x-axis (Figure 3.19). The y-intercept (a) of the line fit through the data will allow you to calculate the age of the sample since the value of ^{235}U/^{238}U is usually assumed to be a constant in meteoritic and terrestrial rocks with a value of 137.88. However recently, this ratio has been found to vary (Richter et al., 2010; Condon et al., 2010) and the value of 137.837 is sometimes used. Tissot et al. (2016) has found that decay of the short-lived radionuclide ^{247}Cm to ^{235}U would affect this value. The slope (b) of the fitted line will depend on the age of the sample and the initial Pb isotopic ratios.

If you assume that the contribution of initial lead is negligible (which is a valid assumption for many minerals) (Bouvier et al., 2007),

$$\frac{^{207}Pb^*}{^{206}Pb^*} = \left(\frac{^{235}U}{^{238}U}\right)\frac{\left(e^{\lambda_{235}t} - 1\right)}{\left(e^{\lambda_{238}t} - 1\right)}$$ (3.23)

where $*$ is used to indicate the radiogenic contribution of the isotope. The y-intercept (a) of *Equation 3.22* will then be equal to ^{207}Pb*/^{206}Pb*.

Since ^{235}U/^{238}U is assumed to have a constant value, the calculated age will not be affected by the loss of uranium (Dickin, 2005). Ages calculated with these equations are called Pb–Pb ages or dates. Pb–Pb dating has been used to calculate the ages of CAIs, which are the oldest dated material in the Solar System (Figure 3.19) (e.g., Bouvier and Wadhwa, 2010; Connelly et al., 2012). Pb–Pb ages are considered more precise than other ages since the isotopic ratios are determined of the same element instead of two geochemically different elements (Dalrymple, 1991), which leads to less analytical uncertainty.

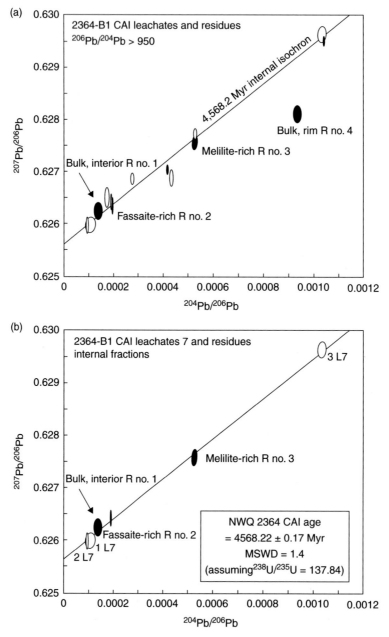

Figure 3.19 $^{207}Pb/^{206}Pb$ versus $^{204}Pb/^{206}Pb$ ratios for (a) residues (filled symbols) and the last three leachates (open symbols) and (b) residues and the last leachate (L7) for bulk samples and mineral fractions from a type B CAI from NWA 2364 (Bouvier and Wadhwa, 2010). The inverse Pb–Pb isochron plotted in (a) and (b) is calculated from the data in (b). Leaching with acid is done to remove the terrestrial lead component. Error ellipses are $\pm 2\sigma$ where σ is the standard deviation. Mean square weighted deviation (MSWD) is a measured of the goodness of the fit. Credit: Audrey Bouvier, Western University, and Nature Publishing Group.

3.9.3 Potassium–Argon and Argon–Argon Dating

Potassium–argon (K–Ar) dating is based on the decay of ^{40}K to either stable ^{40}Ca by emission of a beta particle or stable ^{40}Ar by electron capture or positron emission (which is very rare) (Dickin, 2005). The half-life of the decay to ^{40}Ca is 1.397×10^9 years, while the half-life of the electron capture decay to ^{40}Ar is 1.193×10^{10} years. The half-life for the decay of ^{40}K will be 1.251×10^9 years. The decay to ^{40}Ca occurs 89.1% of the time and to ^{40}Ar occurs 10.9% of the time. Potassium is very common in meteoritic minerals. The most abundant naturally occurring isotope of calcium is ^{40}Ar. Since nonradiogenic ^{40}Ca is usually so much more abundant in the mineral than radiogenic ^{40}Ca, the relative abundances of daughter ^{40}Ar are only used in the dating.

The decay equation of ^{40}K to ^{40}Ar will be

$$^{40}\mathrm{Ar} = \left(\frac{\lambda_{EC}}{\lambda_T} \right) \left(^{40}\mathrm{K} \right) (e^{\lambda_T t} - 1) + {}^{40}\mathrm{Ar}_0 \tag{3.24}$$

where λ_{EC} is the electron capture decay constant, λ_T is the total decay constant ^{40}Ar$_0$ is the initial abundance of ^{40}Ar. The isochron equation is

Since the initial abundance of argon in the mineral is usually assumed to be zero, the equation becomes

$$^{40}\mathrm{Ar}^* = \left(\frac{\lambda_{EC}}{\lambda_T} \right) \left(^{40}\mathrm{K} \right) \left(e^{\lambda_T t} - 1 \right) \tag{3.25}$$

where ^{40}Ar* is the radiogenic ^{40}Ar. However since argon is present in the atmosphere, it is extremely difficult to remove all nonradiogenic argon from the system. If nonradiogenic argon (^{36}Ar) is detected by the mass spectrometer, atmospheric contamination is usually assumed to have occurred (Dickin, 2005). The ratio of atmospheric (*atm*) ^{40}K to ^{36}K is

$$\left(\frac{^{40}\mathrm{Ar}}{^{36}\mathrm{Ar}} \right)_{atm} = 295.5. \tag{3.26}$$

To correct for atmospheric contributions to the abundances of ^{40}Ar, the abundances of ^{36}Ar and the relative ratios of different isotopes of argon in the atmosphere can be used to subtract out the contribution due to the atmospheric ^{40}Ar.

The K–Ar age (t) will be

$$t = \frac{1}{\lambda_T} \ln \left(\frac{^{40}\mathrm{Ar}^*}{^{40}\mathrm{K}} \frac{\lambda_T}{\lambda_{EC}} + 1 \right). \tag{3.27}$$

However, one major issue with K–Ar dating is that K and Ar abundances cannot be determined using the same instrument since potassium is found as a solid part of minerals, while argon is found a gas. So K and Ar isotopes must be measured on separate samples.

Bogard (1995) finds that the isotopic systems (in order of easiest to reset by shock heating due to impact) are K–Ar (easiest), Rb–Sr, Pb–Pb, and Sm–Nd (hardest). The K–Ar is easiest to reset since the daughter isotope is a noble gas and easily lost from the mineral.

To be able to measure the potassium and argon abundances at the same time, argon–argon dating was invented, which determines the age by converting the potassium into argon using neutrons. The reaction is

$$^{39}\text{K} + n \rightarrow {}^{39}\text{Ar} + p. \tag{3.28}$$

Argon–argon dating is done differently than most other forms of isotopic dating since a stable isotope is converted into a radioactive one. [Iodine–xenon (I–Xe) dating is also done by irradiating the sample with neutrons (e.g., Busfield et al., 2008)]. A meteoritic sample containing stable ^{39}K is irradiated with neutrons in a nuclear reactor to convert this potassium isotope into the radioactive noble gas ^{39}Ar. A standard sample (monitor) containing ^{39}K with a known age is also irradiated at the same time with the same neutron flux. To monitor variations in the neutron flux, a number of monitors must be used and the samples placed them. Corrections must be made for argon isotopes produced from calcium, potassium, argon, and chlorine induced by the neutron irradiation.

The irradiation parameter (J) will be

$$J = \frac{e^{\lambda_T t_m} - 1}{\left({}^{40}\text{Ar}^* / {}^{39}\text{Ar}\right)_m} \tag{3.29}$$

where $\left({}^{40}\text{Ar}^* / {}^{39}\text{Ar}\right)_m$ is the isotopic ratio measured for the monitor. The age of the sample will then be

$$t = \frac{1}{\lambda_T} \ln\left[J\left(\frac{{}^{40}\text{Ar}^*}{{}^{39}\text{Ar}}\right) + 1\right]. \tag{3.30}$$

One advantage of the Ar–Ar dating method is that the relative argon abundances can be measured by stepwise heating. During stepwise heating, the least tightly bound argon in the crystals will be released at the lowest temperatures and the more tightly bound argon will be released at higher temperatures. Ages can be determined after each heating step. If the age is the same for each step, the mineral is assumed to not have undergone any argon loss through a heating event (e.g., impact) or any argon diffusion. A plateau age is calculated by determining the "plateau" in the stepwise release spectrum (Figure 3.20) of the calculated age versus the fraction of ^{39}Ar that is cumulatively released (Dickin, 2005). The integrated age is determined from all the ^{39}Ar that is released. Also measured is the K/Ca ratio during the degassing, which helps to determine which phase is being degassed at a particular temperature. For example, the K/Ca ratio will be lower for pyroxenes and olivines than for feldspars.

Argon isotopic data are usually plotted as an inverse isochron, which plots values of $^{36}Ar/^{40}Ar$ versus $^{39}Ar/^{40}Ar$ (Figure 3.20). Since ^{36}Ar is nonradiogenic, the age of the sample can be determined by extrapolating the fitted line to a $^{36}Ar/^{40}Ar$ value of 0 and using that $^{36}Ar/^{40}Ar$ value to calculate the age (Kuiper, 2002). This age will the age when argon started to be "trapped" in the mineral, usually after a shock event that would have liberated the argon.

3.9.4 ^{26}Aluminum–^{26}Magnesium

The isotope ^{26}Al decays to ^{26}Mg with a half-life of 7.17×10^5 years, which allows the dating of events in the first ~5 Ma of Solar System history. The abundances of ^{26}Mg in the mineral will be the initial abundance of ^{26}Mg and the abundances of ^{26}Mg from the decay of ^{26}Al. The equation (Davis et al., 2015) for the abundances of ^{26}Mg relative to ^{24}Mg can be written as

$$\frac{^{26}Mg}{^{24}Mg} = \left(\frac{^{27}Al}{^{24}Mg}\right)\left(\frac{^{26}Al}{^{27}Al}\right)_0 + \left(\frac{^{26}Mg}{^{24}Mg}\right)_0 \tag{3.31}$$

where $^{26}Mg/^{24}Mg$ and $^{27}Al/^{24}Mg$ are the ratios measured today and $(^{26}Al/^{27}Al)_0$ and $(^{26}Mg/^{24}Mg)_0$ are the initial ratios at the time the body was isotopically homogenized.

The first convincing evidence for the presence of ^{26}Al (which would have all decayed due to its short half-life) in meteorites was the discovery of an excess of ^{26}Mg in an Allende chondrule (Lee et al., 1976). An excess is when the ^{26}Mg are enriched, while the abundances of ^{24}Mg and ^{25}Mg are terrestrial in value. A strong correlation was found between the excess of ^{26}Mg and the Al/Mg ratio, implying that the ^{26}Mg excess was related to the aluminum content. Measurements of CAIs find that the initial ("canonical") $^{26}Al/^{27}Al$ ratio was ~5×10^{-5} (e.g., MacPherson et al., 1995; Jacobsen et al., 2008; Wasserburg et al., 2012; Kita et al., 2013). Uncertainties in determining the canonical $^{26}Al/^{27}Al$ value correspond to ±20000 years, which Jacobsen et al. (2008) suggests is the timescale over which Allende CAIs formed.

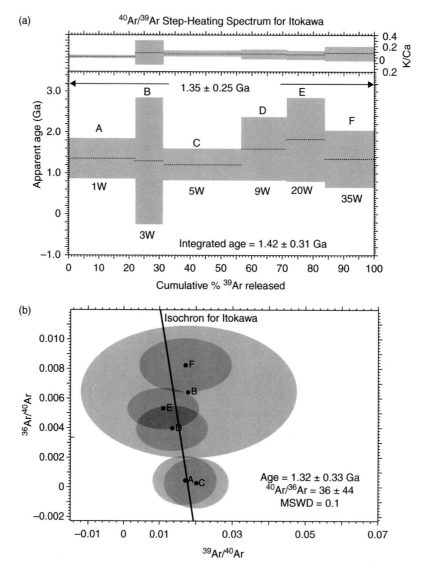

Figure 3.20 A) Stepwise heating spectrum for grains (measured as a single sample) retrieved from asteroid (25143) Itokawa by the Hayabusa sample return mission (Park et al., 2015). Also given is the K/Ca ratio, which helps to determine which phase is being degassed. B) Inverse isochron. The plateau age is 1.35 ± 0.25 Ga, the integrated total fusion age is 1.42 ± 0.31 Ga, and the age from the inverse isochron plot is 1.32 ± 0.33 Ga. This inverse isochron age is interpreted as indicating the age that the Itokawa rubble pile formed. Mean square weighted deviation (MSWD) is a measured of the goodness of the fit. Credit: Jisun Park, Kingsborough Community College, and John Wiley & Sons, Inc.

Magnesium isotopic data are usually presented as $\delta^{25}Mg$ and $\delta^{26}Mg$ (Davis et al., 2015) where

$$\delta^{25}Mg(\%_o) = \left[\frac{\left(^{25}Mg/^{24}Mg\right)_{MEAS}}{\left(^{25}Mg/^{24}Mg\right)_{STD}} - 1 \right] \times 1000 \qquad (3.32)$$

and

$$\delta^{26}Mg(\%_o) = \left[\frac{\left(^{26}Mg/^{24}Mg\right)_{MEAS}}{\left(^{26}Mg/^{24}Mg\right)_{STD}} - 1 \right] \times 1000. \qquad (3.33)$$

MEAS denotes the isotopic data from the mass spectrometer for the measured samples and *STD* denotes the isotopic data for the standard samples. These values are given in terms of parts per thousand. To determine the actual ^{26}Mg excess ($\delta^{26}Mg^*$), the measured data needs to be corrected for mass fractionation, which will almost always occur due to processes such as evaporation, diffusion, and ionization occurring during the mass spectroscopy. The simplest correction (Davis et al., 2015) will be

$$\delta^{26}Mg^*(\%_o) = \delta^{26}Mg - 2\delta^{25}Mg. \qquad (3.34)$$

Dating using the ^{26}Al–^{26}Mg system determines ages relative to CAIs, the oldest dating material in the Solar System. The equation (Kita and Ushikubo, 2012) for determining the age of a chondrule relative to a CAI is

$$\Delta t = \ln\left[\frac{\left(^{26}Al/^{27}Al\right)_{CAI}}{\left(^{26}Al/^{27}Al\right)_{Chondrule}} \right] \times \frac{1}{\lambda_{26}} \qquad (3.35)$$

where $\left(^{26}Al/^{27}Al\right)_{CAI}$ is the initial $^{26}Al/^{27}Al$ ratio of CAIs, $\left(^{26}Al/^{27}Al\right)_{Chondrule}$ is the initial $^{26}Al/^{27}Al$ ratio for the measured chondrule, and λ_{26} is the decay constant for ^{26}Al. The $^{26}Al/^{27}Al$ ratio of CAIs is the "canonical" ratio of 5×10^{-5}. The initial $^{26}Al/^{27}Al$ ratio for the measured chondrule is determined from the slope of the line (as seen from *Equation 3.31*) fit to the $^{26}Mg/^{24}Mg$ [or $\delta^{26}Mg^*(\%_o)$] y-axis) and the $^{27}Al/^{24}Mg$ (x-axis) values plotted (Figure 3.21) for measurements of minerals in the chondrule. A positive Δt indicates a younger age. If the age of the CAI is known from another dating method, an absolute can be determined. To determine the relative age of a CAI, the age is usually relative to angrites, which are the oldest known achondritic meteorites.

Example 3.2

If a chondrule has an initial $^{26}Al/^{27}Al$ ratio of 7.6×10^{-6}, what is its age relative to CAIs?

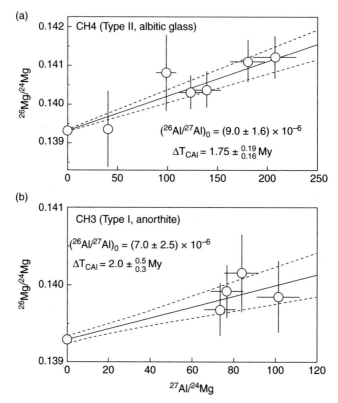

Figure 3.21 ^{26}Mg–^{26}Al isochrons for a (a) Type II and a (b) Type I chondrule (Kita and Ushikubo, 2012). Credit: Noriko Kita, University of Wisconsin, and John Wiley & Sons, Inc.

The initial ^{26}Al/^{27}Al ratio for the CAI is assumed to be the canonical value of 5×10^{-5}. Substituting the values in *Equation 3.35* results in the equation

$$\Delta t = \ln\left[\frac{5 \times 10^{-5}}{7.6 \times 10^{-6}}\right] \times \frac{1}{9.67 \times 10^{-7}} \text{ yrs.} \qquad (3.36)$$

The value of Δt for the chondrule is then 1.95×10^{6} years, which means that the chondrule formed 1.95 Ma after CAIs formed.

Radioactive decay can also heat the material. The decay of ^{26}Al to ^{26}Mg is thought to be the primary source of the heating and melting of meteoritic material. This isotope has a relatively short half-life and gives off energy when it decays (4 MeV or 6.4×10^{-13} J per atom) (Hevey and Sanders, 2006), and was present in significant abundances in meteorites when their parent bodies formed. The degree of heating for a body will be a function of a number of quantities such as the abundance of ^{26}Al, the accretion time, and the water ice content (e.g., Ghosh et al., 2002).

3.9.5 Hafnium–Tungsten Ages

Hafnium–tungsten (Hf–W) dating is often used to date planetary differentiation due to the different chemical affinities of hafnium and tungsten. The isotope ^{182}Hf decays via beta-minus decay to ^{182}W with a half-life of 8.90×10^6 years, which allows the dating of events in the first ~60 Ma of Solar System history (Kleine et al., 2009). Hafnium and tungsten are both very refractory elements; however, hafnium is a lithophile element, while tungsten is a siderophile element (Lee and Halliday, 1995; Halliday, 2000). Lithophile elements bond predominately with oxygen, so these elements will be found in the crust and mantle of a differentiated body. Siderophile elements will preferentially partition into metal phases during metal–silicate segregation. During core formation, the mantle will be enriched in hafnium and the metallic iron core will be enriched in tungsten. The Hf/W ratio in the silicate crust and mantle will be extremely high, while the ratio in the core will be approximately zero. How well W partitions into the core depends on a number of factors such as pressure, temperature, oxygen fugacity, and silicate melt composition (Kleine et al., 2009). After core formation, further fractionation can occur due to the relatively incompatible nature of W relative to Hf in many minerals such as clinopyroxenes and ilmenite.

If core formation take place relatively early in Solar System history (before the ^{182}Hf decays away), crust and mantle material should have an excess of ^{182}W compared to chondrites (the assumed starting material) due to its high Hf/W ratio, while metallic iron should be deficient in ^{182}W compared to chondrites due to its Hf/W ratio of approximately zero (Jacobsen, 2005). The core formation age is determined relative to the formation age of CAIs. The core formation age of a meteorite from a differentiated body is a complex function of ratios of the measured hafnium and tungsten isotopes of the crust/mantle or core, the degree of fractionation of Hf/W, and the ^{182}Hf/^{180}Hf ratio (where ^{180}Hf is a stable isotope) at the time of CAI formation (Kleine et al., 2009).

3.9.6 Ages of Meteorites

Technological advances have increased the precision with which ages can be determined. CAIs tend to have the oldest calculated ages of any analyzed material from our Solar System (e.g., Amelin et al., 2002, 2010; Bouvier and Wadhwa, 2010; Connelly et al., 2012). Bouvier and Wadhwa (2010) measured the Pb–Pb age of one type B CAI in CV3 chondrite NWA 2364 as 4568.22 ± 0.17 Ma and found that the Pb–Pb ages for type B CAIs from CV3 chondrites Allende and Efremovka ranged from 4568.5 ± 0.5 Ma to 4567.11 ± 0.16 Ma. Connelly et al. (2012) found that chondrules formed contemporaneously with CAIs, which they calculated to have formed at 4567.30 ± 0.16 Ma using values for CAIs with measured ^{235}U/^{238}U ratios. Previous researchers have found an age gap of ~1–4 Ma between CAI and chondrule formation (e.g., Kita et al., 2000, 2005; Krot et al., 2009). However, Connelly et al. (2012) found Pb–Pb ages ranging from 4567.32 ± 0.42 Ma for chondrules in Allende to 4564.71 ± 0.30 Ma for chondrules in L3 chondrite NWA 5697. In an earlier study, Amelin and Krot (2007) measured Pb–Pb ages that ranged from 4566.6 ± 1.0 Ma for chondrules in Allende to 4562.7 ± 0.5 Ma for chondrules in CB chondrite Gujba.

Compared to CAIs and chondrules, mineral grains in meteorites tend to have slightly younger crystallization ages. Brennecka and Wadhwa (2012) determined the Pb–Pb age of angrite D'Orbigny as 4563.37 ± 0.25 Ma. Nyquist et al. (2009) calculated a Pb–Pb age of 4558.55 ± 0.15 Ma for the LEW86010 angrite. Lugmair and Shukolyukov (1998, 2001) calculated a ^{53}Mn–^{53}Cr age for HEDs of 4564.8 ± 0.9 Ma, while Trinquier et al. (2008) calculated a very similar ^{53}Mn–^{53}Cr age for HEDs of 4564.9 ± 1.1 Ma. The angrite and HED ages are calculated from when minerals on these parent bodies crystallized. Dolomites in the CM2 chondrite Sutter's Mill have ^{53}Mn–^{53}Cr ages of 4564.8–4562.2 Ma (Jilly et al., 2014), which is dating the time that aqueous alteration occurred on the parent body.

The Pb–Pb ages of chondrules (4562.7 ± 0.5 Ma) in CB chondrites are younger than typically found in carbonaceous chondrites and similar to ages of the metal determined using the ^{182}Hf–^{182}W system (Krot et al., 2005). Krot et al. (2005) argues that CB chondrites formed through the impact of two planetesimals after dust in the protoplanetary disk had dissipated.

Relative ages determined for magmatic iron meteorites (Qin et al., 2008; Kleine et al., 2009) using the Hf–W system are consistent with the iron meteorite parent bodies forming within ~1 Ma after CAI formation. Pb–Pb ages of zircons from eucrites have ages of 4554 ± 7 Ma (Misawa et al., 2005).

The prevailing scenario (Kleine and Rudge, 2011) is that CAIs formed close to the Sun and were then transported a few AU out to where chondrules were forming (Ciesla, 2010). At the same time as chondrules were forming, iron meteorite parent bodies were also forming.

3.9.7 Cosmic-Ray Exposure Ages

Cosmic-ray exposure ages measure how long a meteorite was exposed to galactic cosmic rays in space (Marti and Graf, 1992; Eugster et al., 2006). The galactic cosmic rays are either protons or α particles (helium nuclei). The meteorite is either part of a meteoroid in space or near the surface of a large asteroid. These galactic cosmic rays interact with particles in the upper few meters of the surface and produce cosmogenic radionuclides (isotopes). These isotopes are produced either through spallation or neutron-capture reactions (Jull, 2006). Spallation is due to the collision of a nucleus with a high-energy particle, which results in the disintegration of the nucleus and the production of many neutrons. As these secondary neutrons slow down as they travel through the material, they can be captured by nuclei to produce radionuclides. Commonly used radionuclides for determining cosmic-ray exposure ages are the stable noble gas isotopes ^{3}He, ^{21}Ne, and ^{36}Ar (McSween and Huss, 2010). The low natural abundances of these stable noble gas isotopes in meteorites allow the effects of cosmic-ray irradiation to be easily detected compared to other isotopes. Since these isotopes are stable, the measured excesses of these cosmogenic isotopes will be proportional to the age.

To calculate a cosmic-ray exposure age, a number of assumptions must be made (Eugster et al., 2006). The flux of cosmic rays is assumed to be constant in space and time. The shape and chemical composition of the sample is assumed not to have changed. All non-cosmogenic production of the measured nuclides must also be known. The sample is assumed not to have lost any of the cosmogenic radionuclides that are being measured.

The meteoroids are usually assumed to be spherical and their sizes given in terms of radii (Eugster et al., 2006). The units of depth are given in terms of g/cm^2, which is distance (cm) times density (g/cm^3). These units are used since it better expresses the number of atoms available for nuclear reactions (Eugster et al., 2006).

For a large body, the production of a cosmogenic nuclide will first increase to a depth of ~50 g/cm^2 and then start decreasing by a factor of 2 for 100 g/cm^2 (Jull, 2006). This type of irradiation is called 2π irradiation since the galactic cosmic rays are coming from 180° of solid angle. For much smaller objects in space, the production of a cosmogenic nuclide will increase until the center of the object. This is called 4π irradiation since the galactic cosmic rays are coming from 360° of solid angle.

As expected, CI and CM chondrites tend to have the shortest cosmic-ray exposure ages (usually less than 2 Ma) (Eugster et al., 2006), which is consistent with their fragile nature. Also as expected, iron meteorites can have cosmic-ray exposure ages longer than 1 Ga, which is consistent with the strength of metallic iron. Cosmic-ray exposure ages of ordinary chondrites have been widely studied due to the large number of recovered meteorites. Marti and Graf (1992) have found that the distributions for all three ordinary chondrite groups (H, L, LL) have peaks at specific cosmic-ray exposure ages, which is consistent with impacts at these times on their parent bodies that liberated significant amounts of material. H chondrites peak at ~7 and ~33 Ma ago, L chondrites peak at ~28 and ~40 Ma ago, and LL chondrites peak at ~15 Ma ago. HEDs (Welten et al., 1997b; Eugster et al., 2006) peak at ~22 and ~36 Ma and just the eucrites peak at ~12 Ma. These clusters are thought to be the result of impacts at these times on the HED parent body.

Vokrouhlický and Farinella (2000) have found that the distribution of cosmic-ray exposure ages for H chondrites, L chondrites, and HEDs is consistent with movement in the asteroid belt due to the Yarkovsky effect. Cosmic-ray exposure ages for these meteorites tend to span a range from ~1 to ~100 Ma. However, fragments that enter meteorite-supplying resonances (3:1, 5:2, v_6) have lifetimes of 10 Ma or less with a median lifetime of 2–3 Ma (Morbidelli and Gladman, 1998). The Yarkovsky effect would allow fragments to drift for a significant time period with velocities of 10^{-2} to 10^{-4} AU/Ma before they reach the meteorite-supplying resonance, which would supply fragments to Earth relatively quickly. Drift due to the Yarkovsky effect drift is much more consistent with the long cosmic-ray exposure ages found for ordinary chondrites and HEDs than direct ejection of fragments into a resonance due to an impact on its original parent body.

Example 3.3

How long would it take an ordinary chondrite meteorite fragment to drift 0.3 AU to a meteorite-supplying resonance assuming the quickest Yarkovsky drift rate? Is this consistent with cosmic-ray exposure ages for these meteorites?

The time (t) it would take to enter the resonance can be calculated from the formula

$$t = d / v \qquad (3.37)$$

where v is the velocity and d is the distance. For velocities of 10^{-2} AU/Ma over a distance of 0.2 AU, the time to drift to the resonance would be

$$t = (0.3 / 10^{-2}) \text{ Ma} = 30 \text{ Ma.} \qquad (3.38)$$

This drift time is consistent with the ~1–100 Ma cosmic-ray exposure ages of ordinary chondrites.

3.9.8 Terrestrial Ages

The terrestrial age is the time that a meteorite has spent on the surface of the Earth (Jull, 2006). A meteorite in space is exposed to significant fluxes of galactic cosmic rays. However when a meteorite lands on Earth, this exposure is drastically reduced due to shielding by the Earth's atmosphere. Commonly used radionuclides for determining terrestrial ages include ^{14}C, ^{10}Be, ^{26}Al, ^{36}Cl, ^{41}Ca, and ^{81}Kr. The radionuclide that is used for determining the terrestrial age should have a half-life comparable to the terrestrial age of the meteorite (Herzog et al., 2015).

Meteorites from Antarctica tend to have terrestrial ages of 1 Ma or less (Welten et al., 1997a; Nishiizumi et al., 1989; Jull, 2006). One Antarctic H chondrite (FRO 01149) was found to have a terrestrial age of 3.0 ± 0.3 Ma (Welten et al., 2008). Meteorites in non-Antarctic deserts tend to have terrestrial ages of ~50 000 years or less (Welten et al., 2004; Jull, 2006) due to the accelerated degree of terrestrial weathering in those environments compared to Antarctica.

3.10 Planetary Meteorites

Meteorites are also known to originate from the Moon and Mars. Martian meteorites tend to have much younger crystallization ages than samples believed to originate on asteroids, while lunar meteorites tend to have slightly younger ages.

Meteorites could also possibly be derived from Mercury and Venus, but none have been "conclusively" identified so far. Meteorites appear dynamically possible from Mercury (Gladman and Coffey, 2009), but it is dynamically much more difficult to derive meteorites from Venus (Gladman et al., 1996). Using Messenger spacecraft observations of Mercury, Vaughan and Head (2014) predict that Mercurian meteorites would have younger crystallization ages (~4 Ga) compared to most meteorites. Ungrouped achondrite NWA 7325 has been postulated to be a fragment of Mercury.

3.10.1 Lunar Meteorites

Lunar meteorites are easier to recognize since they can be directly compared to samples returned by the American Apollo missions (11–17) (382 kg of returned material) and the Soviet Luna missions (16, 20, 24) (0.326 kg). The first lunar meteorite to be recognized was Allan Hills 81005 (ALH 81005 or ALHA81005) by Brian Mason (1917–2009) (Score and Mason, 1983). Lunar meteorites and returned lunar samples fall on the terrestrial fractionation line (e.g., Young et al., 2016). Over 110 meteorites are thought to be from the Moon. (Paired samples are counted as one meteorite.)

Lunar meteorites and samples fall on the terrestrial fractionation line. Young et al. (2016) argues that the indistinguishable oxygen isotope values between the Earth and the Moon argues that there must have been rigorous mixing during the giant impact that has been proposed to form the Moon.

3.10.2 Martian Meteorites

Martian meteorites were much harder to initially recognize since Martian samples have never been returned to Earth. A number of meteorites were initially recognized to have crystallization ages of ~1.3 Ga or younger (e.g., Nyquist et al., 2001). These meteorites were much younger than other meteorites. These ages implied that the parent body of these meteorites was molten for a much longer time period than the parent bodies of almost all meteorites. Such an object would be planetary-sized, since large bodies take a much longer time to cool and crystallize minerals than small asteroidal-sized bodies. The most obvious parent body for these meteorites was Mars (e.g., Wood and Ashwal, 1981) since deriving meteorites from Venus is dynamically much more difficult.

The definitive proof was that trapped gases in glasses in these meteorites had elemental and isotopic compositions similar to those found in the Martian atmosphere as measured by Viking 1 and 2 (Bogard and Johnson, 1983; Tremain et al., 2000). The Viking 1 and 2 spacecraft landed on Mars in 1976. The measured glass was formed through impacts, which trapped part of its parent body's atmosphere in the glasses. Of all the planetary bodies with atmospheres, only Mars had measured CO_2, N_2, and noble gas abundances and relative noble gas isotopic ratios comparable to what was measured in the glasses in these meteorites (Tremain et al., 2000). The first meteorite that was definitely shown to be from Mars was EET 79001 (or EETA79001). Oxygen isotopic values of Martian meteorites (Figure 3.9) also fall on a fractionation line that lies slightly above the terrestrial fractionation line, implying their formation on the same parent body.

Over 80 meteorites have been identified as being from Mars. (Paired samples are counted as one meteorite.) Martian meteorites are nicknamed SNCs after the three main Martian meteorite groups: shergottites (named after Shergotty), nakhlites (Nakhla), and chassignites (Chassigny). Shergottites are pyroxene-rich and contain predominately pigeonite and augite. Shergottites also contain a significant amount of plagioclase. Nakhlites are composed predominately of augite with some olivine. Chassigny is composed primarily of olivine. Over 100 meteorites have been identified as being from Mars (Figure 3.22).

The most famous Martian meteorite is ALH 84001. This Antarctic meteorite was argued to contain evidence for extraterrestrial life (McKay et al., 1996). It contains abundant PAHs and carbonate globules that contain single domain magnetite and iron sulfides. This meteorite also has an extremely old crystallization age (~4.51 Ga) (e.g., Nyquist et al., 2001).

3.11 Meteorites on Mars

Over 20 meteorites have also been identified on the surface of Mars (Ashley, 2015). These meteorites have sizes of approximately ~0.3 to ~2 meters and have significant FeNi abundances.

Figure 3.22 Image of Martian meteorite NWA 7034, which is nicknamed "Black Beauty" due to its black fusion crust (Agee et al., 2013). Credit: NASA. (A black and white version of this figure will appear in some formats. For the color version, please refer to the plate section.)

Questions

1) How would you identify if a rock is a meteorite?
2) What is the best continent on Earth to find meteorites? Why?
3) Why are iron meteorites the commonest type of meteorite to find on the surface of the Earth, while ordinary chondrites are the most likely to fall?
4) Why do meteorites from differentiated planetary bodies have much more diverse mineralogies than chondrites?
5) What is the difference between an absolute age and a relative age of a meteorite?
6) The measured values for $^{87}Rb/^{86}Sr$ and $^{87}Sr/^{86}Sr$ for the Piplia Kalan eucrite are given in Table 3.11. What is the calculated age of this meteorite? What can you assume from this value if other dating systems give an age of 4.57 Ga for this meteorite?

Meteorites, Minerals, and Isotopes

Table 3.11 *The $^{87}Rb/^{86}Sr$ and $^{87}Sr/^{86}Sr$ values measured for different samples of the Piplia Kalan eucrite (Kumar et al., 1999)*

Sample	$\dfrac{^{87}Rb}{^{86}Sr}$	$\dfrac{^{87}Sr}{^{86}Sr}$
1	0.0053	0.699348
2	0.0205	0.700225
3	0.0105	0.699624
4	0.0151	0.699885

7) If a chondrule has an initial $^{26}Al/^{27}Al$ ratio of 8.0×10^{-6}, what is its age relative to CAIs?

8) Why would heating due to ^{26}Al only occur at the beginning of Solar System's history? How could you determine if ^{26}Al was important in heating meteoritic bodies?

9) a) Why would you expect CM chondrites to tend to have shorter cosmic-ray exposure ages than ordinary chondrites?

   b) Why would you expect ordinary chondrites to tend to have shorter cosmic-ray exposure ages than iron meteorites?

10) Why are some meteorites thought to originate on Mars? Why were meteorites from Mars harder to initially recognize than ones from the Moon?

4

Reflectance Spectroscopy and Asteroid Taxonomy

4.1 Reflectance Spectra

The most widely used wavelength region to study asteroids is in the visible and near-infrared (~0.4 to ~2.5 μm) because this is where the Sun emits most of its energy, which is either absorbed or reflected by the asteroid. Luckily for interpreting asteroid compositions, many minerals found in meteorites (and, therefore, asteroids) have characteristic absorption bands in this wavelength region. Also, this wavelength region is very easy to observe from Earth, since the atmosphere is relatively transparent in this wavelength region (Figure 1.2).

Reflectance spectroscopy is the study of light that is scattered off a surface back to your telescope and detector. Light that is scattered includes light that is reflected from a surface or refracted through a particle (or particles) and then scattered back to the detector (Clark, 1999). Light that is not scattered is absorbed by the surface.

A reflectance spectrum plots the reflectance of an object (*y*-axis) versus wavelength (*x*-axis) (Figure 4.1). Since how well the body reflects light (its visual albedo) is often not known for an asteroid and measured independently from the reflectances, the reflectance spectrum is usually normalized to unity at 0.55 μm, which is approximately in the middle of the visible wavelength region.

Any observed body will also emit blackbody radiation. However for a reflectance spectrum measured out to ~2.5 μm, this emitted flux does not become significant except for low-albedo near-Earth objects (Figure 4.2). These bodies will have high surface temperatures and, therefore, have blackbody curves that peak at relatively short wavelengths with "significant" blackbody fluxes shortward of 2.5 μm.

4.1.1 Asteroid Reflectance Spectra

A reflectance spectrum plots the reflectance of an object versus wavelength; however, what is actually measured at the telescope is the flux from the body. This

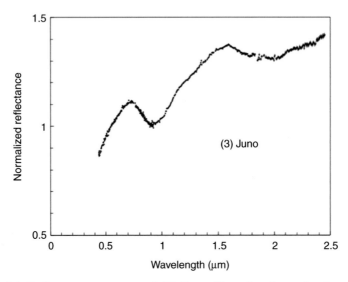

Figure 4.1 Reflectance spectrum of (3) Juno. Note the absorption bands. The spectrum is normalized to unity at 0.55 μm.

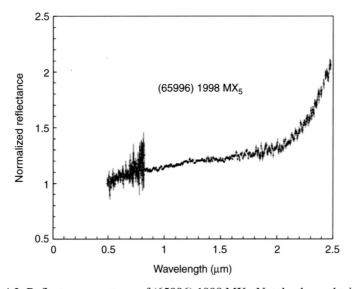

Figure 4.2 Reflectance spectrum of (65996) 1998 MX_5. Not the thermal tail longward of 2 μm. The spectrum is normalized to unity at 0.55 μm.

flux is radiation from the Sun that has interacted with the surface of the object and has been scattered in the direction of the telescope. However for telescopes on the surface of the Earth, the flux from the body will also be affected by the Earth's atmosphere. For example, absorption bands in the 3 region are extremely difficult to fully characterize from Earth-based observations due to telluric absorption features from atmospheric H_2O (e.g., Rivkin et al., 2002).

Since the flux from the Sun is extremely difficult to measure, a standard star with solar-like colors (e.g., G2 spectral type) is measured instead. The formula (Gaffey et al., 2002) to calculate the reflectance R_λ at a particular wavelength is

$$R_\lambda = \frac{F_\lambda(\text{object})}{F_\lambda(\text{star})} \times \frac{F_\lambda(\text{star})}{F_\lambda(\text{Sun})} \tag{4.1}$$

where F_λ is the flux at a particular wavelength for the object, the standard star, and the Sun.

4.1.2 Absorption Bands

Absorption bands are a range of wavelengths where photons are absorbed due to a particular transition. In a reflectance spectrum, an absorption band causes a decrease in reflectance since light is being preferentially absorbed and not scattered. These absorption bands are due primarily to electronic transitions and vibrational bands where photons are preferentially absorbed by minerals. Electronic transitions occur when an electron absorbs a photon and then moves to a higher energy level. These electronic bands are due to crystal field transitions, charge-transfer absorptions, and conduction bands. The names of these transitions refer to what is causing the transition. Crystal field transitions are due to the distortion of the energies of $3d$ (or $4d$) electron orbitals in a crystal field. Charge-transfer transitions are due to the movement of electron from one ion to another, which transfers charge. Conduction bands are due to the movement of electrons from a lower energy region (valence band) to a higher one (conduction band).

Vibrational absorption bands are due to the absorption of photons by molecules where the frequency of the absorbed radiation matches the vibrational frequency of the bond. The vibrational motion must cause a change in the dipole moment of the molecule for photons to be absorbed. Molecules can also have absorption bands associated with rotation and translation (movement of the molecule as a whole).

Absorption bands have often been characterized as a Gaussian distribution. A Gaussian distribution has a symmetric "bell curve" shape. The formula for a Gaussian distribution [$g(x)$] of a mineral absorption band is

$$g(x) = s \times e^{\left[-(x-\mu)^2/2\sigma^2\right]} \tag{4.2}$$

where s is the strength of the band, x is the random variable (with units of energy), μ is the center of the band in energy, and σ is the width of the band (given by the standard deviation) in energy (Sunshine et al., 1990). Sunshine et al. (1990) found that mineral absorption bands could be better fit with a modified Gaussian $[m(x)]$ with the form

$$m(x) = s \times e^{\left[-(x^{-1}-\mu^{-1})^2/2\sigma^2\right]}.$$

(4.3)

This equation is the basis for Modified Gaussian Modeling (MGM) of absorption bands.

4.1.3 *Crystal Field Transitions*

To understand crystal field transitions, the electron structure around an atom must be understood. The electron configuration for an atom derived from Bohr's model is a series of shells, sub-shells, and orbitals around a nucleus where an electron has the highest probability of being found. Each shell is defined by a number (e.g., 1, 2, 3) and each sub-shell is defined by a letter (e.g., *s, p, d, f*). The *s* sub-shells contain two orbitals, the *p* sub-shells contain three orbitals, and the *d* sub-shells contain five orbitals. Each orbital contains electrons with unique energies, angular momentums, and angular momentum vectors. Each orbital can only contain two electrons with one electron having a spin in one direction (spin up) and another in another direction (spin down) due to the Pauli exclusion principle.

The most prominent absorption bands in the visible and near-infrared are primarily due to transitions due to transition metals in different minerals. Transition metals are elements that have incompletely filled *d* or *f* sub-shells or cations with incompletely filled *d* or *f* sub-shells. First transition series metals have incompletely filled 3*d* sub-shells or cations with incompletely filled 3*d* sub-shells. Transition metals include Fe^{2+}, Fe^{3+}, Cr^{3+}, Mn^{2+}, and Mn^{3+}; however, Fe^{2+} is the most abundant of these transition metal ions in meteorites. For example, olivine, pyroxene, and phyllosilicate group minerals are all found in meteorites and can contain Fe^{2+}. These minerals have absorption bands due to crystal field transitions (e.g., Burns, 1993). Crystal field transitions for Fe^{2+}, Fe^{3+}, Cr^{3+}, Mn^{2+}, and Mn^{3+} are due to the absorption of photons by electrons in the partially filled inner (3*d*) orbitals of these transition metal ions.

For an isolated transition metal ion, electrons can reside in any of the d orbitals (Burns, 1993). The baricenter is the energy of the unsplit energy levels. However in a crystal structure, the non-spherical electrostatic field causes the degenerate (same energy) d orbitals to split into different energy levels. The amount of splitting is a function of the type and arrangement of the ligands around the metal ion. Ligands include ions (negatively charged ions) and dipolar molecules. In the non-spherical electrostatic field, the five d orbitals can be broken into two groups (called e_g and t_{2g}) depending on how the orbitals are aligned with the ligands. There are two orbitals in the e_g group and three orbitals in the t_{2g} group. The non-spherical electrostatic

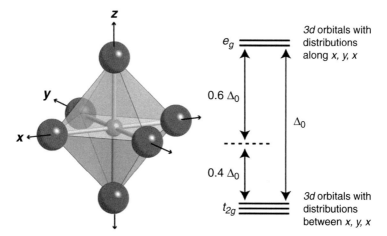

Figure 4.3 Idealized 6-coordinated polyhedron around a hypothetical Fe atom, along with the crystal field splitting diagram. When the Fe atom is in octahedral coordination, the energies of the Fe $3d$ orbitals change as a function of proximity to electronic distributions of the surrounding oxygen anions. When the octahedron is not perfectly symmetrical further splitting occurs among the e_g and t_{2g} orbitals. Such splitting of the energy levels makes possible multiple electronic transitions that can give rise to absorption features such as the bands seen in olivines and pyroxenes. Figure from Klima et al. (2011). Credit: Rachel Klima, JHUAPL and John Wiley & Sons, Inc. (A black and white version of this figure will appear in some formats. For the color version, please refer to the plate section.)

field splits all the orbitals into different energy levels about the baricenter with the two e_g group orbitals either having higher or lower energies than the three t_{2g} group orbitals. If the two e_g group orbitals have energies higher than the barycenter, the three t_{2g} group orbitals must be at lower energies than the barycenter and vice versa. For the crystallographic sites found in olivines and pyroxenes (Figure 4.3), the three t_{2g} group orbitals are at lower energies than the barycenter.

The distribution of the electrons in the d orbitals is controlled by two competing tendencies, which are part of Hund's rules (Burns, 1993). Electrons try be distributed over as many of the $3d$ orbitals as possible with spins in the same direction but also want to populate the orbitals with the lowest energies. The electron configuration for Fe^{2+} is $1s^2\ 2s^2\ 2p^6\ 3s^2\ 3p^6\ 3d^6$. Thus, the five $3d$ orbitals of Fe^{2+} contain only six electrons with the orbital with the lowest energy containing two electrons.

Particular transition metals will occupy sites in minerals if their sizes and charges are appropriate. Since different minerals have different chemical compositions and/or different crystal structures, the crystal field (energy level) splittings will be different for different minerals. Therefore, different minerals will have different characteristic crystal field absorptions and thus absorption bands at different wavelengths and different structures.

These bands are broader than expected due to the thermal vibration of atoms (Putnis, 1992). These vibrations change the distances between atoms and, therefore, the crystal field splittings will also oscillate around some average value. A number of bands may also overlap, which will also broaden an absorption feature.

A charge-transfer absorption is where an electron absorbs a photon and then the electron transfers from one ion to another ion. Conduction bands are present in some minerals, such as sulfides. Electrons can reside in a higher energy level called the "conduction band" where the electrons can move freely and a lower energy region called the "valence band" where electrons are part of individual atoms. The energy that is needed to excite an electron between the two bands is called the "band gap."

Absorption bands are usually characterized by a band center and a band depth. The band center is the center of the band after a linear continuum is removed. The continuum is the slope of the spectrum without the absorption band. Usually two points slightly outside the band are chosen and a straight line fit between these two points. The band depth can be calculated by the formula

$$\text{band depth} = 1 - \left(\frac{R_b}{R_c} \right) \tag{4.4}$$

where R_b is the reflectance at the band center and R_c is the reflectance of the continuum at the band center wavelength (Clark and Roush, 1984). The depth of an absorption band is related to the abundance of the absorber and also the grain size (Clark, 1999). Decreasing the grain size tends to decrease the band depth since smaller grain sizes have more surfaces where light can be scattered off. Larger grain sizes increase the probability that the material will absorb the photons.

Example 4.1

What will be the band depth of an absorption band that has a reflectance value of 0.8 at the band center wavelength of 0.9 μm? The continuum has a reflectance of 1.0 at 0.8 μm and a reflectance of 1.1 at 1.05 μm.

The slope of the continuum will be

$$\text{slope} = \left(\frac{1.1 - 1.0}{1.05 - 0.8} \right) \frac{1}{\mu m} = \frac{0.4}{\mu m}. \tag{4.5}$$

The reflectance at the continuum (R_c) can then be calculated from the formula

$$\left(\frac{R_c - 1.0}{0.9 - 0.8} \right) \frac{1}{\mu m} = \frac{0.4}{\mu m}. \tag{4.6}$$

The reflectance at the continuum at 0.9 μm will then be 1.04. The band depth can then be calculated from

$$\text{band depth} = 1 - \left(\frac{0.8}{1.04}\right) = 0.23.$$ (4.7)

The band depth will be 23%.

4.1.4 Absorption

Photons are absorbed by a material (e.g., Clark, 1999) according to Beer's law where

$$I = I_0 e^{-\alpha x}$$ (4.8)

where I is the observed intensity, I_0 is the original intensity, α is the absorption coefficient, and x is the distance traveled through the medium. The absorption coefficient usually has units of inverse length (m^{-1} or cm^{-1}) so x will have units of length (m or cm).

The absorption coefficient is related to a material's complex index of refraction ($\underline{n}$), which measures the speed of light in that material. The formula for the complex index of refraction ($\underline{n}$) is

$$\underline{n} = n + ik$$ (4.9)

where n is the real part of the complex index of refraction, k is the extinction coefficient (imaginary or complex index of refraction), and i is the imaginary unit that solves the equation $x^2 = -1$. The quantities n and k are called optical constants. These optical constants are an inherent property of the material and not dependent on particle size.

The real part of the index of refraction is a measure of how much the light is refracted, while the imaginary part is a measure of how much the light is absorbed. The real part of the complex index of refraction is the speed of light in a vacuum divided by the speed of light through the material. The absorption and extinction coefficients for a material vary versus wavelength. Spectral bands are due to variations in the extinction coefficient over a wavelength region (Lucey, 1998). The extinction coefficient varies by many orders of magnitude more than the real part of the index of refraction over different wavelengths.

The absorption coefficient (α) is calculated from the extinction coefficient (k) by the formula

$$\alpha = \frac{4\pi k}{\lambda}$$ (4.10)

where λ is the wavelength of light.

The reflectance (R) at a particular wavelength of a flat surface in a vacuum with light normally incident on it will just be a function of the optical constants of the material at those wavelengths (e.g., Hapke, 1993, 2012b; Clark, 1999) and is given by the Fresnel equation

$$R = \frac{(n-1)^2 + k^2}{(n+1)^2 + k^2}.$$ (4.11)

If $k = 0$, then R will equal $\dfrac{(n-1)^2}{(n+1)^2}$. If $k \gg 0$, then R will equal 1. However, calculating the reflectance of a material is much more complicated since the incident and emission almost always occurs at angles away from the normal on particulate surfaces.

From *Equation 4.9*, surfaces with large extinction coefficients (e.g., metallic iron) will have high reflectances (e.g., Hapke, 2012b). This is due to materials with large extinction coefficients also having large conductivities. These large conductivities allow currents to be produced at the surface due to the incident radiation, which does not allow the propagation of light through the material. Therefore, light will tend to be reflected from the material.

Pyroxenes tend to have larger absorption coefficients than olivine so pyroxenes tend to dominate the spectral properties of pyroxene–olivine mixtures since they are more absorbing. So for minerals in the visible and near-infrared, the strength of the absorption bands does not directly correspond to their actual abundances in the mixture.

4.2 Radiative Transfer Modeling

When light strikes a regolith, it will be absorbed or scattered by individual particles. Absorption is when a particle absorbs a photon. Scattering includes a number of processes that affect how a photon interacts with a particle. Light can either be scattered back to the observer (backward scattering) or forward (forward scattering). Forward scattered light will interact with another particle. Backward scattering includes the photon being reflected from the surface of the grain or being transmitted through the grain until it reaches the surface on the other side of the grain and then reflected back. Forward scattering includes the photon being transmitted all the way through the grain or being diffracted by the particle. Diffraction is the slight bending of light as it passes by a particle. Hapke (1999) found that diffracted light could be treated as unscattered light when modeling the interaction of light with a regolith.

Radiative transfer is the physical phenomenon of energy transfer when light interacts with a material. Radiative transfer modeling of asteroid surfaces allows the reflectance of a surface to be modeled by simulating the interaction of light with a particulate surface. Since the composition of the asteroid's surface is unknown, this modeling allows the surface mineralogy to be estimated by trying to reproduce the measured reflectance spectrum using the spectral properties of known minerals. However as expected, the modeling of how light interacts with a particulate surface is extremely complicated since you are dealing with sextillions (10^{21}) of photons interacting with trillions ($\sim 10^{12}$) of particles on the surface each second. In a series of papers (Hapke, 1981, 1984, 1986, 1999, 2001, 2002, 2008, 2012a; Hapke and Wells, 1981) and books (Hapke, 1993, 2012b), Bruce Hapke developed a way to model these interactions, which is called Hapke theory. The following discussion of Hapke theory is based on the papers and books written by Hapke plus numerous other papers (e.g., Mustard and Pieters, 1989; Hiroi and Pieters, 1994; Lawrence and Lucey, 2007; Ciarniello et al., 2014; Li et al., 2015) and theses (e.g., Sklute, 2014). A discussion of possible shortcomings to Hapke theory can be found in Shkuratov et al. (2012) and Hapke (2013). Hapke theory assumes that the particle size must be much larger than the wavelength, the particles must be touching, and the particles must be randomly oriented.

The light interacting with a surface is usually given in terms of power (P). Power is the energy per unit time and is given in terms of watts (joules per second). Two quantities that are often used to model how radiation interacts with a material are irradiance and radiance. Irradiance measures the incident flux of radiation on a surface, while radiance measures the emitted flux of radiation from that surface. The irradiance (J) is the power per unit area (flux) of a collimated (where light rays are parallel) beam received by a surface. The radiance (I), also known as intensity or brightness, is the uncollimated (where the light rays are not parallel) power per unit solid angle per unit area that is radiated by a surface. The measured radiance will be a combination of light that has been scattered only once off the surface and light that has been scattered off multiple surfaces. In the metric system, the units of irradiance are watts per square meter, while the units of radiance are usually given as watts per steradian (sr) per square meter. Steradian is the unit of solid angle (two-dimensional angle in three-dimensional space) and is dimensionless. The maximum solid angle that can be subtended is 4π steradians.

Two angles are needed to define the geometry of the interaction of light on the surface. These angles are the incident angle (i) and the emission angle (e) of the light interacting with the surface. The incident light on the surface will make an angle i with a line normal to the surface. The reflected light makes an angle e with a line normal to the surface. The incident and emission angles are usually given as $\mu_o = \cos i$ and $\mu = \cos e$. Cosines are used since the irradiance striking a surface is proportional to the cosine of the incident angle. This relation is called Lambert's cosine law after Johann Lambert (1728–1777) who first saw this relationship between incident light and angle. The phase angle (g) is the angle formed between the incident and emission directions.

The ratio of the power of the incident radiation that is affected by a single particle to the total irradiance is the extinction cross section (σ_E). Extinction includes the processes of absorption and scattering, which both remove energy from a beam of light incident on a surface in a particular direction. This extinction cross section (σ_E) is $\dfrac{P_E}{J}$, the scattering cross section (σ_S) is $\dfrac{P_S}{J}$, and the absorption cross section (σ_A) is $\dfrac{P_A}{J}$ where P_E is the extincted power, P_A is the absorbed power, and P_S is the scattered power. These cross sections are theoretical areas that are related to the probability that a photon interacts with a particle. Since energy is conserved ($P_E = P_S + P_A$), $\sigma_E = \sigma_S + \sigma_A$.

The geometrical cross section (σ) of a particle is πa^2 where a is the radius of the particle. The ratio of the cross section for the particular interaction (extinction, scattering, absorption) to the geometrical cross section is the efficiency (Q) for that particular interaction. The extinction efficiency is $Q_E = \dfrac{\sigma_E}{\sigma}$, the scattering efficiency is $Q_S = \dfrac{\sigma_S}{\sigma}$, and the absorption efficiency is $Q_A = \dfrac{\sigma_A}{\sigma}$. Therefore, $Q_E = Q_S + Q_A$.

4.2.1 Bidirectional Reflectance for Mixture of Single Type of Particle

Hapke theory calculates the bidirectional reflectance, which is the ratio of the radiance (I) scattered by a surface at a particular angle to the irradiance (J) incident on the surface. The units of bidirectional reflectance are sr^{-1}. The bidirectional reflectance [$r_\lambda(i,e,g)$] of a surface at a particular incident angle (i) and emission angle (e) relative to the normal at a surface and at a particular phase angle (g) is approximated by the formula

$$r_\lambda(i,e,g) = \frac{w}{4\pi} \frac{\mu_0}{\mu_0 + \mu} \left[\left(1 + B(g)\right) p(g) + H(\mu_0)H(\mu) - 1 \right] sr^{-1}. \quad (4.12)$$

The measured bidirectional reflectance is a function of the incident and emission angles. This bidirectional reflectance equation accounts for single and multiple scattering and also backscatter.

Each part of the equation will be discussed. Let us first assume that surface is composed of one type of particle (usually a mineral). The single scattering albedo (*w*) is the probability that a photon is scattered by a particle. The single scattering albedo is the ratio of the scattered power to the total power that is affected (called extinction) by a single particle. The single scattering albedo is equal to

$$w = \frac{P_S}{P_E} = \frac{\sigma_S}{\sigma_E}.$$

(4.13)

The single scattering albedo is a function of wavelength. The $\frac{1}{4\pi}$ term adjusts the reflectance so it has units of sr^{-1}. The $\frac{w}{4\pi} \frac{\mu_0}{\mu_0 + \mu}$ term is the reflectance per solid angle that is just due to single scattering. This reflectance term is the pretty much the Lommel–Seeliger law, which models isotropic scattering off a surface. The bidirectional reflectance is a function of wavelength since the single scattering albedo is also a function of wavelength.

The single particle phase function [$p(g)$], which is a function of the phase angle, describes the angular pattern of the scattered power off a particle. For isotropic scattering, $p(g)$ is unity since the particle scatters equally in all directions. For minerals, $p(g)$ is usually approximated by

$$p(g) = 1 + b\cos(g) + c[1.5\cos^2(g) - 0.5]$$

(4.14)

where *b* and *c* are constants that determine the shape of the scattering function for a particular mineral. The constant *b* describes the degree of forward and backward scattering and the constant *c* describes the degree of side scattering. These constants are functions of wavelength (Mustard and Pieters, 1989). However for simplicity, Lucey (1998) uses values of −0.4 for *b* and 0.25 for *c* for his radiative transfer modeling of olivine and pyroxene spectra. These values of *b* and *c* are estimates of the average values for the minerals studied by Mustard and Pieters (1989) that tended to forward scatter.

As mentioned earlier, backscatter is the sharp surge in brightness when an object is observed near zero phase angle. $B(g)$ is the backscatter function, which is a function of phase angle. The backscatter term can be approximated by the formula

$$B(g) = \frac{B_0}{1 + \frac{1}{h}\tan\left(\frac{g}{2}\right)}$$

(4.15)

where B_0 is the amplitude of the opposition effect, *h* is the angular width parameter of the opposition effect, and *g* is the phase angle. Mustard and Pieters (1989) found that the contribution due to backscatter for phase angles 15° or greater is negligible, so this term can be set to zero for these phase angles.

$H(\mu_0)$ and $H(\mu)$ are both Chandrasekhar H functions, which are used to approximate multiple isotropic scattering of radiation in a material (Chandrasekhar, 1960). A Chandrasekhar H function solves the two-stream solution for radiative transfer, which models the passage of radiation as fluxes in two directions (forward and backward). A thin layer of particles is assumed to exist over a much thicker layer of particles. The "−1" term that is added in the bidirectional reflectance equation removes the component of the first surface isotropic scattering from the $H(\mu_0)H(\mu)$ term, which has already been accounted for with the $p(g)$ term. The Chandrasekhar H function can be approximated by

$$H(x) \approx \left\{ 1 - [1 - \gamma]x \left[r_0 + \left(1 - \frac{1}{2}r_0 - r_0 x \right) \ln \left(\frac{1+x}{x} \right) \right] \right\}^{-1} \tag{4.16}$$

where

$$r_0 = \frac{1 - \gamma}{1 + \gamma} \tag{4.17}$$

and

$$\gamma = \sqrt{1 - w}. \tag{4.18}$$

Either μ_0 or μ will be substituted for x. Since the Chandrasekhar H functions are a function of the single scattering albedo, they are also a function of wavelength.

4.2.2 Bidirectional Reflectance for Mixture of Particles

The single scattering albedo in the bidirectional reflectance equation is affected if the surface is a mixture of particles. The single scattering albedo (w) can be calculated for a mixture using the equation

$$w = \frac{\sum\limits_j \dfrac{M_j}{\rho_j d_j} w_j}{\sum\limits_j \dfrac{M_j}{\rho_j d_j}} \tag{4.19}$$

where w_j is the probability that a photon is scattered by a particle for each mineral (where the subscript j indicates a different mineral), M_j is the mass fraction for each mineral, ρ_j is the single particle density for each mineral, and d_j is the average

Table 4.1 *Prominent absorption features identified in meteorite and asteroid spectra. Features are listed by their approximate band center except for the UV feature shortward of ~0.7 μm.*

Feature (μm)	Origin	Reference
0.5	Fe^{2+} in oldhamite	Burbine et al. (2002)
0.506	Fe^{2+} in pyroxene	Hiroi et al. (2001a); Cloutis et al. (2010c)
<0.7	Fe^{2+}	Clark (1999); Vilas and Hendrix (2015)
0.7	Fe^{3+} in phyllosilicates	Vilas and Gaffey (1989)
0.9	Fe^{2+} in pyroxene	Burns (1993)
1.0	Fe^{2+} in magnetite	Yang and Jewitt (2010); Cloutis et al. (2011a)
1.1	Fe^{2+} in olivine	Burns (1993)
1.9	Fe^{2+} in pyroxene	Burns (1993)
2.2	Fe^{2+} in spinel	Sunshine et al. (2008)
2.7	–OH in phyllosilicates	Takir et al. (2013)
3.0	H_2O	Gaffey et al. (1989)
3.4	C–H in organics	Rivkin and Emery (2010)

effective particle size of each mineral. This equation weights the average single scattering albedo of each type of particle by its geometrical cross section. Hapke theory has been used extensively to model the spectra of meteorites and asteroids (e.g., Clark, 1995; Lawrence and Lucey, 2007) and has been found to be able to duplicate very well their reflectance spectra with plausible mineral assemblages and compositions.

4.3 Mineral Absorption Features

A number of different absorption features have been identified in meteorite and asteroid spectra. A listing of some of the most prominent absorption features are given in Table 4.1.

4.3.1 Olivine

Olivine has three bands at ~0.9, ~1.1, and ~1.25 μm that form an asymmetric 1 μm feature (Figure 4.4). Forsterite (Fo) is the magnesium end member of the olivine solid solution series, while fayalite (Fa) is the iron-rich end member. The bands move to slightly longer wavelengths as Fe^{2+} substitutes for Mg.

These absorptions are due to Fe^{2+} in two different sites in olivine, which are called M1 and M2. M1 and M2 are two distorted octahedral coordination sites in the olivine crystal structure. The M1 site is more distorted than the M2 site. The

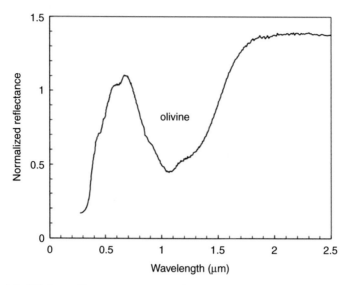

Figure 4.4 Olivine reflectance spectrum from Clark et al. (2007) in the visible and near-infrared. Note the asymmetric ~1 μm band. The spectrum is normalized to unity at 0.55 μm.

absorptions at ~0.9 and ~1.25 μm are due to transitions of $3d$ electrons in the M1 site, while the absorption at ~1.1 μm is due to transitions of $3d$ electrons in the M2 site.

Olivine also has a strong absorption shortward of ~0.7 μm, which is due to a charge-transfer absorption between oxygen and Fe^{2+}. This is called an oxygen–metal charge transfer. This charge transfer produces high intensity bands that are generally centered in the ultraviolet but do extend into the visible wavelength region.

4.3.2 Pyroxene

Pyroxenes tend to have two symmetric bands at ~0.9–1.0 (called Band I) and ~1.9–2.0 μm (Band II) (Figure 4.5). These absorptions are due to Fe^{2+}. Band I is due to transitions of $3d$ electrons in the M1 and M2 sites, while Band II is due to transitions of $3d$ electrons in the M2 site. The bands move to longer wavelengths as Fe^{2+} substitutes for Mg^{2+} (e.g., Klima et al., 2007) and/or Ca^{2+} substitutes for Fe^{2+} (e.g., Klima et al., 2011). FeO-free pyroxenes (enstatite) do not have these absorption bands due to the absence of Fe^{2+} and, therefore, have featureless, flat spectra in the visible and near-infrared (Figure 4.6). High-Ca pyroxenes (such as hedenbergite) have Ca occupying the M2 site so Fe^{2+} in the M1 site causes absorption bands at ~0.93–0.98 and ~1.18 μm and no band at ~1.9–2.0 μm (Figure 4.7). Pyroxenes also

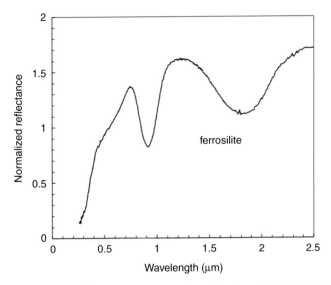

Figure 4.5 Ferrosilite reflectance spectrum from Clark et al. (2007) in the visible and near-infrared. Note the symmetric bands at ~0.9 and ~1.9 μm. The spectrum is normalized to unity at 0.55 μm.

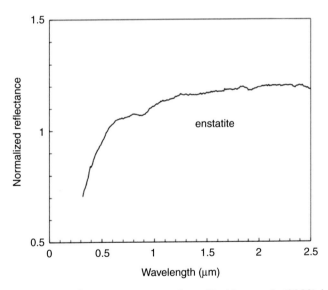

Figure 4.6 Enstatite reflectance spectrum from Burbine et al. (2002) in the visible and near-infrared. Enstatite is from the Peña Blanca Spring aubrite. Weak absorption features are due to terrestrial weathering. The spectrum is normalized to unity at 0.55 μm.

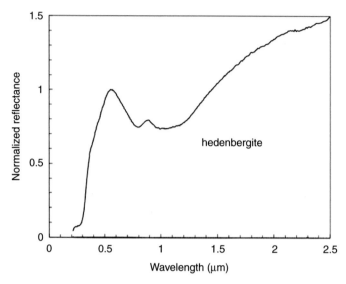

Figure 4.7 Hedenbergite reflectance spectrum from Clark et al. (2007) in the visible and near-infrared. Note the absence of a ~2 μm band. The spectrum is normalized to unity at 0.55 μm.

have a strong absorption shortward of ~0.7 μm, which is due to a charge-transfer absorption between oxygen and Fe^{2+}.

4.3.3 Phyllosilicates

Phyllosilicates have a number of vibrational absorption bands, which primarily occur in the near- to mid-infrared. The positions of these vibrational bands are a function of mineralogy and temperature (e.g., Henning, 2010). Water-bearing and hydroxyl-bearing minerals (Figure 4.8) can also have very strong vibrational absorption bands that are present in asteroid spectra (e.g., Rivkin et al., 2002). H_2O is a V-shaped molecule with hydrogen atoms at the ends of each of the arms and an oxygen atom at the apex of the V. H_2O can have three types of vibrations (Gaffey et al., 1989): a bend where the V opens and closes at a fundamental frequency corresponding to a wavelength of ~6.1 μm, a symmetric stretch where the arms both lengthen and shorten together at ~3.05 μm, and an asymmetric stretch where one arm lengthens or shortens at ~2.9 μm. The bend and symmetric stretch fundamentals combine with the first overtone of the bend fundamental to produce a strong absorption feature in the 3 μm wavelength region. (An overtone is any resonant frequency above the fundamental frequency.) OH in minerals can only undergo an asymmetric stretch and causes a sharp absorption band at ~2.7 μm

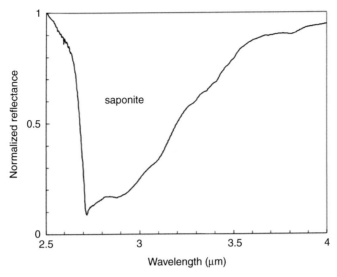

Figure 4.8 Saponite reflectance spectrum from Clark et al. (2007) in the 3 μm region. Spectrum normalized to unity at 2.5 μm.

(e.g., Rivkin et al., 2002). The structure of the 3 μm feature depends on what water-bearing and hydroxyl-bearing minerals are present as evident by the different 3 μm band shapes for different meteorite types (e.g., Sato et al., 1997; Takir et al., 2013). Mg-rich serpentines have a feature at ~2.72 μm, while the Fe-rich cronstedtite has a feature at ~2.85 μm (Takir et al., 2013). An absorption band at ~1.4–1.5 μm is caused by an overtone of the O-H stretch and absorption band at ~1.9–2.0 μm is caused by a combination of the O-H stretch and the H_2O bend (Cloutis, 2001).

A feature centered at ~0.7 μm in the spectra of some phyllosilicates (e.g., Vilas and Gaffey, 1989; Cloutis et al., 2011b) is attributed to an intervalence charge transfer ($Fe^{2+} \rightarrow Fe^{3+}$) transition in in phyllosilicates. This feature at ~0.7 μm has been found in the spectra of cronstedtite, which contains Fe^{3+}.

4.3.4 Spinel

If aluminous spinel contains some Fe^{2+} in its crystal structure, the spinel will have a strong 2 μm absorption feature (Sunshine et al., 2008) (Figure 4.9). Reflectance spectra of CAIs show this strong 2 μm absorption band. The spectrum of a spinel can be differentiated from an orthopyroxene spectrum because the spinel spectrum has a 2 μm band that is stronger than its 1 μm band. Orthopyroxenes will have a stronger 1 μm band than its 2 μm band.

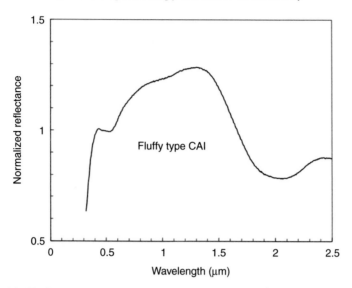

Figure 4.9 Fluffy type A CAI reflectance spectrum from Allende from Sunshine et al. (2008). Note how the ~2 μm band is stronger than the ~1 μm band. The spectrum is normalized to unity at 0.55 μm.

4.3.5 Sulfides

Conduction bands are present in some sulfides. Oldhamite (CaS), which is found in aubrites and enstatite chondrites (Watters and Prinz, 1979), has a strong conduction band that is centered at ~0.5 μm (Burbine et al., 2002) plus a feature at ~0.9 that is due to a Fe^{2+} crystal field transition band (Figure 4.10).

4.3.6 Metallic Iron

Common minerals found in meteorites that do not have absorption bands are kamacite and taenite. Both minerals are composed of Fe and Ni. Kamacite has a lower concentration of Ni than taenite plus a different crystal structure. Cloutis et al. (2010a) found that metallic iron powders had slightly red spectral slopes (Figure 4.11) and that there was no systematic effect due to Ni concentrations on the spectral slopes of the metallic iron spectra.

4.4 Reflectance Spectra of Meteorites

The reflectance spectra of meteorites can be measured in the laboratory and then directly compared to asteroid spectra. Samples are usually made into a powder and then measured with a spectrometer. The reflectance spectra is usually measured on samples of tens to hundreds of milligrams of meteoritic material, which will be

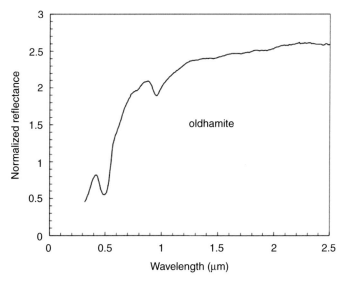

Figure 4.10 Oldhamite reflectance spectrum from the Norton County aubrite from Burbine et al. (2002) in the visible and near-infrared. The spectrum is normalized to unity at 0.55 μm.

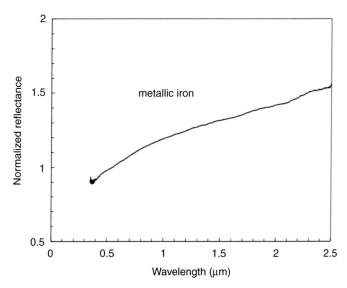

Figure 4.11 Odessa (iron) meteorite reflectance spectrum from Cloutis et al. (2010a). The meteorite sample has been ground to a powder and has a grain size of less than 45 μm. Note the absence of absorption features. The spectrum is normalized to unity at 0.55 μm.

compared to the reflectance spectra of asteroids that are usually tens to hundreds of kilometers in diameter. One of the most widely used spectroscopy facilities is the Keck/NASA Reflectance Experiment Laboratory (RELAB) facility located at Brown University. RELAB has two spectrometers. One is a near-ultraviolet, visible, and near-infrared bidirectional spectrometer and the other is a near- and mid-infrared FTIR (Fourier transform infrared) spectrometer.

A large number of meteorites have had their spectra measured. Johnson and Fanale (1973) did a spectroscopic study of carbonaceous chondrites from ~0.3 to 2.5 μm, while Salisbury and Hunt (1974) and Salisbury et al. (1975) did a spectroscopic study of stony meteorites in the same wavelength region. Michael Gaffey did one of the most extensive studies of the reflectance spectra of a wide range of meteorite groups as part of his PhD thesis (Gaffey, 1974). His spectra were published in Gaffey (1976). Dunn et al. (2010) did a coordinated spectral, mineralogical, and compositional study of ordinary chondrites. Spectra were obtained at RELAB, X-ray fluorescence data were obtained to determine mineralogies, and thin sections were probed to determine mineral chemistries. In a series of papers, Edward Cloutis did a spectroscopic study of a large number of carbonaceous chondrites (Cloutis et al., 2011a, 2011b, 2012a, 2012b, 2012c, 2012d, 2012f) and ureilites (Cloutis et al., 2010b).

4.4.1 Ordinary Chondrites

Ordinary chondrites have absorption features due to olivine and pyroxene (Gaffey, 1976; Dunn et al., 2010) (Figure 4.12). The absorption features of olivine and pyroxene overlap in the ~0.9–1.0 μm region (Band I), while only the pyroxene band is present around ~1.9 μm (Band II). Since H chondrites have approximately equal abundances of olivine and pyroxene and the pyroxenes are more absorbing, Band I is primarily due to the pyroxenes. The relatively low-Fe contents of their pyroxenes causes the Band II to have a band center at ~1.92 μm. Since L chondrites have higher abundances of olivine to pyroxene and higher Fe contents in their silicates, Band I and Band II move to longer wavelengths and Band II has more of a spectral contribution of olivine, which distorts the symmetric pyroxene feature. LL chondrites have the highest abundance of olivine among ordinary chondrites and the highest Fe-contents in their silicates, Band I and Band II are at the longest wavelengths and Band I has a shape most similar to a pure olivine mixture. Metallic iron and opaques will suppress the band depths relative to olivine–pyroxene mixtures.

4.4.2 Carbonaceous Chondrites

In the visible and near-infrared (Figure 4.13), carbonaceous chondrites have low visual albedos and suppressed absorption bands due to their high abundances of

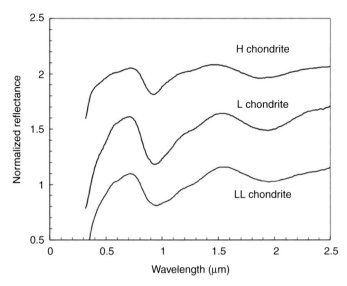

Figure 4.12 Reflectance spectra of H, L, and LL chondrites in the visible and near-infrared from Burbine et al. (2003). The meteorites are H5 chondrite Allegan, L6 chondrite Girgenti, and LL4 chondrite Greenwell Springs. Note how the LL chondrite has a stronger absorption due to olivine than the H and L chondrite. The spectra are taken of powders that have had the metal sieved out of them. The spectra are normalized to unity at 0.55 μm and offset in reflectance.

opaques. CI chondrites are relatively featureless and red-sloped. CM chondrites have a 0.7 μm band and are also red-sloped. CV chondrites have an absorption band centered at ~1 μm due to olivine.

In the ~3 μm wavelength region (Figure 4.14), carbonaceous chondrites (e.g., CI, CM) contain phyllosilicates that have strong absorptions due to –OH. Beck et al. (2010) and Takir et al. (2013) revolutionized the study of this 3 μm region by showing that phyllosilicate-bearing meteorites (CI and CM chondrites) measured under dry (elevated temperature) conditions and under vacuum have a strong feature at ~2.7 μm that tends to be distinctly different from the features found for these meteorites when their spectra are measured at room temperature and at ambient pressure. Spectra that were measured at room temperature and ambient pressure have extremely wide 3 μm bands due to absorbed terrestrial water. The presence of a strong feature at ~2.7 μm is consistent with meteorites containing significant abundances of Mg-rich phyllosilicates. In Figure 4.14, the feature at ~3.4 μm in the CI chondrite spectrum is due to organics.

4.4.3 HEDs

HEDs (howardites, eucrites, diogenites) all have prominent pyroxene bands (e.g., Gaffey, 1976; Beck et al., 2011a) (Figure 4.15). Since the bands move to longer

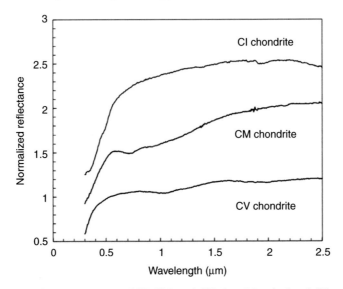

Figure 4.13 Reflectance spectra of CI, CM, and CV chondrites in the visible and near-infrared. The meteorites are the CI1 chondrite Orgueil, CM2 chondrite Murchison, and CV3 chondrite Allende. The spectra of the CI and CM chondrites are from Hiroi et al. (1993), while the CV spectrum is from Sunshine et al. (2008). Note the ~0.7 µm band in the CM chondrite spectra and the 1 µm band in the CV chondrite spectra. The spectra are normalized to unity at 0.55 µm and offset in reflectance.

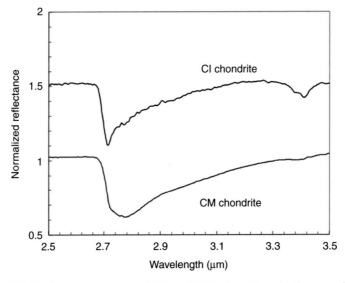

Figure 4.14 Reflectance spectra of CI and CM chondrites in the near-infrared from Takir et al. (2013). Note the strong band between 2.7–2.8 µm. The meteorites are CI1 chondrite Ivuna and CM2 chondrite LAP 03786. The spectra are offset in reflectance.

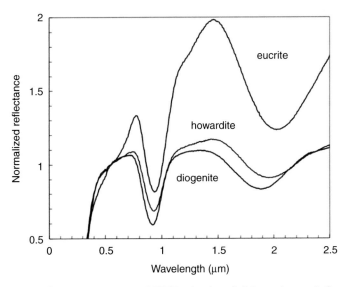

Figure 4.15 Reflectance spectra of HEDs in the visible and near-infrared. The meteorites are eucrite Bouvante, howardite EET 87503, and diogenite Johnstown. The eucrite spectrum is from Burbine et al. (2001), the howardite spectrum is from Hiroi et al. (1994), and the diogenite spectrum is from Hiroi et al. (1995). Note the strong pyroxene absorption bands. Spectra are normalized to unity at 0.55 μm.

wavelengths as Fe^{2+} substitutes for Mg (e.g., Klima et al., 2007) and/or Ca substitutes for Fe^{2+} (e.g., Klima et al., 2011), eucrites have band centers at longer wavelengths than howardites, and howardites have band centers at longer wavelengths than diogenites.

4.5 Space Weathering

Space weathering is the alteration of the regolith of a planetary body without an atmosphere as it is exposed to space (e.g., Zeller and Ronca, 1967; Chapman, 2004). The regolith undergoes optical, chemical, and physical changes. Regolith is the granular material on the surface of a body.

Space weathering was first evident from the analysis of returned lunar samples (e.g., Hapke et al., 1975; Pieters et al., 2000; Hapke, 2001; Chapman, 2004). Lunar soils have different spectral properties than lunar rock fragments in the visible and near-infrared. Comminution is the breaking down of larger fragments into smaller fragments. Lunar soils have redder continuum slopes, lower visual albedos, and weaker absorption bands than lunar rock fragments. These types of spectral changes are called lunar-style space weathering (LSSW). The spectral changes on the Moon

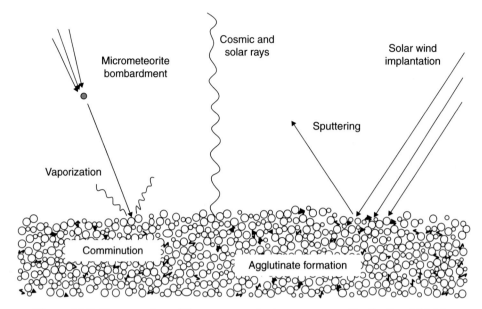

Figure 4.16 Cartoon of possible space weathering processes. Credit: Sarah Noble, NASA.

were due to the production of microscopic iron grains during agglutination, which is the production of small glassy breccias after micrometeorite impacts. These microscopic iron particles are called nanophase iron particles or npFe^0.

Space weathering appears to be due to a variety of processes (Figure 4.16). Originally micrometeorite impacts were proposed to produce these nanophase iron particles. Sasaki et al. (2001) carried out analog experiments in the laboratory by irradiating olivine samples with a nanosecond-pulse laser, and these experiments indeed produced these nanophase iron particles. They concluded that such phases could actually have been produced on asteroid surfaces by micrometeorite impacts that could redden a spectrum and darken the surface. This experiment was reproduced by Kurahashi et al. (2002) and Loeffler et al. (2016). However, alteration by the solar wind (e.g., Brunetto, 2009; Loeffler et al., 2009; Bennett et al., 2013) has been argued by Vernazza et al. (2009) to be the prime mechanism of space weathering since the timescale for space weathering seems to be relatively rapid (~10^6 years). Nanophase iron atoms will be ejected from the surface due to the bombardment of ions, which is a process called sputtering, and coat the surface. Cosmic and solar rays striking the surface will also create different isotopes on the surface.

When the visible colors of the most commonly observed asteroids (S-complex asteroids) were compared to the visible reflectance spectra of the

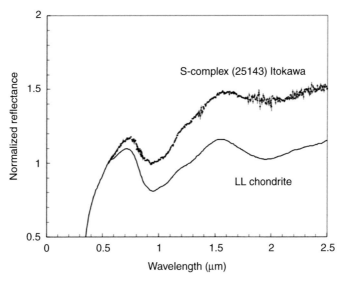

Figure 4.17 Reflectance spectra of an ordinary chondrite (LL4 chondrite Greenwell Springs) (Burbine et al., 2003) versus S-complex [(25143) Itokawa] (Binzel et al., 2001b). Note the similarity in spectral shape but the large difference in spectral slope. Itokawa was the target of the Hayabusa sample return mission. Spectra are normalized to unity at 0.55 μm.

commonest meteorites (ordinary chondrites) starting in the early 1970s, both types of bodies had absorption bands consistent with mixtures of olivine and/or pyroxene. However, the asteroids had redder spectra with shallower absorption bands than the ordinary chondrites. These spectral differences were also apparent when near-infrared spectra started to become more readily available in the 1980s. These spectral differences are readily apparent when comparing the reflectance spectrum of Hayabusa target (25143) Itokawa and an LL chondrite (Figure 4.17).

4.6 Mineralogical Formulas

Gaffey et al. (2002) developed a number of formulas for determine the pyroxene chemistry from a reflectance spectrum with absorption bands just due to pyroxene. The formulas (with uncertainties) for calculating the Fs (ferrosilite) and Wo (wollastonite) contents are

$$\mathrm{Wo}(\pm 3) = \left(347.9 \times \frac{\mathrm{BI\,Center}}{\mu m} - 313.6 \right) \mathrm{mol\%} \quad (Fs < 10; \mathrm{Wo}_{\sim 5\text{-}35} \text{ excluded}), \quad (4.20)$$

$$Wo(\pm 3) = \left(456.2 \times \frac{BI\,Center}{\mu m} - 416.9 \right) mol\%$$

$$(Fs = 10 - 25; Wo_{\sim 10-25} \text{ excluded}),$$

(4.21)

$$Wo(\pm 4) = \left(418.9 \times \frac{BI\,Center}{\mu m} - 380.9 \right) mol\% \quad (Fs = 25 - 50),$$

(4.22)

$$Fs(\pm 5) = \left(268.2 \times \frac{BII\,Center}{\mu m} - 483.7 \right) mol\% \quad (Wo < 11),$$

(4.23)

$$Fs(\pm 5) = \left(57.5 \times \frac{BII\,Center}{\mu m} - 72.7 \right) mol\% \quad (Wo = 11 - 30; Fs_{<25} \text{ excluded}),$$

(4.24)

$$Fs(\pm 4) = \left(-12.9 \times \frac{BII\,Center}{\mu m} + 45.9 \right) mol\% \quad (Wo = 30 - 45),$$

(4.25)

and

$$Fs(\pm 4) = \left(-118.0 \times \frac{BII\,Center}{\mu m} + 278.5 \right) mol\% \quad (Wo > 45).$$

(4.26)

The band centers are measured in μm. Excluded mineral compositions are not found among natural pyroxenes. The En (enstatite) content can be calculated from the formula

$$En = 100 - Fs - Wo.$$

(4.27)

Since neither the Fs nor the Wo content is known initially, the equations are used in an iterative fashion.

Example 4.2

The spectrum of a pyroxene has band centers at 0.92 and 1.91 μm. Estimate its ferrosilite and wollastonite contents.

Using the Band I center, the Wo content from *Equations 4.20, 4.21,* and *4.22* are

$$Wo(\pm 3) = \left(347.9 \times \frac{0.92\,\mu m}{\mu m} - 313.6 \right) mol\% = 6.5\,mol\%$$

$$(Fs < 10; Wo_{\sim 5-35} \text{ excluded}),$$

(4.28)

$$\text{Wo}(\pm 3) = \left(456.2 \times \frac{0.92\,\mu m}{\mu m} - 416.9 \right) \text{mol}\% = 2.8\,\text{mol}\% \tag{4.29}$$

$$(\text{Fs} = 10 - 25; \text{Wo}_{\sim 10-25} \text{ excluded}),$$

and

$$\text{Wo}(\pm 4) = \left(418.9 \times \frac{0.92\,\mu m}{\mu m} - 380.9 \right) \text{mol}\% = 4.5\,\text{mol}\% \ (\text{Fs} = 25 - 50). \tag{4.30}$$

Since all the Wo contents are below 11 mol%, *Equation 4.23* must be used to determine the Fs content. The Fs content will be

$$\text{Fs}(\pm 5) = \left(268.2 \times \frac{1.91\,\mu m}{\mu m} - 483.7 \right) \text{mol}\% = 28.6\,\text{mol}\% \ (\text{Wo} < 11). \tag{4.31}$$

For a Fs content of 28.6 mol%, *Equation 4.30* must be used for the Wo content, which results in a content of 4.5 mol%. So the composition of the pyroxene is $\text{Fs}_{28.6\pm 5}\text{Wo}_{4.5\pm 4}$.

Cloutis et al. (1986) developed a formula for determining the relative abundance of orthopyroxene in an olivine–orthopyroxene mixture. For spectra of mixtures of olivine and low-Ca pyroxene, Cloutis et al. (1986) found that the ratio of the areas of Band II to Band I (the Band Area Ratio or *BAR*) is proportional to the relative abundances of olivine and low-Ca pyroxene. The Band I area and Band II area are determined (Figure 4.18) by drawing a straight line tangent to the peaks on both sides of the bands (~0.7 and ~1.4–1.5 μm, respectively, for Band I and ~1.4–1.5 and ~2.4–2.5 μm, respectively, for Band II) and determining the area between the straight line and the spectrum. To determine the actual area of each band, the area under the polyhedron defined by the peaks is first calculated. This area then has the area under the spectrum subtracted from it. The area under a curve can be calculated using a mathematical procedure such as the trapezoidal rule or Riemann sums. The Band II area is then divided by the Band I area to calculate the Band Area Ratio.

From the work of Cloutis et al. (1986), the formula for determining the mineralogy for the spectrum of an orthopyroxene–olivine mixture can be written (Gastineau-Lyons et al., 2002) as

$$\text{opx} / (\text{opx} + \text{ol}) = (0.417 \times BAR) + 0.052. \tag{4.32}$$

This formula is valid for opx/(opx + ol) ratios of 10–90% (Gaffey et al., 2002). The spectra of low-Ca pyroxenes will have a large *BAR*, while the spectra of olivines

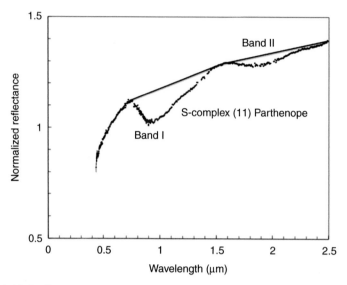

Figure 4.18 Reflectance spectra of S-complex (11) Parthenope. Band I and II are labeled. The tangent lines that define the areas of Band I and Band II, respectively, are shown. Due to a weak Band II, this object will have a small Band Area Ratio. The spectrum is normalized to unity at 0.55 µm.

will have a *BAR* with a value near 0. This formula can also be written to calculate the ol/(ol + opx) ratio, which will be

$$ol / (ol + opx) = -0.417 \times BAR + 0.948. \tag{4.33}$$

One problem with determining Band Area Ratios is that it may not be possible to accurately determine the Band II area since the whole Band II is often not fully covered with the near-infrared spectrum. So the calculated *BAR* may be a lower limit since the Band II area may be underestimated. This is called "The Red Edge Problem" (Lindsay et al., 2016). Lindsay et al. (2016) has developed equations for determining ol/(ol + opx) ratios for reflectance spectra that cut off at 2.40 and 2.45 µm, respectively.

Dunn et al. (2010) developed a number of formulas for determining the Fa and Fs contents and ol/(ol + px) ratio for ordinary chondrite assemblages. The formulas were developed by analyzing the reflectance spectra of ordinary chondrite powders (Jarosewich, 1990) with known mineralogies. The formulas with uncertainties are

$$Fa(\pm 1.3) = \left[-1284.9 \times \left(\frac{BI\,Center}{\mu m} \right)^2 + 2656.5 \times \frac{BI\,Center}{\mu m} - 1342.3 \right] mol\%, \tag{4.34}$$

$$Fs(\pm 1.4) = \left[-879.1 \times \left(\frac{BI\,Center}{\mu m} \right)^2 + 1824.9 \times \frac{BI\,Center}{\mu m} - 921.7 \right] mol\%, \quad (4.35)$$

and

$$ol\,/\,(ol + px)(\pm 0.03) = (0.242 \times BAR) + 0.728. \qquad (4.36)$$

These formulas were calculated from analyzing only ordinary chondrite spectra and may not be valid for other types of olivine–pyroxene mixtures.

Example 4.3

The spectrum of an ordinary chondrite assemblage has a Band I center at 0.93 and a Band Area Ratio of 0.80. Estimate its fayalite and ferrosilite contents and ol/(ol + px) ratio.

With these values, the fayalite and ferrosilite contents and ol/(ol + px) ratio will be

$$Fa(\pm 1.3) = \left[-1284.9 \times \left(\frac{0.93\,\mu m}{\mu m} \right)^2 + 2656.5 \times \frac{0.93\,\mu m}{\mu m} - 1342.3 \right] mol\% \quad (4.37)$$

$$= 16.9\,mol\%,$$

$$Fs(\pm 1.4) = \left[-879.1 \times \left(\frac{0.93\,\mu m}{\mu m} \right)^2 + 1824.9 \times \frac{0.93\,\mu m}{\mu m} - 921.7 \right] mol\% \quad (4.38)$$

$$= 15.1\,mol\%,$$

and

$$(ol + px)\,(\pm 0.03) = (0.242 \times 0.80) + 0.728 = 0.92.$$

The composition of the olivine will be $Fa_{16.9\pm 1.3}$, the composition of the pyroxene will be $Fs_{15.1\pm 1.4}$, and the ol/(ol + px) ratio will be 0.92.

4.7 Temperature

One complication in analyzing band centers and Band Area Ratios for asteroid spectra is that the positions and strengths of mineral absorption bands are a function of temperature (Burns, 1993). The reflectance spectra of most minerals and meteorites were measured at room temperature, while most main-belt asteroids are at much lower temperatures.

As the temperature decreases, the amplitude of the thermal vibrations of the cation about the center of its coordination site also decreases, which will reduce the width of the band. Since the coordination site is contracting and the distance between the cations and oxygen is decreasing, the energy level splitting will therefore increase, which usually results in a band that moves to shorter wavelengths. The band positions of olivines and low-Ca pyroxenes tend to decrease in width and move to shorter wavelengths as the temperature of the sample decreases (Singer and Roush, 1985; Roush and Singer, 1987; Hinrichs et al., 1999; Schade and Wäsch, 1999; Moroz et al., 2000; Reddy et al., 2012c; Sanchez et al., 2012).

4.8 Taxonomy

Asteroid taxonomies are based on the hope that asteroids that reflect light similarly over a particular wavelength range also have similar surface mineralogies. The asteroid taxonomic system is somewhat similar in design to the widely used taxonomy that is used for stellar classification. The hope for all asteroid taxonomies is to be as robust as the one that is currently used for stars.

Both stars and asteroids have bands due to absorptions due to electrons; however, stars have much narrower bands than asteroids since their electrons are not found in minerals, which tends to widen the bands. Starting in the 1880s, stars were initially classified according to the strength of their hydrogen Balmer absorption lines (in the visible wavelength region) with A-type stars having the strongest lines and O-type stars having the weakest lines. Annie Jump Cannon (1863–1941) realized that the spectral types could be better rearranged in order of surface temperature. The sequence of O, B, A, F, G, K, and M ("Oh, Be A Fine Guy/Girl, Kiss Me") spectral types goes from hottest (O) to coolest surface temperature (A). A number is used to indicate temperature subdivisions with 0 as the hottest in a spectral type and 9 as the coolest. Since some stars have the same temperatures but very different luminosities, luminosity classes used with the spectral types were developed to indicate stars of different sizes. Luminosity classes uses Roman numerals to indicate different types of stars primarily above the main sequence (luminosity class V). Additional letters are used to identify other absorption or emission features in their spectra. The Sun is classified as a G2 V star.

The physics of black body radiation and absorption and emission by electrons in a hot ball of gas are relatively well understood and this understanding has allowed a stellar classification to be developed over the last hundred years that groups objects with different temperatures, luminosities, and spectral features (absorption and emission). However, the difficulty in truly modeling the physics of mineral absorption features in a particulate regolith have resulted in taxonomies that may not group objects together with similar mineralogies.

4.8.1 Asteroid Taxonomy

Similarities in how different asteroids reflect light over a wide range of wavelength have been used to group asteroids into different classes. Different asteroids have long been known to reflect light differently at different wavelengths. The hope is that asteroids that reflect light similarly have similar surface compositions. Asteroid taxonomies tend to use an upper case letter (or a series of upper case and lower case letters) to indicate a specific spectral type.

Currently used asteroid taxonomies are all based on the works of Zellner (1973) who classified asteroids as either stony or carbonaceous and Chapman et al. (1975) who classified most asteroids as either S- (silicaceous) or C- (carbonaceous) types due to their spectral similarity to stony-iron meteorites and carbonaceous chondrites, respectively. These classifications were based on visible spectral reflectance properties and visual albedo. S-types have redder spectra, higher albedos, and absorption bands at ~1 μm, while C-types have more neutral colors, lower albedos, and weak to absent absorption bands. Chapman et al. (1975) designated a few objects [e.g., (4) Vesta, (16) Psyche, (349) Dembowska] as U (Unclassified) since they had parameters outside the S and C groupings. Zellner and Gradie (1976) introduced the M-type for bodies with featureless visible spectra with a slight positive spectral slope and moderate visual albedos and the E-type for bodies with featureless visible spectra and high visual albedos. M-types were interpreted as metal-rich, while E-types were interpreted as pure-enstatite.

Tholen (1984) classified asteroids into 14 separate classes (Figure 4.19) using Eight-Color Asteroid Survey (ECAS) data and visual geometric albedos (when available). The ECAS survey (Tedesco et al., 1982; Zellner et al., 1985) observed asteroids using eight filters from 0.337 to 1.041 μm. The filters were chosen to allow for the identification of characteristic absorption bands found in both meteorite and asteroid spectra (Tedesco et al., 1982). These classes included the S-types, C-group (B-types, C-types, F-types, G-types), X-group (E-types, M-types, P-types), A-types, D-types, Q-types, R-types, T-types, and V-types.

When available, near-infrared colors (past 1.1 μm) were also used to create new asteroid classes or subdivide existing classes. The K-class was proposed by Bell (1988) for objects in the (221) Eos family (including Eos) that have visible and near-infrared spectral properties and visual albedos intermediate between S-types and C-types. The K-class was later incorporated into future taxonomies. Also using visible and near-infrared spectra, Gaffey et al. (1993) developed a classification system for S-types based on Band Area Ratios and Band I centers to create

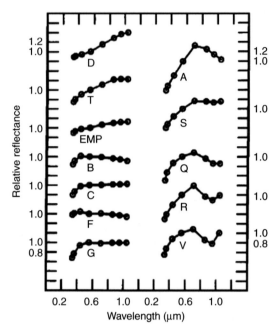

Figure 4.19 Reflectance spectra of the Tholen (1984) classes from Howell et al. (1994b). Credit: Ellen Howell, University of Arizona and John Wiley & Sons, Inc.

the S(I), S(II), S(III), S(IV), S(V), S(VI), and S(VII) subtypes, which have varying interpreted ratios of olivine to pyroxene. The Band Area Ratio tends to increase and the Band I center tends to decrease as the Roman numeral increases in value, which tends to indicate an increasing pyroxene concentration. Rivkin et al. (1995) suggested the W-class ("wet M") for M-types with 3 μm absorption features.

CCD spectral observations also identified asteroids with distinct spectroscopic properties. Phase I of the Small Main-belt Asteroid Spectroscopic Survey (SMASS) observed over 300 asteroids (Xu et al., 1995) in the visible using a CCD. Using data from this survey, Binzel and Xu (1993) found that a number of asteroids in the Vesta family and between Vesta and the 3:1 and v_6 meteorite-supplying resonances had V-type visible spectra and appeared to be fragments of Vesta. Some objects also had spectra that resembled HEDs but with stronger 1 μm absorption features. Binzel and Xu (1993) called these objects J-types for the diogenite Johnstown; however, near-infrared spectra of these bodies out to 1.65 μm did not confirm a spectral similarity to diogenites (Burbine et al., 2001). The J-type classification was not used in later taxonomies. Binzel et al. (1993) also found that (3628) Božněmcová had a unique visible spectrum that resembled LL chondrites. The letter O (for ordinary chondrites) was used to identify this type of visible spectrum.

However, more recent near-infrared spectra do not appear consistent with ordinary chondrites (Burbine et al., 2011).

Bus and Binzel (2002b) classified asteroids into 26 classes using spectra from SMASS. This survey observed ~1500 asteroids in the visible using a CCD. The classes included the S-complex (S, Sa, Sl, Sq, Sr, Sv), C-complex (B, C, Cb, Cg, Cgh, Ch), X-complex (X, Xc, Xe, Xk), A, D, K, L, Ld, O, Q, R, T, and V. The S-complex broke up S-asteroids into S-types and objects spectrally intermediate between S-types and A-types (Sa), L-types (Sl), Q-types (Sq), R-types (Sr), and V-types (Sv). The C-complex included C- and B-types plus objects intermediate between C- and B-types (Cb), those with deep UV features [Cg where the g refers to the G-class of Tholen (1984)], those with a 0.7 μm absorption feature (Ch), and those with a strong UV feature and 0.7 μm absorption feature (Cgh). The X-complex included X-types, objects intermediate between X- and C-types (Xc), objects with an absorption feature shortward of 0.55 μm (Xe where the e refers to E-types, which typically have this feature), and objects intermediate between X- and K-types (Xk). The A-, D-, K-, O-, Q-, R-, T-, and V-types all had spectral properties to how they were defined previously. L-types had spectra with a very strong UV feature and a flat reflectance past 0.75 μm and appear to form a spectral continuum between K- and S-types. Ld-types have spectra intermediate between L- and D-types.

The current Bus–DeMeo taxonomy (Figure 4.20) (DeMeo et al., 2009) classifies asteroids into 25 classes on the basis of both visible and near-infrared data. The classes include the S-complex (S, Sa, Sq, Sr, Sv), C-complex (B, C, Cb, Cg, Cgh, Ch), X-complex (X, Xc, Xe, Xk, Xn), A, D, K, L, O, Q, R, T, and V. S-complex bodies (excluding the Sa-types) and V-types with very red spectral slopes were also given the notation "w" to indicate reddening due to interpreted "space weathering." The spectral properties of each taxonomic class are listed in Table 4.2. An Xn-class was added later after the original taxonomy was published to identify bodies with spectra similar to (44) Nysa, which have low to medium spectral slopes and a weak and narrow feature at 0.9 μm.

The Sloan Digital Sky Survey (SDSS) was a multi-filter imaging system and spectroscopic redshift survey that is primarily used to obtain spectra of galaxies, quasars, and stars. SDSS uses a 2.5-meter telescope at the Apache Point Observatory in New Mexico. SDSS obtained colors of ~500 000 moving objects. These bodies were observed using five filters (u', 0.3543 μm; g', 0.4770 μm; r', 0.6231 μm; i', 0.7625 μm; z', 0.9134 μm). SDSS was not able to measure colors for the brightest asteroids (H < 12) (DeMeo and Carry, 2013).

Originally SDSS colors were used to define three broad types (S-, C-, and V-types) (Ivezić et al., 2001; Jurić et al., 2002) but each of these three types contained a number of different taxonomic classes. Using SDSS colors for over 60 000 asteroids, Carvano et al. (2010) defined nine classes (V_p, O_p, Q_p, S_p, A_p, L_p, D_p, X_p,

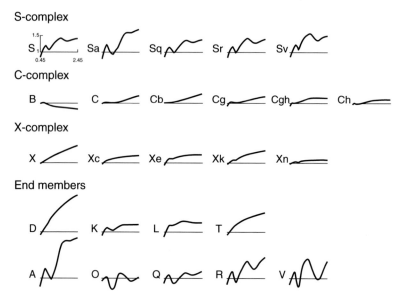

Figure 4.20 Reflectance spectra of the Bus–DeMeo taxonomic classes (DeMeo et al., 2009). Credit: Francesca DeMeo, Massachusetts Institute of Technology.

C_p) where the lowercase p stands for photometric. Except for the L_p class (which includes the K, L, and Ld classes), the classes are intended to be directly comparable to the Bus and Binzel (2002b) classes. SDSS colors can also be used to identify interesting objects that can be designated for later higher resolution spectral observations. For example, SDSS colors have been very useful for identifying objects with V-type visible spectra (e.g., Roig and Gil-Hutton, 2006).

4.9 Taxonomic Classes

The spectral properties of each taxonomic type are defined slightly differently in each taxonomy. The following sections discuss the basic properties and interpreted mineralogies of each type. Each meteorite group has been linked to particular taxonomic types (Table 4.3). The terms "S-complex," "C-complex," and "X-complex" are used to refer to taxonomic types with roughly similar spectral properties but do have more relatively subtle spectral differences. The term "complex" for taxonomic types was first coined by Bus (1999).

4.9.1 S-complex

The commonest asteroids in the inner main belt are S-complex asteroids. S-complex asteroids tend to have moderately strong UV features due to Fe^{2+} and

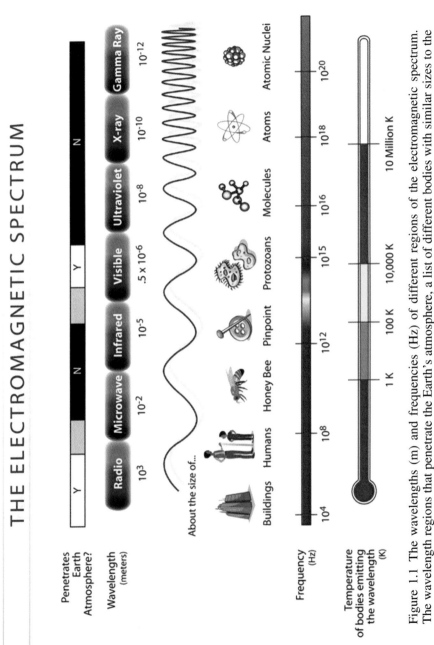

Figure 1.1 The wavelengths (m) and frequencies (Hz) of different regions of the electromagnetic spectrum. The wavelength regions that penetrate the Earth's atmosphere, a list of different bodies with similar sizes to the wavelengths of each region, and the temperatures of different bodies emitting black body radiation primarily at that wavelength. Credit: NASA.

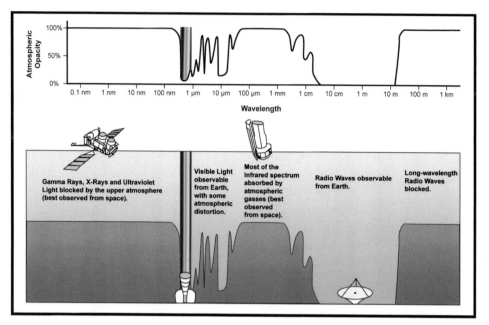

Figure 1.2 The transmittance of electromagnetic radiation through the atmosphere. Microwaves are included as part of the short wavelength radio waves. Atmospheric opacity is the amount of light absorbed by the atmosphere with 100% opacity indicating total absorption and 0% opacity indicating total transmission. Credit: NASA.

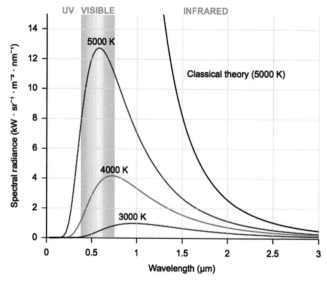

Figure 1.3 Black body curves for bodies at temperatures at 5000 K, 4000 K, and 3000 K plus the classical theory prediction for a body at 5000 K. Credit: Dark Kule.

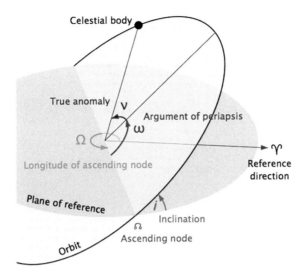

Figure 2.1 Five of the six orbital elements of a celestial body orbiting the Sun. The semi-major axis of the orbit, which is one-half of the major axis, is not shown. The orbital plane is in yellow. The ecliptic is in light gray. The reference direction points toward the vernal equinox (♈). Credit: Lasunncty.

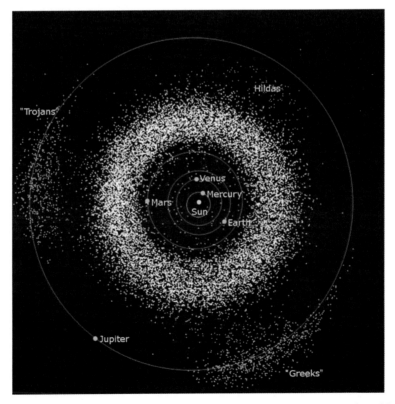

Figure 2.2 The orbits of the terrestrial planets and minor planets out to the orbit of Jupiter. The white points are asteroids in the main asteroid belt, the orange points are the Hilda asteroids, and the Trojan asteroids are in green. The orbits are for bodies that were discovered as of August 14, 2006. Credit: Mdf.

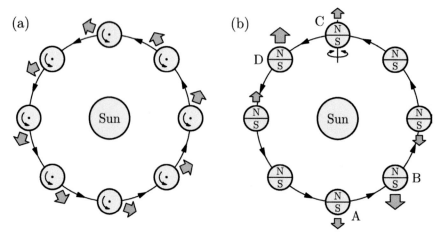

Figure 2.5 Plot of (a) the diurnal Yarkovsky effect and (b) the seasonal Yarkovsky effect. A circular orbit is assumed for both bodies. The obliquity is assumed to be 0° on the left side of the figure and 90° on the right side of the figure. For (a) the diurnal Yarkovsky effect, the body is heated by the Sun at "noon" (when the Sun is directly overhead). The body has prograde rotation. But due to thermal inertia, the maximum temperature will occur on the "afternoon" side, which "pushes" the object in the opposite direction. The body will accelerate in the direction of the gold arrow, which is in the same direction as orbital motion. This acceleration causes the semi-major axis of the body to increase. A body with retrograde motion will accelerate in the opposite direction to orbital motion and will have its semi-major axis decrease. For (b) the seasonal Yarkovsky effect during the "summer," the northern hemisphere will be hotter than the southern hemisphere. During "June" (A), the northern hemisphere will receive the most direct sunlight but due to thermal inertia, the body will be hottest in "July–September" (B). During "December" (A), the southern hemisphere will receive the most direct sunlight but due to thermal inertia, the body will be hottest in "January–February" (B). The body will accelerate in the opposite direction to orbital motion, which causes the semi-major axis of the body to decrease. Credit: David Vokrouhlický, Southwest Research Institute.

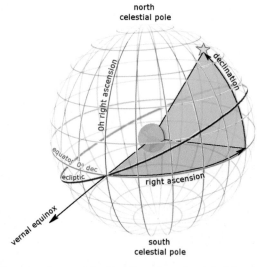

Figure 2.6 Right ascension and declination of a celestial object as seen from outside the celestial sphere. The celestial equator has a declination of 0°. The right ascension is 0 hours on the hour circle (great circle) that passes through the vernal equinox and the celestial poles. Credit: Tfr000.

Figure 3.4 A meteorite being collected in Antarctica by the ANSMET team. A picture is being taken of the meteorite in the ice. Credit: US Antarctic Search for Meteorites program/Katie Joy.

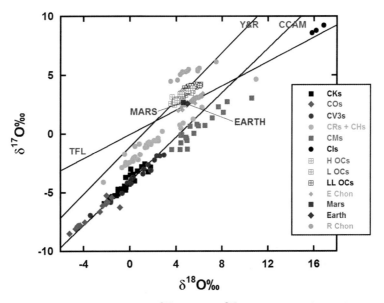

Figure 3.8 Oxygen isotope plot ($\delta^{18}O$ versus $\delta^{17}O$) for chondrites. The average oxygen isotopic values for the Earth and Mars are also plotted. TFL is the terrestrial fractionation line, CCAM is the carbonaceous chondrite anhydrous mineral line, and Y&R is the Young and Russell (1998) slope 1 line. Credit: Richard Greenwood, Open University.

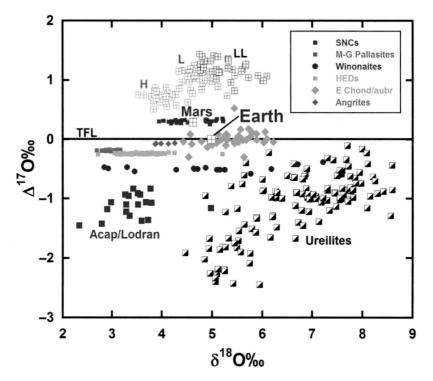

Figure 3.9 Oxygen isotope plot ($\Delta^{17}O$ versus $\delta^{18}O$) for achondrites. The average oxygen isotopic values for the Earth, Mars (from Martian meteorites), and ordinary chondrites (H, L, LL) are also plotted. SNCs (Shergottites, Nakhlites, Chassignites) are Martian meteorites. Enstatite chondrites (E Chond) and aubrites (aubr) have indistinguishable oxygen isotopic compositions and are plotted using the same symbols. M-G pallasites are main-group pallasites. HEDs are howardites, eucrites, and diogenites. Credit: Richard Greenwood, Open University.

Figure 3.17 Slice of the Esquel pallasite. Note the greenish olivine grains. Credit: Doug Bowman.

Figure 3.22 Image of Martian meteorite NWA 7034, which is nicknamed "Black Beauty" due to its black fusion crust (Agee et al., 2013). Credit: NASA.

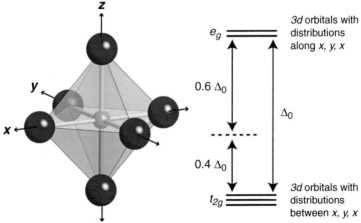

Figure 4.3 Idealized 6-coordinated polyhedron around a hypothetical Fe atom, along with the crystal field splitting diagram. When the Fe atom is in octahedral coordination, the energies of the Fe $3d$ orbitals change as a function of proximity to electronic distributions of the surrounding oxygen anions. When the octahedron is not perfectly symmetrical further splitting occurs among the e_g and t_{2g} orbitals. Such splitting of the energy levels makes possible multiple electronic transitions that can give rise to absorption features such as the bands seen in olivines and pyroxenes. Figure from Klima et al. (2011). Credit: Rachel Klima, JHUAPL and John Wiley & Sons, Inc.

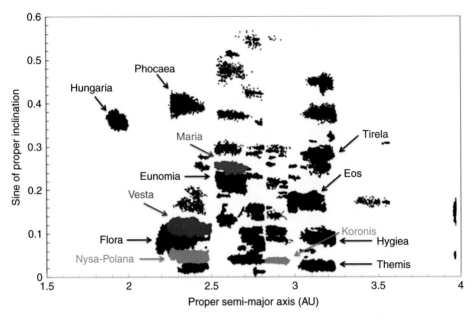

Figure 5.1 Plot of sine of proper inclination versus semi-major axis for family members from the ~120 families defined by Nesvorný et al. (2015).

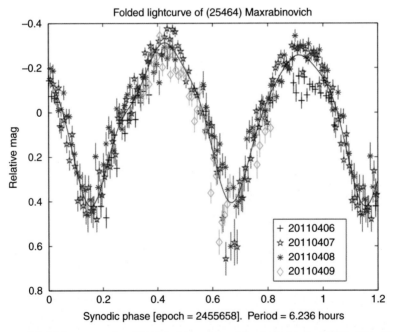

Figure 5.3 Light curve for (25464) Maxrabinovich (Polishook, 2012). The light curve plots relative magnitude versus phase. Note the two maxima and two minima. Credit: David Polishook, Weizmann Institute of Science, and The Minor Planet Bulletin.

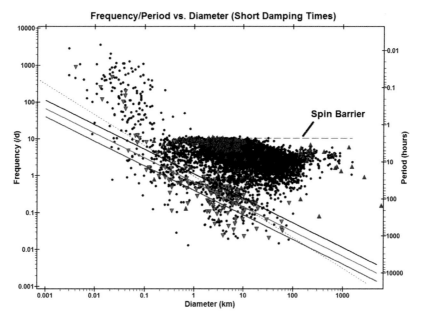

Figure 5.4 Frequency (cycles per day) (or period in hours) versus diameter (km) for asteroids (Warner et al., 2009). Red triangles are binary objects. Green triangles are tumblers. The dashed horizontal line is the spin barrier. The three solid diagonal lines represent the "tumbling barrier" based on damping times of 200 Ma (black line), 1 Ga (green line), and 4.5 Ga (red line) where an object below the line is a good candidate for being a tumbler. The thin dashed diagonal line is based on the collisional lifetime of the objects. Credit: Brian D. Warner, Center for Solar System Studies.

Figure 5.5 Image of Arecibo Observatory. The telescope is 305 meters in diameter. Credit: National Oceanic and Atmospheric Administration.

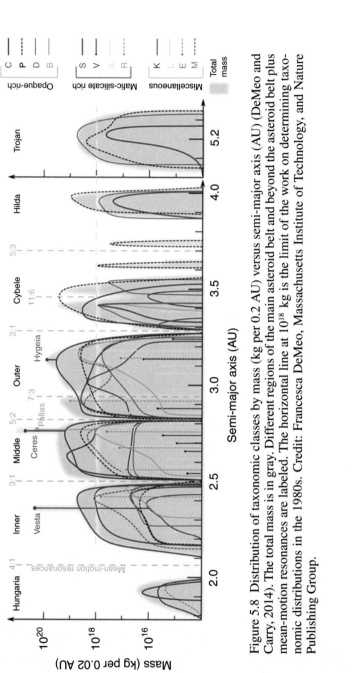

Figure 5.8 Distribution of taxonomic classes by mass (kg per 0.2 AU) versus semi-major axis (AU) (DeMeo and Carry, 2014). The total mass is in gray. Different regions of the main asteroid belt and beyond the asteroid belt plus mean-motion resonances are labeled. The horizontal line at 10^{18} kg is the limit of the work on determining taxonomic distributions in the 1980s. Credit: Francesca DeMeo, Massachusetts Institute of Technology, and Nature Publishing Group.

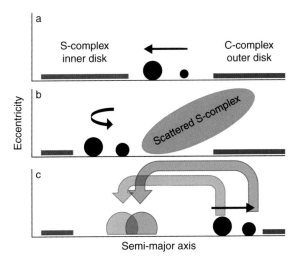

Figure 5.9 Cartoon of the Grand Tack during the first 5 Ma of Solar System history. Eccentricity versus semi-major axis is plotted. (a) Initially S-complex bodies form interior to Jupiter (large circle) and Saturn (small circle) and C-complex bodies form exterior to these giant planets. (b) As Jupiter and Saturn migrate towards the Sun, S-complex bodies are scattered outward from the Sun. (c) As Jupiter and Saturn migrate away from the Sun, the scattered S-complex bodies are scattered again towards the dynamically stable asteroid belt region and the previously undisturbed C-complex bodies are scattered into the asteroid belt region. The S-complex bodies will be preferentially found in the inner main belt while the C-complex bodies will be preferentially found in the outer main belt. Credit: Kevin Walsh, Southwest Research Institute.

Figure 6.2 Image of comet Hale–Bopp taken in April 1997. The blue gas tail is on the left of the image and the white dust tail is on the right. Credit: E. Kolmhofer and H. Raab, Johannes-Kepler-Observatory.

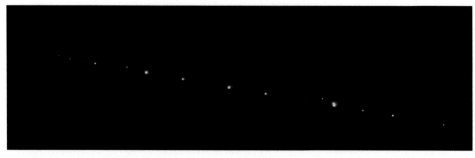

Figure 6.4 A Hubble Space Telescope image of comet Shoemaker–Levy 9, taken in May 1994 with the Wide Field Planetary Camera 2 (WFPC2). Credit: NASA.

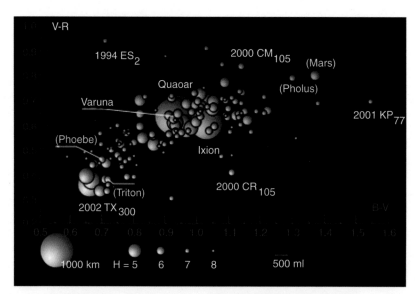

Figure 6.8 Color–color plot of trans-Neptunian objects for bodies discovered as of 2006. Larger objects are plotted with larger circles. For reference, Mars, Phoebe (moon of Saturn), Pholus (centaur), and Triton (moon of Neptune) are also plotted. Mars and Triton are not plotted using the same size scale as the trans-Neptunian objects. Credit: Drbogdan.

Figure 6.11 This image was produced from four images of (134340) Pluto from New Horizons' Long Range Reconnaissance Imager (LORRI) that were combined with color data from the Ralph/Multispectral Visual Imaging Camera (MVIC). Pluto has a diameter of 2374 km. Note the relatively craterless heart-shaped region called Tombaugh Regio. Credit: NASA/JHUAPL/SwRI.

Figure 6.12 This image of Charon combines blue, red and infrared images taken by the New Horizon's Ralph/Multispectral Visual Imaging Camera (MVIC). Charon as a diameter of 1212 km. The colors are processed to best highlight the variation of surface properties across Charon. Note the large canyon across Charon's surface and the reddish color apparently due to tholins around the north pole. Credit: NASA/JHUAPL/SwRI.

Figure 7.3 Image of Meteor Crater in Northern Arizona taken by the NASA Earth Observatory. Meteor Crater is ~1.2 km in diameter. Layers of exposed limestone and sandstone are visible just beneath the crater rim. Credit: NASA.

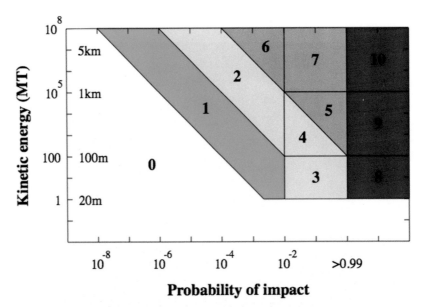

Figure 7.7 Torino Scale values for an impacting object. Kinetic energy in megatons is plotted versus impact probability. The estimated diameters in km are also given on the y-axis. White (number"0") indicates no hazard, green ("1") indicates a routine discovery, yellow ("2," "3," "4") indicates objects meriting more attention from astronomers, orange ("5," "6," "7") indicates objects that will have a close encounter with the Earth and could possibly cause regional or global devastation, and red ("8," "9," "10") indicates a certain collision. Credit: Richard Binzel, Massachusetts Institute of Technology.

THE TORINO SCALE
Assessing Asteroid/Comet Impact Predictions

No Hazard	**0**	The likelihood of collision is zero, or is so low as to be effectively zero. Also applies to small objects such as meteors and bolides that burn up in the atmosphere as well as infrequent meteorite falls that rarely cause damage.
Normal	**1**	A routine discovery in which a pass near the Earth is predicted that poses no unusual level of danger. Current calculations show the chance of collision is extremely unlikely with no cause for public attention or public concern. New telescopic observations very likely will lead to re-assignment to Level 0.
Meriting Attention by Astronomers	**2**	A discovery, which may become routine with expanded searches, of an object making a somewhat close but not highly unusual pass near the Earth. While meriting attention by astronomers, there is no cause for public attention or public concern as an actual collision is very unlikely. New telescopic observations very likely will lead to re-assignment to Level 0.
	3	A close encounter, meriting attention by astronomers. Current calculations give a 1% or greater chance of collision capable of localized destruction. Most likely, new telescopic observations will lead to re-assignment to Level 0. Attention by the public and by public officials is merited if the encounter is less than a decade away.
	4	A close encounter, meriting attention by astronomers. Current calculations give a 1% or greater chance of collision capable of regional devastation. Most likely, new telescopic observations will lead to re-assignment to Level 0. Attention by the public and by public officials is merited if the encounter is less than a decade away.
Threatening	**5**	A close encounter posing a serious, but still uncertain threat of regional devastation. Critical attention by astronomers is needed to determine conclusively whether or not a collision will occur. If the encounter is less than a decade away, governmental contingency planning may be warranted.
	6	A close encounter by a large object posing a serious. but still uncertain threat of a global catastrophe. Critical attention by astronomers is needed to determine conclusively whether or not a collision will occur. If the encounter is less than three decades away, governmental contingency planning may be warranted.
	7	A very close encounter by a large object, which if occurring this century, poses an unprecedented but still uncertain threat of a global catastrophe. For such a threat in this century, international contingency planning is warranted, especially to determine urgently and conclusively whether or not a collision will occur.
Certain Collisions	**8**	A collision is certain, capable of causing localized destruction for an impact over land or possibly a tsunami if close offshore. Such events occur on average between once per 50 years and once per several 1000 years.
	9	A collision is certain, capable of causing unprecedented regional devastation for a land impact or the threat of a major tsunami for an ocean impact. Such events occur on average between once per 10,000 years and once per 100,000 years.
	10	A collision is certain, capable of causing a global climatic catastrophe that may threaten the future of civilization as we know it, whether impacting land or ocean. Such events occur on average once per 100,000 years, or less often.

Figure 7.8 Discussion of each of the Torino Scale values (Morrison et al., 2004). Credit: NASA.

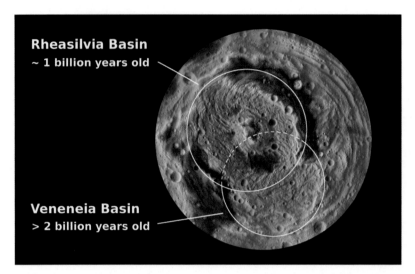

Figure 8.11 Topographic map of the southern hemisphere of (4) Vesta made by the Dawn mission. The map is color-coded by elevation with red showing the higher areas and blue showing the lower areas. Rheasilvia, the largest impact basin on Vesta, is ~500 km in diameter. The other basin, Veneneia, is ~400 kilometers across and lies partially beneath Rheasilvia. The topographic map was constructed from analyzing images from Dawn's framing camera that were taken with varying Sun and viewing angles. Credit: NASA/JPL-Caltech/UCLA/MPS/DLR/IDA/PSI.

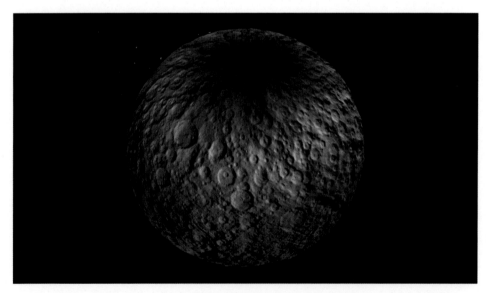

Figure 8.14 Map of neutron counts acquired by the GRaND instrument on Dawn for a portion of the northern hemisphere of (1) Ceres. The color scale of the map is from blue (lowest neutron count) to red (highest neutron count). The color information is based on the number of neutrons detected per second by GRaND. These data reflect the concentration of hydrogen in the upper meter of the regolith. Counts decrease with increasing hydrogen concentration. Lower neutron counts near the pole suggest the presence of water ice within a meter of the surface at high latitudes. Credit: NASA/JPL-Caltech/UCLA/MPS/DLR/IDA/PSI.

Table 4.2 *Spectral properties of Bus–DeMeo taxonomic classes*

A	Deep and extremely broad absorption band with a minimum near 1 μm, may or may not have shallow 2 μm absorption band, very highly sloped.
B	Linear, negatively sloping often with a slight round bump around 0.6 μm and/ or a slightly concave up curvature in the 1 to 2 μm region.
C	Linear, neutral visible slope often a slight rough bump around 0.6 μm and low but positive slope after 1.3. May exhibit slight feature longward of 1 μm.
Cb	Linear with a small positive slope that begins around 1.1 μm.
Cg	Small positive slope that begins around 1.3 μm and a pronounced UV dropoff.
Cgh	Small positive slope that begins around 1 μm and a pronounced UV dropoff similar to Cg also includes a broad, shallow absorption band centered near 0.7 μm similar to Ch.
Ch	Small positive slope that begins around 1.1 μm and slightly pronounced UV dropoff also includes a broad, shallow absorption band centered near 0.7 μm.
D	Linear with very steep slope, some show slight curvature or gentle kink around 1.5 μm.
K	Wide absorption band centered just longward of 1 μm, the left maximum and the minimum are sharply pointed and the walls of the absorption are linear with very little curvature.
L	Steep slope in visible region leveling out abruptly around 0.7 μm. There is often a gentle concave down curvature in the infrared with a maximum around 1.5 μm. There may or may not be a 2 μm absorption feature.
O	Very rounded and deep, "bowl" shape absorption feature at 1 μm as well as a significant absorption feature at 2 μm.
Q	Distinct 1 μm absorption feature with evidence of another feature near 1.3 μm, a 2 μm feature exists with varying depths between objects.
R	Deep 1 and 2 μm features, the 1 μm feature is much narrower than a Q-type, but slightly broader than a V-type.
S	Moderate 1 and 2 μm features. The 2 μm feature may vary in depth between objects.
Sa	Has a deep and extremely broad absorption band at 1 μm, has similar features to A-types but is less red.
Sq	Has a wide 1 μm absorption band with evidence of a feature near 1.3 μm like the Q-types, except the 1 μm feature is more shallow for the Sq.
Sr	Has a fairly narrow 1 μm feature similar to but more shallow than an R-type as well as a 2 μm feature.
Sv	Has a very narrow 1 μm absorption band similar to but more shallow than a V-type as well as a 2 μm feature.
T	Linear with moderate to high slope and often gently concaving down.
V	Very strong and very narrow 1 μm absorption and as well as a strong 2 μm absorption feature.
X	Linear with medium to high slope.
Xc	Low to medium slope and slightly curved and concave downward.
Xe	Low to medium slope similar to either Xc- or Xk-types, but with an absorption band feature shortward of 0.55 μm.
Xk	Slightly curved and concave downward similar to Xc-type but with a faint to feature between 0.8 to 1 μm.
Xn	Low to medium slopes with a weak narrow feature at 0.9 μm.

Table 4.3 *Postulated best taxonomic class analogs for each meteorite type in order of decreasing fall percentage. The fall percentages are calculated using data for classified meteorites from the Meteoritical Bulletin Database.*

Type	Fall percentage	Best taxonomic class analogs
L	37.2	Q-type (Binzel et al., 2004), S-complex (Gaffey et al., 1993)
H	33.8	Q-type (Binzel et al., 2004), S-complex (Gaffey et al., 1993)
LL	8.2	Q-type (Binzel et al., 2004), S-complex (Nakamura et al., 2011)
HED	5.8	V-type (McCord et al., 1970; Consolmagno and Drake, 1977)
iron	4.6	M-type (Cloutis et al., 1990; Shepard et al., 2010, 2011, 2015)
CM	1.5	C-complex (Vilas and Gaffey, 1989; Cloutis et al., 2011b)
L/LL	1.0	Q-type (Binzel et al., 2004), S-complex (Gaffey et al., 1993)
aubrite	0.8	E-type (Zellner, 1975; Zellner et al., 1977; Clark et al., 2004)
EH	0.8	M-type (Chapman and Salisbury, 1973; Shepard et al., 2010)
EL	0.7	M-type (Gaffey and McCord, 1978; Shepard et al., 2010)
CV	0.7	K-type (Bell, 1988)
mesosiderite	0.7	M-type (Shepard et al., 2010), S-complex (Gaffey et al., 1993)
CO	0.6	K-type (Bell, 1988; Clark et al., 2009)
ureilite	0.6	C-complex (Jenniskens et al., 2009), S-type (Gaffey et al., 1993)
CI	0.5	C-complex (Johnson and Fanale, 1973; Cloutis et al., 2011a)
Martian	0.5	Mars (Bogard and Johnson, 1983)
pallasite	0.4	A-type (Cruikshank and Hartmann, 1984; Sunshine et al., 2007)
C2-ungrouped	0.3	D-type (Hiroi et al., 2001b), T-type (Hiroi and Hasegawa, 2003)
CR	0.3	C-complex (Hiroi et al., 1996; Sato et al., 1997)
H/L	0.3	Q-type (Binzel et al., 2004), S-complex (Gaffey et al., 1993)
acapulcoite/ lodranite	0.2	S-complex (Gaffey et al., 1993)
CK	0.2	K-type (Clark et al., 2009; Cloutis et al., 2012e)
angrite	0.1	S-complex (Rivkin et al., 2007)
C3-ungrouped	0.1	K-type (Clark et al., 2009)
CB	0.1	M-type (Shepard et al., 2010)

Type	Fall percentage	Best taxonomic class analogs
K	0.1	C-complex (Gaffey, 1980)
R	0.1	A-type (Sunshine et al., 2007)
winonaite	0.1	S-complex (Gaffey et al., 1993)
brachinite	all finds	A-type (Cruikshank and Hartmann, 1984; Sunshine et al., 2007)
CH	all finds	M-type (Shepard et al., 2010)
lunar	all finds	Moon (Marvin, 1983)

absorption features in the visible and near-infrared that are usually due to olivine and/or pyroxene (Figure 4.20). It has long been argued whether S-types are primarily bodies with ordinary chondrite-like mineralogies (e.g., Feierberg et al., 1982) or are bodies that have experienced melting and differentiation. Many questions concerning the mineralogy of S-complex asteroids were answered by the Hayabusa sample return mission (Section 8.10). S-complex asteroids are subdivided on the basis of a number of relatively subtle spectral differences.

The S-type designation of DeMeo et al. (2009) is the typical classification for objects with absorption features due to pyroxene and olivine (Figure 4.21). Approximately 40% of asteroids classified by DeMeo et al. (2009) are designated as S-types. In the visible (Bus and Binzel, 2002b), S-types have moderately strong UV features and moderately deep 1 μm bands. In the near-infrared, S-types have moderately strong 1 and 2 μm features.

S-type (Bus and Binzel, 2002a) and S(IV)-asteroid (Gaffey et al., 1993) (6) Hebe has been postulated as the parent body of the H chondrites (Farinella et al., 1993b; Gaffey and Gilbert, 1998). S(IV)-types have Band Area Ratios and Band I centers consistent with ordinary chondrites. The Band Area Ratios and Band I centers of rotational spectra of Hebe are consistent (Gaffey and Gilbert, 1998) with H chondrites. Hebe is also located near the 3:1 mean motion and v_6 secular resonances and is argued to be a major potential source of meteorites because it should be supplying relatively large number of fragments into meteorite-supplying fragments (Farinella et al., 1993a). Bottke et al. (2010) finds that the distribution of cosmic-ray exposure ages for H chondrites is consistent with derivation from Hebe. Rivkin et al. (2001) also detected a weak 3 μm band on Hebe, consistent with aqueous alteration products that have been detected (Rubin et al., 2002) in some H chondrites. Hebe's calculated bulk density (~3800 kg/m³) (Baer et al., 2011) is consistent with the bulk density of H chondrites (~3710 kg/m³) (Macke, 2010).

Gaffey and Gilbert (1998) proposed that impacts melted the surface and caused it to differentiate into layers of silicates overlaying layers of metallic iron. These metallic iron sheets were exposed by later impacts. However, other researchers believe that asteroid sized impacts on the surface of Hebe would not produce enough melt and it would solidify too quickly to produce large sheets of metal (Kerr, 1996). Evidence that some type of metal segregation can occur can be found in the H chondrite Portales Valley meteorite (Rubin et al., 2001), an impact melt breccia with large regions of coarse metal with a Widmanstätten pattern. A Widmanstätten pattern indicates slow cooling.

Sa-types have intermediate spectra between S- and A-types in the visible (Bus and Binzel, 2002b) and near-infrared (DeMeo et al., 2009) (Figure 4.20). Sa-types have distinctive olivine-like bands but are less red than A-types. Sa-types are relatively rare with less than 1% of bodies classified by DeMeo et al. (2009) having this designation. Rivkin et al. (2007) observed Mars Trojan Sa-type (5261) Eureka and found that it had a very broad 1 µm but no 2 µm band and resembled the spectra of angrites (Burbine et al., 2006).

Sq-types have intermediate spectra between S- and Q-types (Figure 4.20). Sq-types are relatively common (Binzel et al., 2004) among near-Earth asteroids but relatively rare among main-belt objects. Binzel et al. (2004) groups the Sq- and Q-types into the Q-complex since Q-types are also relatively common among near-Earth asteroids but also extremely rare among main-belt objects. Mothé-Diniz et al. (2010) found that a vast majority of Sq- and Sk-bodies had visible spectra that matched ordinary chondrites.

Main-belt objects classified as Sq by DeMeo et al. (2009) include (3) Juno and (11) Parthenope. Both of these objects were classified as S(IV)-types by Gaffey et al. (1993). Gaffey et al. (1993) and Gaffey (1995) identified Juno as being a leading candidate as an ordinary chondrite parent body. Parthenope was identified by Farinella et al. (1993a) as a body that may be supplying relatively large number of fragments into meteorite-supplying resonances.

Sr-types have intermediate spectra between S- and R-types with a narrow 1 µm band that is similar to but shallower than the bands found in R-types (Figure 4.20). Using MGM on visible and near-infrared spectra, Sunshine et al. (2004) found that a number of S- or Sr-types, including (17) Thetis and members of the (808) Merxia and (847) Agnia families, had silicate mineralogies consistent with low- and high-CA pyroxene with minor amounts of olivine (<20%). Approximately 40% of the total pyroxene on their surfaces is high-Ca pyroxene, which is only consistent with bodies that have undergone melting since partial melting of chondritic material tends to produce melts enriched in low-Ca pyroxene.

Sv-types have intermediate spectra between S- and V-types (Figure 4.20). Sv-types have a very narrow 1 µm band that is similar to but shallower than the band found in V-types. Sv-types are relatively rare with only 2 (< 1%) of objects classified by DeMeo et al. (2009) have this designation.

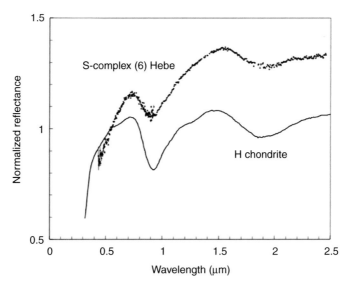

Figure 4.21 Reflectance spectra of an H5 chondrite (Allegan) (Burbine et al., 2003) versus S-complex (6) Hebe in the visible and near-infrared. Note the similarity in spectral shape but the difference in spectral slope. Spectra are normalized to unity at 0.55 μm.

4.9.2 C-complex

The commonest asteroids in the outer main belt are the C-complex asteroids. C-complex asteroids tend to have weak absorption bands in the UV and visible wavelength regions and relatively featureless spectra from ~1 to ~2.5 μm that have blueish to slight reddish spectral slopes (Figure 4.20).

The C-complex bodies of DeMeo et al. (2009) include the B, C, Cb, Cg, and Cgh objects. Approximately 13% of classified asteroids by DeMeo et al. (2009) are part of the C-complex. Approximately 60% of observed C-complex objects (Rivkin et al., 2002) have 3 μm bands, indicating hydrated silicates on their surfaces. Using SDSS colors to estimate that 30 ± 5% of C-complex objects have 0.7 μm bands (and, therefore, have 3 μm bands) and assuming half of the remaining C-complex bodies have 3 μm bands, Rivkin (2012) estimates that approximately two-thirds of C-complex asteroids have hydrated minerals on their surfaces. Traditionally, C-complex objects have been linked with carbonaceous chondrites (e.g., Johnson and Fanale, 1973, Feierberg et al., 1985) or thermally metamorphosed carbonaceous chondrites (Hiroi et al., 1993; Ostrowski et al., 2010, 2011). The two largest objects in the main belt are both C-complex bodies: dwarf planet (1) Ceres (diameter of ~950 km) and (2) Pallas (diameter of ~530 km).

B-type asteroids have bluish spectral slopes longward of ~0.5 μm (Figure 4.20). Yang and Jewitt (2010) noted that a number of asteroids that have orbits consistent with comets or associated with meteor showers have been classified as B-types, implying that these objects may have originally incorporated significant amounts of water ice. This is consistent with the idea that there is a continuum between dark asteroids and comets (Gounelle et al., 2008).

On the basis of visible and near-infrared spectra, Clark et al. (2010) broke up B-type spectra into three groups. One is the Pallas group, which have blue spectral slopes like Pallas. Another is the Themis group, which have concave curve shapes like (24) Themis. The Other group has spectra unlike Pallas or Themis.

Pallas has long been known (Larson et al., 1979, 1983; Feierberg et al., 1981; Jones et al., 1990) to have a 3 μm band that was similar in structure to the bands found for CM chondrites. Beck et al. (2010) noted the Pallas should even be a better match to CM chondrites in the 3 μm region if terrestrial adsorbed water is removed from CM chondrite spectra. Pallas' estimated bulk density of 2400–2600 kg/m^3 (Schmidt et al., 2009; Zielenbach, 2011; Baer et al., 2011) and the presence of a 3 μm band are both consistent with a body that formed from water-rich material.

Yang and Jewitt (2010) found that the concave shapes of a number of B-types in the near-infrared are consistent with a broad 1 μm absorption feature that is consistent with the mineral magnetite. One object, (335) Roberta, a Themis group B-type (Clark et al., 2010), had a spectrum consistent with a mixture of a CI chondrite and magnetite. Both Roberta and the CI-magnetite mixture have a broad 1 μm band and a 3 μm feature.

Themis was found to have spectral features in the 3 μm wavelength region consistent with water ice (band at ~3.1 μm) and organics (band at ~3.4 μm) (Campins et al., 2010; Rivkin and Emery, 2010). However, Beck et al. (2011b) have found that goethite also has absorption features similar to those found for Themis. Due to the absence of detectable CN emission lines, which is used as tracer of water in comets, and assuming cometary H_2O/CN mixing ratios, an upper limit for surface water ice was determined for Themis by Jewitt and Guilbert-Lepoutre (2012) to be less than ~10%.

Ceres is the most studied C-complex body due to its large size; however, the mineralogy of Ceres has been considerably debated (e.g., Rivkin et al., 2011a; McCord et al., 2011). In the visible and near-infrared (~0.4–2.5 μm), Ceres has moderate strength UV features and a broad absorption band centered around 1.2 μm that has been attributed to magnetite (Larson et al., 1979; Rivkin et al., 2011a). With a spatial resolution of ~75 km, Carry et al. (2012) found that Ceres was very homogeneous spectrally in the ranges 1.17–1.32 μm and 1.45–2.35 μm. However, the most prominent absorption bands for Ceres are in the 3 μm region (Figure 4.22), but these bands are unlike those found in carbonaceous chondrites (Figure 4.14). King et al. (1992) believed that the feature at ~3.06 μm was due to an NH_4-bearing phyllosilicate, which has an absorption feature due to NH_4^+ that is approximately at

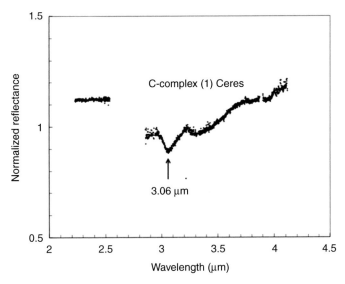

Figure 4.22 Reflectance spectrum of C-complex (1) Ceres (Milliken and Rivkin, 2009) in the 3 μm region. Note the band at 3.06 μm.

~3.07 μm and has a similar shape. Vernazza et al. (2005) also observed this feature at ~3.06 μm and attributed it to water ice in a mixture of ion-irradiated organics. Milliken and Rivkin (2009) argued that the 3 μm absorption features for Ceres are consistent with the presence of the hydroxide brucite, magnesium carbonates, and serpentines. This type of assemblage is not found in known carbonaceous chondrites. For an object with Ceres' estimated bulk density of ~2100 kg/m^3 (Thomas et al., 2005; Zielenbach, 2011; Baer et al., 2011), thermal modeling of Ceres by Castillo-Rogez (2011) finds that a significant amount of water ice must exist in the interior to produce Ceres' density for the temperatures that Ceres must have experienced. Hopefully, many of these questions concerning Ceres' mineralogy will be answered by the Dawn mission (Section 8.13.2).

C-complex bodies with a feature at ~0.7 μm are given the "h" designation (Figure 4.20). This band tends to be found in the spectra of CM2 chondrites (Cloutis et al., 2011b). Objects with a ~0.7 μm band almost always have a 3 μm band (Vilas, 1994; Howell et al., 2001; Rivkin et al., 2002; Rivkin et al., 2015), which is consistent with this linkage with CM chondrites. Ch-types (13) Egeria and (19) Fortuna were originally noted by Burbine (1998) to be spectrally similar to CM chondrites in the visible and near-infrared plus have 3 μm band strengths (Jones et al., 1990) consistent with CM2 chondrites. Egeria and Fortuna are both located near the 3:1 resonance, which may make it easier for the relatively weak CM fragments to make it into a meteorite-supplying resonance. CM fragments may not be strong enough to drift considerable distances in the belt and still survive. Cgh-types have stronger UV features than Ch-types.

Cruikshank and Brown (1987) identified a ~3.4 μm absorption band on Ch-type (130) Elektra, which was originally classified as a G-type by Tholen (1984). This band was argued to be due to a C–H stretching mode in organics (e.g., Moroz et al., 1998). However, this feature was not found (Cruikshank et al., 2002) in a later spectrum of Elektra or in the spectra of a number of observed G-, P-, or D-types. Cruikshank et al. (2002) argues that this band may not be present in asteroid spectra due to space weathering, which could dehydrogenate C–H-bearing hydrocarbons over time through interaction with the solar wind.

4.9.3 X-complex

X-complex asteroids have reddish spectral slopes in the visible and in the near-infrared (Figure 4.20). Tholen (1984) subdivided X-complex members on the basis of visual albedo into E-, M-, and P-types. E-types have high visual albedos, M-types have moderate visual albedos, and P-types have low visual albedos. Historically, E-types have been linked with enstatite-rich material, M-types with metallic iron, and P-types with organic-rich material due to their respective visible albedos and their relatively flat spectra in the visible and near-infrared. Since the Bus–DeMeo taxonomy does not use visual albedo, the E-, M-, and P-types are not defined in that taxonomy and instead X-complex bodies are subdivided according to spectral shape.

The E-types have been typically linked with the aubrites (e.g., Zellner, 1975; Zellner et al., 1977) since both types of objects have high albedos and relatively flat spectra longward of 0.55 μm. The aubrites are achondritic enstatite-rich meteorites (Watters and Prinz, 1979). Enstatite is an FeO-poor pyroxene that is whitish in appearance. Among meteorites (Gaffey, 1976), only aubrites have such high albedos and no prominent 1 and 2 μm absorption bands. Approximately 60% of identified E-types (Clark et al., 2004) are found in the Hungaria region near 2 AU where (434) Hungaria is located. Approximately two-thirds of the observed E-types were found to have 3 μm absorptions (Rivkin et al., 1995; Rivkin et al., 2002), but none of these objects were located in the Hungaria region. If E-types are primarily igneous bodies, the presence of a 3 μm band due to hydrated silicates is surprising because these objects would be expected to have been heated to temperatures during differentiation too high to retain hydrated silicates.

E-types do not have entirely featureless visible and near-infrared spectra. For example, a number of weak bands have been found in the reflectance spectra of E-types. A high percentage (Clark et al., 2004) of E-types were classified as Xe-types by Bus and Binzel (2002b) and DeMeo et al. (2009) (Figure 4.20) due to the presence of a ~0.5 μm feature (Bus and Binzel, 2002a) that was not present in the lower resolution ECAS data. Xe-type (64) Angelina was noted by Kelley and Gaffey (2002) to have a number of weak bands in the visible and near-infrared, which they argue were due to accessory phases mixed with the dominant mineral enstatite. Near-Earth asteroid (3103) Eger, classified as an E-type by Veeder et al. (1989) and an Xe-type by Binzel et al. (2004), was postulated by Gaffey et al. (1992) to be the source body of the aubrites and also to be derived from the Hungaria region.

The ~0.5 μm absorption band found in the spectra of Xe-types was argued to be due to oldhamite (CaS) by Burbine et al. (2002), a mineral commonly found in aubrites and enstatite chondrites but usually in very low abundances (Watters and Prinz, 1979). Aubrites have much higher visual albedos (~0.42–0.48) (excluding the chondritic inclusion-rich Cumberland Falls) (Cloutis and Gaffey, 1993) than enstatite chondrites (~0.05–0.18) since enstatite chondrites have significant abundances of metallic iron and opaques, which reduces the visual albedo. Due to these low concentrations in aubrites, Shestopalov et al. (2010) proposes that the feature is instead due to Ti^{3+} crystal field bands or Fe^{2+} to Ti^{4+} charge-transfer transitions in pyroxenes. However, oldhamite weathers very easily on Earth when exposed to atmospheric moisture (Okada et al., 1981) so the abundances found in meteorites may not be representative of the abundances on the surfaces of enstatite-rich asteroids.

Asteroid (44) Nysa, classified as an E-type by Zellner et al. (1977) and Tholen (1984) and now as an Xn-type (Figure 4.20), has long been known to have weak bands at ~0.9 and ~1.8 μm, which have been attributed to a low-FeO pyroxene (Gaffey et al., 1989; Clark et al., 2004). Cloutis and Gaffey (1993) argued that these weak bands could be due to carbonaceous chondritic inclusions that are found in some aubrites. Nysa's 3 μm band (Rivkin et al., 1995) could be consistent with carbonaceous chondritic inclusions.

M-type asteroids have been historically linked with iron meteorites due to both types of objects having relatively flat spectra in the visible and near-infrared and moderate visual albedos; however, enstatite chondrites are also spectrally similar to M-types and have similar visual geometric albedos. However, most of these asteroids (e.g., Hardersen et al., 2005, 2011; Fornasier et al., 2010, 2011; Ockert-Bell et al., 2010) have absorption features at ~0.9 μm, which are consistent with the presence of Fe-bearing silicates. Birlan et al. (2007) observed a number of different M-types in the near-infrared and found them to be featureless with spectral slopes consistent with metallic iron surfaces. Rivkin et al. (1995) found that a number of M-type asteroids had 3 μm absorptions (W-types). Approximately 40% of observed M-types had 3 μm bands (Rivkin et al., 2002), which is consistent with hydrated silicates on the surface and not consistent with metallic iron-dominated assemblages.

Psyche, classified as an M-type by Tholen (1984) and as an Xk-type (Figure 4.20) by DeMeo et al. (2009), is one of the most studied asteroids due to its large diameter (~250 km). Psyche has long been thought to be a metallic iron core due to its extremely high radar albedo (0.42 ± 0.10) (Ostro et al., 1985; Shepard et al., 2010). Radar albedos are discussed in Section 5.4.2. However, a feature at ~0.9 μm (Hardersen et al., 2005) in the spectra of Psyche is consistent with Fe-bearing silicates on its surface. Binzel et al. (1995) did not find any significant variation in spectral properties in the visible versus rotation for Psyche, implying that there are no significant surface variations in silicate abundances. Rivkin et al. (1995, 2000) did not identify a 3 μm band in Psyche's spectrum, which is consistent with a metallic iron-rich surface. Psyche's estimated bulk density, ~6700 kg/m³ (Baer et al., 2011), is consistent with a metallic iron mineralogy with some porosity.

P-types are commonly found in the Hilda group and among Jupiter Trojans. About a quarter of observed Trojans (Roig et al., 2008b) have spectral slopes compatible with P-type asteroids. Most P-types do not have a 3 μm band (Rivkin et al., 2002). The spectral properties of P-types asteroids are usually thought to be due to organics, which would redden and darken the surface. Hiroi et al. (2004) was able to duplicate the spectral properties of P-types by spectrally mixing unaltered CI/CM and Tagish Lake (C2-ungrouped) material with CI/CM and Tagish Lake (C2-ungrouped) material that has undergone thermal metamorphism and/or space weathering. Schaefer et al. (2010) found that the opposition surges calculated from the phase curves (brightness versus phase angle) of Jupiter Trojans were consistent with the opposition surges of outer Solar System bodies that have lost their ices (e.g., centaurs, extinct comets) and not with main-belt C- and P-type asteroids, implying that Jupiter Trojans and main-belt C- and P-types have different surface compositions and/or textures.

Main-belt P-type (65) Cybele has a sem-major axis of 3.43 AU. Cybele has a visual albedo of 0.050 ± 0.005 (Müller and Blommaert, 2004). This body was classified as an Xk by DeMeo et al. (2009) (Figure 4.20). Cybele was found to have an absorption featured at ~3.1 μm (Licandro et al., 2011) that is consistent with water ice. However, previous observations of Cybele (Jones et al., 1990) did not find a 3 μm band. Due to the absence of detectable CN emission lines, which is used as tracer of water in comets, and assuming cometary H_2O/CN mixing ratios, an upper limit for surface water ice was determined for Cybele by Jewitt and Guilbert-Lepoutre (2012) to be less than ~10%. Other absorption features for Cybele between 3.2 and 3.6 μm appear consistent with organics (Licandro et al., 2011) with no features detected that were consistent with hydrated silicates. The emissivity spectrum of Cybele (Licandro et al., 2011) in the 5–14 μm range is similar to that found for Trojan asteroids (Emery et al., 2006). Cybele has an estimated bulk density of ~1000 kg/m^3 (Baer et al., 2011), which is consistent with an extremely porous material.

4.9.4 A-types

A-type asteroids have distinctive absorption features in the visible and near-infrared that are characteristic of olivine (Figure 4.20). Meteoritic analogs (Cruikshank and Hartmann, 1984; Sunshine et al., 2007) for A-types are olivine-dominated meteorites such as the pallasites, the brachinites, and the R chondrites. The fayalite content varies among these possible analogs with most pallasites containing olivine with Fa$_{12\pm1}$ (Mittlefehldt et al., 1998), brachinites containing Fa$_{\sim30-35}$ (Mittlefehldt et al., 1998), and R chondrites containing Fa$_{\sim37-40}$ (Brearley and Jones, 1998). A few pallasites have fayalite compositions as high as Fa$_{18}$. Postulated linkages between A-types and pallasites tend to assume surface compositions that do not have a significant abundance of metallic iron. A-types tend to have much redder near-infrared spectra when compared to laboratory measurements of olivine, which is consistent with space weathering reddening their surfaces (Lucey et al., 1998; Hiroi and Sasaki, 2001; Burbine and Binzel, 2002).

Sunshine et al. (2007) found that a number of A-types [(354) Eleonora, (446) Aeternitas, (863) Benkoela, (984) Gretia, (2501) Lohja, (3819) Robinson] had very magnesian (forsteritic) olivine compositions from MGM of A-type visible and near-infrared spectra. Eleonora was classified by Gaffey et al. (1993) as an S(I)-subtype, which have interpreted silicate mineralogies that consist of almost entirely of olivine. However, two A-types, (246) Asporina and (289) Nenetta, were found to be more ferroan (fayalitic) by Sunshine et al. (2007). The forsteritic A-types have mineralogies consistent with differentiation from an ordinary chondrite-like body, while the fayalitic A-types are consistent with the more oxidized R chondrites or melting from such a body. A number of these objects have a Band II, which is consistent with 5–10% pyroxene on their surfaces. Lucey et al. (1998) had previously argued that the reflectance spectra of (246) Asporina, (289) Nenetta, (446) Aeternitas, and (863) Benkoela were consistent with low FeO-bearing olivines measured at asteroidal temperatures.

Most A-type bodies are commonly assumed to be remnants of mantle material from disrupted differentiated bodies. However, only ~2% of bodies classified by DeMeo et al. (2009) were found to be A-types. If differentiated bodies formed metallic iron cores, olivine-dominated mantles, and basaltic crusts, the disruption of these objects should produce significant numbers of olivine-dominated fragments. The low number of identified A-types in the main belt that are currently present in the belt appears inconsistent (Chapman, 1986) with the large number of differentiated bodies that should have disrupted to produce the number of iron meteorites that are present in our meteorite collections. One scenario is that these olivine-dominated objects have been "battered to bits" (Burbine et al., 1996) and are currently at sizes not observable in the main belt.

4.9.5 D-types

D-types have very red spectral slopes and extremely low albedos (Figure 4.20). D-types are the dominant taxonomic type among Jupiter Trojans and are also prevalent in the outer main belt and among the Hildas. D-types have also been observed in the inner main-belt (DeMeo et al., 2014).

Due to their very red spectral slopes and low albedos, D-types are usually linked with organic-rich material (Gradie and Veverka, 1980) that may or not be present in our meteorite collections due to its fragile nature. Hiroi et al. (2001b) has found spectral similarities between C2 chondrite Tagish Lake and D-type asteroids. However since Jupiter Trojan D-types have a lack of distinctive absorption features (e.g., 3.4 µm band due to organics) in the visible and near-infrared out to 4 µm, Emery and Brown (2003, 2004) argue that the red spectral slope is due to space-weathered anhydrous silicates (e.g., pyroxenes, olivines).

4.9.6 K-types

K-type asteroids have a moderately strong UV feature, a broad 1 µm feature, and a red spectral slope (Figure 4.20). K-types have typically been linked to CO and CV chondrites (Bell, 1988; Clark et al., 2009) due to their similar spectral properties in the visible and near-infrared and visual albedos. Clark et al. (2009) found that the best meteorite matches to K-types were CO and CK chondrites. However, Mothé-Diniz and Carvano (2005) noted the spectral similarity of Eos to the differentiated meteorite Divnoe. Mothé-Diniz et al. (2008) using radiative transfer modeling found that a number of K-types in the Eos family had calculated mineralogies consistent with forsteritic olivine ($Fa_{~20}$), which they argue is consistent with either partial differentiation of an ordinary chondrite parent body or a CK chondritic mineralogy.

4.9.7 L-types

L-type asteroids have a strong UV feature, a weak Band I, and a Band II much stronger than Band I (Figure 4.20). A number of L-types have absorption features consistent (Sunshine et al., 2008) with iron oxide-bearing aluminous spinel found in calcium–aluminum inclusions (CAIs). Two of these L-types [(387)

Aquitania and (980) Anacostia] were previously identified by Burbine et al. (1992) as possibly having spinel-rich surfaces. Sunshine et al. (2008) argue that these objects contain extremely high (30 ± 10 vol%) CAI abundances compared to meteorites and may be more ancient than anything found in our meteorite collections. They believe that these objects may have formed before the injection of the ^{26}Al into the Solar System since these objects would have melted with such high CAI abundances and canonical initial ^{26}Al/^{27}Al ratios. However, Hezel and Russell (2008) argue that the actual CAI abundances are lower (10 vol%) and the objects would not have melted if they had canonical ^{26}Al/^{27}Al ratios. Then, these bodies would not necessarily be more ancient than anything found in our meteorite collections.

4.9.8 O-types

The O-type designation was proposed by Binzel et al. (1993) of (3628) Božněmcová due to its spectral similarity to ordinary chondrites in the visible. O-types have moderately steep UV features shortward of 0.54 μm, becoming less steep over the interval from 0.54 to 0.7 μm, and a relatively deep 1 μm band (Bus and Binzel, 2002b). DeMeo et al. (2009) only classified one object (Božněmcová) as an O-type and found that it has a very rounded and deep, "bowl-shaped" absorption feature at 1 μm as well as a 2 μm band (Figure 4.20).

Using visible and near-infrared spectra out to ~1.65 μm, Cloutis et al. (2006) interpreted the reflectance spectrum of Božněmcová as being consistent with an assemblage of clinopyroxene and plagioclase feldspar that would be mineralogically similar to angrites. Angrites contain a clinopyroxene that typically does not have a 2 μm feature (Burbine et al., 2006). However, near-infrared spectra out to ~2.5 μm of Božněmcová (DeMeo et al., 2009; Burbine et al., 2011) have a distinctive 2 μm feature that is not consistent with an angritic composition. Burbine et al. (2011) found that even though the shapes of the absorption bands for Božněmcová appeared characteristic of pyroxenes, the Band I and II centers were offset from the typical values found for pyroxenes. No known meteorite has a spectrum similar to that of Božněmcová.

4.9.9 Q-types

Q-type asteroids have absorption features in the visible and near-infrared due to olivine and pyroxene that tend to be stronger than S-complex objects but less red-sloped (Figure 4.20). Near-Earth asteroid (1862) Apollo was the first identified Q-type (Tholen, 1984) and has long been known to be spectrally similar to LL chondrites in the visible (McFadden et al., 1985) and near-infrared

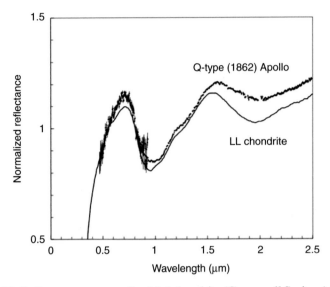

Figure 4.23 Reflectance spectra of an LL4 chondrite (Greenwell Springs) (Burbine et al., 2003) versus Q-type (1862) Apollo in the visible and near-infrared. Note the similarity in spectral shape. Spectra are normalized to unity at 0.55 μm.

(Binzel et al., 2006) (Figure 4.23). Q-types are relatively rare in the main belt but a few (e.g., Vernazza et al., 2006a; Mothé-Diniz and Nesvorný, 2008; Chapman et al., 2009) have been tentatively identified. The prevailing opinion is that Q-types are "unweathered" ordinary chondrite assemblages, while most S-complex objects are "weathered" ordinary chondrite assemblages (e.g., Binzel et al., 2015).

4.9.10 R-types

The first identified R-type was (349) Dembowska. R-types have a deep Band I and Band II (Figure 4.20). The Band II for an R-type is narrow than a Q-type's band but slightly broader than a V-type's band. DeMeo et al. (2009) only classified one object, (349) Dembowska, as an R-type because of its deep 1 and 2 μm features.

Gaffey et al. (1989) interpreted the reflectance spectra of Dembowska as indicating a partially melted pyroxene–olivine assemblage with little to no metallic iron. Using Band Area Ratios and Band I centers, Abell and Gaffey (2000) also interpreted the rotational spectra of Dembowska as indicating a partially melted orthopyroxene–olivine assemblage. Hiroi and Sasaki (2001) find that Dembowska's reflectance spectrum is consistent with a space-weathered orthopyroxene–olivine assemblage. Both Abell and Gaffey (2000) and Hiroi and Sasaki (2001) find Dembowska's orthopyroxene to olivine ratio to be ~45:55.

4.9.11 T-types

In the visible (Tholen, 1984; Bus and Binzel, 2002b), T-types have moderately strong UV features shortward of 0.75 μm and then flatten out longward of 0.85 μm. In the near-infrared (DeMeo et al., 2009), T-types have slopes redder than X-types but not as red as D-types (Figure 4.20). Most T-types were found to have 3 μm bands (Rivkin et al., 2002).

Britt et al. (1992) noted the red near-infrared spectral slopes and low albedos of T-types were similar to troilite. Currently, the spectral properties of T-type asteroids are usually thought to be due to organic-rich silicates that would redden and darken the surface. Hiroi and Hasegawa (2003) noted the spectral similarity between T-type (308) Polyxo and the C2-ungrouped chondrite Tagish Lake.

4.9.12 V-types

The first recognized V-type was (4) Vesta, which has long been known to have an unusual spectrum that was similar to HEDs in the visible and near-infrared (Figures 4.20 and 4.24). McCord et al. (1970) measured and analyzed the visible reflectance spectrum of Vesta and found it was a spectral match to a eucrite. Larson and Fink (1975) then noted the spectral similarity of Vesta to the eucrites in the near-infrared. Vesta had rotational spectral variations (e.g., Gaffey, 1997; Binzel et al., 1997; Carry et al., 2010) consistent with HED mineralogies. Spectra of Vesta in the UV into the visible (0.22–0.95 μm) (Li et al., 2011) matched a fine-grained howardite. The postulated relationship of Vesta with the HEDs was tested by the Dawn spacecraft's encounter with Vesta (Section 8.13.1).

Vesta is commonly assumed to be the HED parent body (e.g., Consolmagno and Drake, 1977) due its presence as the only large (~525 km diameter) body with an HED-like visible and near-infrared spectrum. Originally Vesta was thought to dynamically unable (Wetherill, 1987) to supply large number of fragments to meteorite-supply resonances; however, the discovery of small V-type bodies (commonly called Vestoids) (Binzel and Xu, 1993) in the Vesta family and also between Vesta and the 3:1 meteorite-supplying resonance on the basis of visible observations and the ability of Yarkovsky effect to move small bodies over distances of tenths of an AU in the asteroid belt (e.g., Bottke et al., 2006) argues that Vesta could supply fragments to these resonances. Numerous V-types have been identified in the Vesta family (e.g., Mothé-Diniz et al., 2005), the inner main-belt but outside the Vesta family (e.g., Florczak et al., 2002; Carruba et al., 2005), and the near-Earth population (e.g., Cruikshank et al., 1991; Burbine et al., 2009).

Near-infrared of inner-belt V-types (Burbine et al., 2001; Kelley et al., 2003; Moskovitz et al., 2010; Mayne et al., 2011; Hardersen et al., 2014, 2015) confirm

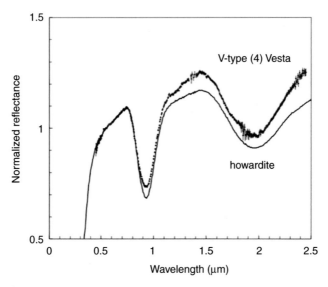

Figure 4.24 Reflectance spectra of a howardite (EET 87503) (Hiroi et al., 1994) versus V-type (4) Vesta in the visible and near-infrared. Note the spectral similarity. Vesta is a target of the Dawn mission. The spectra are normalized to unity at 0.55 μm.

their spectral similarity to eucrites and howardites. A number of these inner-belt Vestoids also have a feature at ~0.506 μm (Vilas et al., 2000), which has been also seen in the spectra of HEDs (e.g., Hiroi et al., 2001a). This feature has been attributed to a spin-forbidden Fe^{2+} absorption band that is commonly found in the spectra of low-Ca pyroxenes (Cloutis et al., 2010c).

Binzel and Xu (1993) also designated some objects with visible HED-like spectra but very deep 1 μm bands as J-types, which they interpreted as having diogenitic mineralogies based on just visible spectra. However, near-infrared observations (out to 1.65 μm) of a number of J-types (Burbine et al., 2001) appeared not to be consistent with diogenite spectra. However, a number of V-types with near-infrared spectra consistent with diogenites have recently been identified in the inner belt (Mayne et al., 2011; De Sanctis et al., 2011; Hardersen et al., 2015). Lim et al. (2011) found that V-type (956) Elisa had a Spitzer Space Telescope emissivity (~5–35 μm) spectrum that was consistent with an olivine–diogenite-like mineralogy. Besides identifying a number of diogenite V-types, Mayne et al. (2011) also identified one V-type, (2511) Patterson, as having a ferroan (Fe-rich) diogenite composition.

However, Vesta appears not to be the only body that formed in the asteroid belt with a basaltic crust. The first identified object was outer-belt body (1459) Magnya (Lazzaro et al., 2000; Hardersen et al., 2004), which has an *a* of 3.14 AU and is observed to have an HED-like spectrum in the visible and near-infrared. Band positions derived for Magnya were interpreted (Hardersen et al., 2004) as indicating a pyroxene mineralogy that is slightly less Fe-rich than Vesta. Magnya is located relatively far from Vesta ($a = 2.36$ AU) and far past the 3:1 resonance, which makes it dynamically difficult (Michtchenko et al., 2002) to derive Magnya from Vesta. Magnya's diameter (17 ± 1 km) (Delbó et al., 2006) is larger than inner-belt Vestoids, but Magnya's geometric visible albedo (0.37 ± 0.06) is similar to Vesta.

A number of V-types bodies have been identified in the middle and outer main-belt (e.g., Roig and Gil-Hutton, 2006) on the basis of SDSS colors. Asteroid (21238) Panarea ($a = 2.54$ AU), which is located past the 3:1 resonance, was found to have a typical visible V-type spectrum (Roig et al., 2008a) and band positions in the near-infrared consistent with diogenites (Moskovitz et al., 2010; De Sanctis et al., 2011). Asteroid (40521) 1999 RL_{95} is another identified V-type body (Roig et al., 2008a) that is also located past the 3:1 resonance at 2.54 AU. Bodies like Panarea and 1999 RL_{95} could be fragments of Vesta that "luckily" were able to drift past the 3:1 resonance (Roig et al., 2008a) or fragments of another differentiated body (Carruba et al., 2007).

Outer-belt V-types have been also identified on the basis of SDSS colors. Two with visible and near-infrared spectra (Duffard and Roig, 2009) are (7472) Kumakiri ($a = 3.02$ AU) and (10537) 1991 RY_{16} ($a = 2.85$ AU). Both objects have visible spectra similar to other V-types except for an absorption band at ~0.65 μm, which was interpreted as possibly being evidence of chromium in the pyroxenes. However, their near-infrared spectra (Moskovitz et al., 2008b; Burbine et al., 2011) have been identified as unusual compared to other V-types. Kumakiri has 1 and 2 μm bands that appear consistent with pyroxenes (Burbine et al., 2011), but the calculated band positions are offset from the values found for HEDs. Possible explanations for this anomalous behavior is that Kumakiri contains an unusual type of pyroxene not typically found in meteorites or this object is a mixture of olivine with high-Ca pyroxene and minor low-Ca pyroxene. No known meteorite has a spectrum similar to that of Kumakiri. Asteroid (10537) 1991 RY_{16} has a near-infrared spectrum (Moskovitz et al., 2008b) consistent with a substantial olivine component, which is unlike the spectral properties of other observed V-types. Scott et al. (2009) argue that these "anomalous" eucrites are evidence for a number of distinct Vesta-like parent bodies besides Vesta.

Questions

1) What are the fundamental causes of the different electronic and vibrational transitions found in meteorites and asteroids?

2) What will be the band depth of an absorption band that has a reflectance value of 0.8 at the band center wavelength of 1.5 μm? The continuum has a reflectance of 1.1 at 1.4 μm and a reflectance of 1.3 at 1.7 μm.

3) Explain the different parts of the formula for bidirectional reflectance developed by Bruce Hapke.

4) How is an olivine reflectance spectrum in the visible and near-infrared different from an orthopyroxene reflectance spectrum in the same wavelength region?

5) Why and how will the reflectance spectrum of a phyllosilicate change in the 3 μm wavelength region if the sample is measured under dry (elevated temperature) conditions and under vacuum?

6) The spectrum of a pyroxene has band centers at 0.94 and 2.00 μm. Estimate its ferrosilite and wollastonite contents.

7) The spectrum of an ordinary chondrite assemblage has a Band I center at 0.99 and a Band Area Ratio of 0.50. Estimate its fayalite and ferrosilite contents and ol/(ol + px) ratio.

8) What are the best meteoritic analogs for the S-complex asteroids? Why?

9) What are the best meteoritic analogs for the C-complex asteroids? Why?

10) What are the best meteoritic analogs for the V-type asteroids? Why?

5

Physical Properties and Families

5.1 Diameters

Only a relatively few asteroids have had their diameters (D) measured directly from Earth since asteroids have relatively small angular sizes as observed from most Earth-based telescopes. Ceres has a maximum angular size of only ~0.8 arc-seconds (Drummond et al., 2014). Most asteroids have angular sizes less than 0.01 arcseconds (Kim et al., 2003). As expected, it is much easier to directly measure asteroid diameters from spacecraft missions.

The commonest direct way to measure the diameter of an asteroid from Earth is by measuring the occultation of a star by the asteroid (Millis and Elliot, 1979). By measuring the time that an asteroid blocks the star from view and knowing the distance to the star and its rate of motion, the diameter can be calculated. Approximately 900 different asteroids have had their diameters measured from over 2000 occultations.

Direct imaging has also been used to measure asteroid diameters. However, the telescopes need to be relatively large and equipped with adaptive optics or be based in space (e.g., Hubble Space Telescope).

5.1.1 Calculating Diameters Using Infrared Measurements

The absolute magnitude is related to the diameter and the visual albedo by the relation

$$H \propto D^2 p_v \tag{5.1}$$

since $D = \left(1329 / \sqrt{p_v}\right) 10^{-0.2H}$ km (*Equation 1.30*). The D^2 term is due to the projected cross-sectional area ($\pi D^2/4$) of the asteroid. The larger body will reflect more light than a smaller body. The higher albedo body will also reflect more light than a lower albedo body. If only the absolute magnitude of an object is known, you cannot determine the visual albedo unless the diameter is known and vice versa.

While the visual brightness is a function of the cross-sectional area times the visual albedo, the emitted infrared flux is proportional to the absorbed visible light with

$$F_{ir} \propto D^2 (1 - p_v).$$

(5.2)

So for the same cross-sectional area, lower albedo objects will be brighter in the infrared than higher albedo objects since lower albedo bodies will have higher surface temperatures and therefore emit more thermal radiation.

At the same distance from the Sun, a high-albedo small body can have the same visual magnitude as a low-albedo large object. However, the low-albedo large object will be much brighter in the infrared than the high-albedo small object since that body is absorbing more radiation and hotter. So by using photometry in both the visual and infrared wavelengths, the diameter and albedo of an object can be determined that fits both the values of the H magnitude and the infrared flux.

5.1.2 Thermal Inertia

The thermal inertia controls the temperature distribution over an asteroid and its resulting infrared flux (Delbó and Tanga, 2009). The higher the thermal inertia, the more resistant a material is to temperature changes. Bare rock has a higher thermal inertia than fine dust of the same material. Metal tends to have a higher thermal inertia than silicates.

The formula for the thermal inertia (Γ) is

$$\Gamma = \sqrt{\kappa \rho C}$$

(5.3)

where κ is the thermal conductivity, ρ is the density, and C is the specific heat capacity. The thermal conductivity measures the ability of a material to conduct heat. The specific heat capacity is the energy involved in changing the temperature of a material (Consolmagno et al., 2013). The units for thermal conductivity are W/(m K), for density are kg/m^3, and for specific heat capacity are J/(kg K). The units of thermal inertia are $J\,m^{-2}s^{-0.5}\,K^{-1}$, which is often known as a thermal inertia unit (TIU).

The thermal conductivity is a function of temperature and porosity (Opeil et al., 2012). It is measured by imparting a pulse of heating to a sample and measuring the temperature change (Opeil et al., 2010). Newly measured thermal conductivity values of meteorites at 200 K range from 0.45 W/(m K) for an L6 chondrite to 5.51 W/(m K) for an EL6 chondrite (Opeil et al., 2012). Meteorite densities are given in Section 3.7 and tend to be ~3000–4000 kg/m³ for chondrites and achondrites. Asteroid densities tend to be lower due to higher porosities (e.g., Carry, 2012). Meteorite densities for irons are ~8000 kg/m³. Specific heat capacity is also a function of temperature. The specific heat capacity is measured by placing the sample in liquid nitrogen, measuring the mass of nitrogen boiled off due to the heat within the sample, and calibrating against measurements of pure quartz (Consolmagno et al., 2013). For asteroidal surface temperatures, Consolmagno et al. (2013) found that the average specific heat capacities for meteorites groups varied from 342 J/(kg K) for IIAB irons to 535 J/(kg K) for CR chondrites. H and L chondrites had average specific heat capacities of 494 J/(kg K). The meteorite thermal conductivities and specific heat capacities are measured on small intact samples and not particulate material as would be found in an asteroid regolith.

The thermal inertia is very dependent on the physical properties of the regolith (e.g., Delbó et al., 2007b). Fine dust (~ 30 J m^{-2}s$^{-0.5}$K^{-1}) and the lunar regolith (~ 50 J m^{-2}s$^{-0.5}$K^{-1}) have very low thermal inertias. Coarse sand (~ 400 J m^{-2}s$^{-0.5}$K^{-1}) has a higher thermal inertia. Bare rock (~ 2500 J m^{-2}s$^{-0.5}$K^{-1}) is much higher. Ness and Emery (2014) argue that asteroids with thermal inertias less than 500 J m^{-2}s$^{-0.5}$K^{-1} likely have a fine-grained regolith. An asteroid rotating slowly with a high thermal inertia will have a similar temperature distribution to an object rotating much more rapidly with a lower thermal inertia (Delbó et al., 2007b).

Example 5.1

What is the thermal inertia of an asteroid with a thermal conductivity of 0.5 W/(m K), a bulk density of 1500 kg/m³, and a specific heat capacity of 500 J/(kg K)? Would you expect the regolith to be fine-grained?

The thermal inertia can be calculated from *Equation 5.3* and will be

$$\Gamma = \sqrt{(0.5)(1500)(500)} \; J \; m^{-2} \; s^{-0.5} \; K^{-1} = 612 \; J \; m^{-2} \; s^{-0.5} \; K^{-1}. \tag{5.4}$$

Since this value is above 500 J m^{-2}s$^{-0.5}$K^{-1}, the surface is expected not to be fine-grained but more similar to bare rock.

Thermal inertias for asteroids are actually calculated by comparing measurements of their infrared fluxes to synthetic fluxes generated using a thermophysical model. Most calculated thermal inertias of asteroids tend to be less than 200 J m^{-2}s$^{-0.5}$K^{-1} (Delbó and Tanga, 2009) for asteroids with diameters of 100 km or less, implying fine-grained regoliths. The thermal inertia of km-sized near-Earth asteroids was estimated to be 200 ± 40 J m^{-2}s$^{-0.5}$K^{-1} (Delbó et al., 2007b). Thermal inertia tends to increase with decreasing diameter (Delbó and Tanga, 2009).

5.1.2 Standard Thermal Model

The Standard Thermal Model (STM) is a relatively simple model (Lebofsky and Spencer, 1989; Delbó and Harris, 2002; Harris and Lagerros, 2002; Delbó, 2004) that has been used to calculate the temperature distribution on the surface of main-belt asteroids. The asteroid is assumed to be spherical. STM assumes that there is instantaneous equilibrium between radiation striking the body's surface (insolation) and thermal emission. At the terminator (the imaginary "moving" line that separates the day and night sides of an object), the temperature is assumed to be 0 kelvin and that there is no thermal emission on the night side. The phase angle is assumed to be 0°. Since observations are never made at exactly 0°, a correction of 0.01 magnitudes/degree is made to the observed infrared magnitude. This correction appears to work for phase angles less than 30°.

The differential incident flux (dF_i) on a surface element of the differential surface area dS will be

$$dF_i = \frac{S_0}{(r^2 / \text{AU}^2)} \cos(\theta) dS \qquad (5.5)$$

where θ is the angle between the incident flux and the normal direction of the surface element. According to Lambert's cosine law, the intensity of observed light will be a function of the cosine of the angle between the incident flux and the surface normal.

The differential absorbed flux (dF_a) will be equal to the differential incident flux (dF_i) that is not reflected $(1 - A)$ and is given by the formula

$$dF_a = (1 - A)dF_i \qquad (5.6)$$

and the differential emitted flux (dF_e) will be

$$dF_e = \eta \varepsilon \sigma T^4 dS. \qquad (5.7)$$

The beaming factor η is a constant that modifies the temperature distribution to account for factors such as the enhancement of thermal emission at small phase angles (beaming) (Harris and Lagerros, 2002). This beaming factor can be thought of as a normalization factor (Harris, 1998). In STM, the beaming factor is often set to 0.756 (Lebofsky and Spencer, 1989). This value was based on fits to ground-based thermal observations of Ceres and Pallas to match diameters determined from stellar occultations (Lebofsky et al., 1986).

The subsolar temperature T_{ss} is the temperature at the subsolar point, which is the point on an asteroid where the Sun is directly overhead ($\theta = 0$). If you equate the absorbed and emitted differential flux for the surface element at the subsolar point,

$$\frac{(1-A)S_0}{(r^2 \, / \, \text{AU}^2)}dS = \eta\varepsilon\sigma T_{ss}^{\,4}dS. \tag{5.8}$$

Integrating both sides results in

$$\frac{(1-A)S_0}{(r^2 \, / \, \text{AU}^2)} = \eta\varepsilon\sigma T_{ss}^{\,4}. \tag{5.9}$$

Solving for the subsolar temperature produces the formula

$$T_{ss} = \sqrt[4]{\frac{(1-A)S_0}{\eta\varepsilon\sigma(r^2 \, / \, \text{AU}^2)}}. \tag{5.10}$$

The visual geometric albedo (p_v) is easier to measure and can be related to the Bond albedo by

$$A \approx qp_v \tag{5.11}$$

where q is the phase integral. The subsolar temperature becomes

$$T_{ss} = \sqrt[4]{\frac{(1-qp_v)S_0}{\eta\varepsilon\sigma(r^2 \, / \, \text{AU}^2)}}. \tag{5.12}$$

The phase integral is related to the slope parameter G by the equation

$$q = 0.290 + 0.684G. \tag{5.13}$$

Example 5.2

What is the subsolar temperature for an asteroid if the visual geometric albedo is 0.20, slope parameter is 0.15, beaming factor is 0.756, emissivity is 0.9, and the distance from the Sun is 2.5 AU?

The phase integral will be

$$q = 0.290 + (0.684)(0.15) = 0.393. \tag{5.14}$$

The subsolar temperature will then be

$$T_{ss} = \sqrt[4]{\frac{[1-(0.393)(0.20)(1366)]}{4(0.756)(0.9)(5.67\times10^{-8})(2.5)^2}} = 190 \text{ K.} \tag{5.15}$$

The subsolar temperature will be 190 K.

The surface temperature distribution is assumed to be

$$T(\theta) = T_{ss}[\cos(\theta)]^{1/4} \tag{5.16}$$

where θ is the angle from the subsolar point. This equation is valid for angles less than 90°. For angles greater than 90° (beyond the terminator), the temperature is assumed to be ~0 K. STM works well if the asteroid has a low thermal inertia, rotates slowly, observed at a small phase angle, and is not heavily cratered or irregularly shaped (Harris and Lagerros, 2002).

5.1.3 Calculating Diameters and Albedos Using STM

The procedure for calculating the diameter and albedo using STM was discussed in detail in Delbó (2004). The absolute magnitude H must be known and the infrared flux must have been measured in at least one wavelength. The total number of measurements is N. First, an estimate is made for the visual geometric albedo p_v. Then an iterative procedure is used to calculate the best estimates for the visual geometric albedo and diameter using one wavelength (λ) or a series of wavelengths (λ_i).

- Using the known absolute magnitude H, a diameter D can be calculated from the formula that equates the absolute magnitude, diameter, and visual geometric albedo (*Equation 1.30*), which is

$$D = \frac{1329}{\sqrt{p_v}} 10^{-0.2H} \text{ km.} \tag{5.17}$$

- The temperature distribution on the surface of the body is calculated from *Equation 5.16*.
- The theoretical flux at the infrared wavelength that was measured is then calculated using the formula

$$F_{\lambda_i} = \frac{\varepsilon D^2}{2\Delta^2} \int_0^{\pi/2} B[\lambda_i, T(\theta)] \cos(\theta) \sin(\theta) d\theta \tag{5.18}$$

where Δ is the geocentric distance of the asteroid and $B[\lambda_i, T(\theta)]$ is the Planck function (Equation 1.2).
- The observed flux (f_{λ_i}) is then scaled to a phase angle g of 0° using the formula

$$f_{0,\lambda_i} = f_{\lambda_i} 10^{-(\beta g/2.5)} \tag{5.19}$$

where the mean phase coefficient β has a value of 0.01 magnitudes/degree.

- A χ^2 value is then determined using the formula

$$\chi^2 = \sum_{i=1}^{N} \left[\frac{(F_{\lambda_i} - f_{0,\lambda_i})^2}{\sigma_i^2} \right] \tag{5.20}$$

where σ_i is the standard deviation for the ith measurement.
- The visual geometric albedo is then adjusted, and these steps are repeated until the value for the χ^2 is minimized.

5.1.4 Near-Earth Asteroid Thermal Model

However, Veeder et al. (1989) found that STM did not work very well with near-Earth asteroids. Calculated visual geometric albedos tended to be too high and, therefore, estimated diameters would be too small. Compared to main-belt asteroids, NEAs tend to have more irregular shapes, less loose surface material, and faster rotation rates. Harris (1998) developed the near-Earth asteroid thermal model (NEATM) to better model the albedos and diameters of near-Earth asteroids. NEATM can also be used with main-belt asteroids.

As with STM, the asteroid is assumed to be spherical. Fluxes must be measured at two or more infrared wavelengths. The beaming factor is allowed to vary in the modeling to produce the best fit to the infrared data. Since NEAs are often observed at phase angles greater than 30° (Harris and Lagerros, 2002), the infrared fluxes are calculated at the phase angle that the object is observed at. The emitted infrared flux is calculated for the part of the asteroid that is illuminated by the Sun and also visible to the observer (Harris and Lagerros, 2002; Delbó, 2004).

5.1.5 Infrared Observations

A number of infrared surveys have been done to determine diameter and albedos of asteroids (Mainzer et al., 2015). NASA's Infrared Astronomical Satellite (IRAS) was the first space-based observatory to do an all-sky survey at infrared wavelengths. IRAS observed the sky at wavelengths of 12, 25, 60, and 100 µm. Launched in January of 1983, the mission lasted 10 months. Diameters and albedos from IRAS were calculated using STM. Diameters calculated from IRAS data varied from diameters calculated from occultations had a root-mean-square (RMS) fractional difference of ~11% (Harris and Lagerros, 2002). Revised IRAS diameters and albedos for over 2200 asteroids were published in Tedesco et al. (2002).

NASA launched the Spitzer Space Telescope (SST), which was originally named the Space Infrared Telescope Facility (SIRTF), in August 2003. Instruments

Table 5.1 *NEOWISE visual geometric albedos (Mainzer et al., 2011) for Bus–DeMeo classes and Tholen E, M, and P classes. Taxonomic types with a standard deviation of 0.000 have only had one member observed.*

Taxonomic type	Visual geometric albedo
A	0.191 ± 0.034
C-complex	0.058 ± 0.028
D	0.048 ± 0.025
E	0.430 ± 0.229
K	0.130 ± 0.058
L	0.149 ± 0.066
M	0.125 ± 0.037
P	0.044 ± 0.014
O	0.339 ± 0.000
Q	0.227 ± 0.000
R	0.148 ± 0.000
S-complex	0.223 ± 0.073
T	0.042 ± 0.004
V	0.362 ± 0.100

on Spitzer are the IRAC (Infrared Array Camera), IRS (Infrared Spectrograph), and MIPS (Multiband Imaging Photometer for Spitzer). After all the liquid helium became depleted, the IRS and MIPS became inoperable. The two shortest wavelength bands of the IRAC camera could still be used with the same sensitivity as before, while the two long wavelength bands could not. Observations at 3.6 and 4.5 μm by the IRAC camera are called the Spitzer Warm Mission.

The AKARI (meaning light in Japanese) was launched by Japan in 2006. AKARI contained two instruments, the InfraRed Camera (IRC) and the Far-Infrared Surveyor (FIS). IRC covered the wavelength range of 2–26 μm and FIS covered the wavelength range of 50–180 μm. The telescope and instruments were cooled to ~6 K using superfluid helium. The "cold" mission phase occurred between 2006 and 2007 and did an all-sky survey at six bands: 9, 18, 65, 90, 140, and 160 μm. Over 5000 asteroids were observed. When the helium coolant became depleted, the "warm" mission phase occurred between 2008 and 2011. Low-resolution spectroscopy was done of targeted objects with the IRC in the range ~2–5 μm.

The WISE/NEOWISE mission (Section 1.8.3) determined diameters and visual albedos for over 150 000 asteroids (Mainzer et al., 2015). Average NEOWISE visual geometric albedos determined for Tholen (1984) classes are given in Table 5.1.

The Herschel Space Observatory was launched in 2009 and placed in Earth's L2 Lagrange point. Herschel is named after William Herschel, who discovered infrared radiation. Herschel was the largest infrared telescope that has ever been launched. It was cooled using superfluid liquid helium and could observe in the wavelength range of 55–671 μm. Due to its observations at long infrared wavelengths, Herschel was used to observe distant minor planets. These minor planets would have extremely cold surfaces and, therefore, emit more radiation in the far-infrared (Mainzer et al., 2015). Herschel stopped operating in 2013 when it ran out of coolant.

5.2 Asteroid Families

Asteroid families are clusterings of asteroids in proper element space (semi-major axis, eccentricity, inclination) that are thought to be due to the collisional disruption of a parent body or a cratering event. Asteroid families were originally named after their lowest numbered member; however, families have recently been named after their largest member (e.g., Masiero et al., 2013), which is often the lowest numbered object. The importance of studying asteroid families is that it allows the interior of an asteroid to be studied.

During the cratering or disruption of the original parent body, the ejected fragments of the asteroid accumulate into larger bodies (Nesvorný et al., 2015). The velocity of the ejected fragments must exceed the escape velocity. The formula for the escape velocity (v_e) from a body is

$$v_e = \sqrt{\frac{2GM}{R}} \tag{5.21}$$

where M is the mass of the original parent body and R is its radius. Since the escape velocity is much smaller than the orbital velocity, the orbits of all the fragments are initially very similar. However, gravitational perturbations will cause the orbits to start to diverge from the original orbit.

Example 5.3

What is the escape velocity from the 525-km diameter Vesta, which has a mass of 2.59×10^{20} kg (Russell et al., 2012)?

The escape velocity will be

$$v_e = \sqrt{\frac{2(6.67 \times 10^{-11})(2.59 \times 10^{20})}{(525000/2)}} \text{ m/s} = 362 \text{ m/s}. \tag{5.22}$$

(Remember that the diameter of the asteroid must be converted to a radius in meters.) The escape velocity will be 362 m/s.

The diameter (D_{PB}) of the original parent body for an asteroid family can be roughly estimated from the formula

$$D_{PB} = (D_{LM}^3 + D_{frag}^3)^{1/3} \qquad (5.23)$$

where D_{LM} is the diameter of the largest member and D_{frag} is the diameter of a sphere with a volume equivalent to all the fragments (Nesvorný et al., 2015). If $D_{LM} > D_{frag}$, the family was most likely created during a cratering event and if $D_{LM} < D_{frag}$, the family was most likely created during a catastrophic collision. This parent body diameter is only a rough estimate since it ignores the smallest, unobserved family members and may also include some interlopers. Interlopers are asteroids that are not believed to be fragments of the original parent body.

5.2.1 Number of Asteroid Families

These families were first recognized by Kiyotsugu Hirayama (1874–1943) in a series of papers (e.g., Hirayama, 1918, 1928, 1933). He found that there were clusterings of asteroids with respect to their orbital elements, most notably mean motion (related to semi-major axis from Kepler's third law), inclination, and eccentricity. Hirayama postulated that these families were due to the breakup of a larger body. He initially recognized the Koronis, Eos, and Themis families and then later identified the Flora and Maria families. These families are called Hirayama families.

The number of asteroid families and the members of each family have changed over time. Up to the late 1980s, the memberships of asteroid families were usually determined visually by looking at plots of asteroid orbital elements with very little statistical analysis to determine how robust the grouping were (Bendjoya and Zappalà, 2002). In the 1990s, Vincenzo Zappalà and coworkers revolutionized asteroid family determinations with their work (Zappalà et al., 1990, 1994, 1995) using the Hierarchical Clustering Method (HCM) on large datasets of asteroid proper elements. HCM uses a clustering method that defines a cutoff distance (d_{cut}), which is calculated from the proper orbital elements. Iterating over each body in the population, the distance in velocity between that body and all other asteroids is calculated (Masiero et al., 2013). The distance between the orbits of two neighboring asteroids must be less than the cutoff distance to be clustered into the asteroid family. The distances are given in meters/second because the defined distances are related to the relative velocities of the fragments. The most widely used distance metric had the form

$$d = na_p \sqrt{\frac{5}{4}\left(\frac{\delta a_p}{a_p}\right)^2 + 2\left(\delta e_p\right)^2 + 2\left(\delta \sin i_p\right)^2} \qquad (5.24)$$

where a_P, e_P, and i_P are the proper elements and n is the mean motion. The mean motion will be equal to $2\pi/p$ in radians or $360°/p$ in degrees where p is the period of the orbit. Therefore, na will be the orbital velocity since it is equal to $2\pi a/p$. For the distance to be in meters/second, a_P must be given in meters and p must be given in radians/second or degrees/second. Cutoff distances used today for defining asteroid families for different regions of the asteroid belt tend to be less than 100 meters/second (Nesvorný et al., 2015).

Example 5.4

What is the distance metric between (158) Koronis and (243) Ida using the Zappalà et al. (1990, 1994) distance metrics? The proper elements for Koronis and Ida and the mean motion of Koronis are given in Table 5.2.

Table 5.2 *Proper elements for (158) Koronis and (243) Ida and mean motion for (158) Koronis*

Name	a_p (AU)	e_p	$\sin i_p$	n (degrees/year)
(158) Koronis	2.869	0.0452	0.0375	74.1
(243) Ida	2.862	0.0429	0.0371	

The orbital velocity (na_p) will be equal to

$$na_p = \left(\frac{74.1°}{\text{yr}}\right)\left(\frac{2\pi}{360°}\right)\left(\frac{\text{yr}}{31540000 \text{ s}}\right)(2.869 \text{ AU})$$

$$\left(\frac{149600000000}{\text{AU}}\right) = 17590 \text{ m/s}. \tag{5.25}$$

Therefore, the distance metric will be

$$d = 17590\frac{m}{s}\sqrt{\frac{5}{4}\left(\frac{2.869-2.862}{2.869}\right)^2 + 2(0.0452-0.0456)^2 + 2(0.0375-0.0364)^2} = 56 \text{ m/s} \tag{5.26}$$

between Koronis and Ida. Remember that when using a distance metric to define an asteroid family, the distance metric is calculated between all objects in the population.

Recent work in identifying families also incorporates albedos. Masiero et al. (2013) identified 76 families in the main belt using HCM analysis of proper elements plus albedos from the NEOWISE mission. Albedos from the IRAS mission were used for 150 objects. IRAS data was used because large bodies tend to saturate the WISE images. There were over 100 000 asteroids in their study. Masiero et al. (2013) divided the belt into three regions and then divided each region into two groups by albedo (low and high). The low-albedo group contains members with albedos lower than 0.155, while the high-albedo group contains members with albedos greater than 0.065.

It had long been postulated (e.g., Granahan and Bell, 1991; Granahan, 1993) that spectroscopic observations would find families that contain M-types (historically argued to be the fragments of iron cores), A-types (olivine lower mantles), R-types (peridotite upper mantles), and V-types (crustal fragments). However, spectroscopic measurements of family members have found them to be typically very spectrally homogeneous. Family members that have spectra that are not consistent with other family members are called interlopers. Michel et al. (2002, 2003) argue that the spectral homogeneity of asteroid families is due to families being the breakup of a pre-fragmented parent body, which results in large family members being the products of the gravitational accumulation of smaller bodies. This reaccumulation results in spectrally homogenizing the family members.

Nesvorný et al. (2015) identified 122 notable asteroid families (Figure 5.1). A list of the largest asteroid families are given in Table 5.3. These families range from over 19 000 members for the (44) Nysa–(142) Polana family to a number of families with fewer than 10 members. The complicated structure of the Nysa–Polana family appears to be the result of the breakup of multiple parent bodies (Dykhuis and Greenberg, 2015).

The age of an asteroid family can be estimated from the spread in proper semi-major axis due to the Yarkovsky effect (Nesvorný et al., 2015). Some asteroid families have estimated ages less than 1 Ma (e.g., Vokrouhlický et al., 2016). No asteroid family appears to have an estimated age greater than ~4 Ga (Brasil et al., 2016).

Nesvorný et al. (2015) argue that list of families with parent bodies greater than 200 km in size is relatively complete, while those from parent bodies of 100–200 km in size is largely incomplete. They find that the estimated ages of the families derived from parent bodies greater than 200 km is randomly distributed over the lifetime of the Solar System, while those derived from parent bodies 100–200 km in size tend to have ages less than 1 Ga. Nesvorný et al. (2015) assume that there are many unidentified families with ages greater than 1 Ga. A few of the most spectrally studied of these asteroid families are discussed in the next sections.

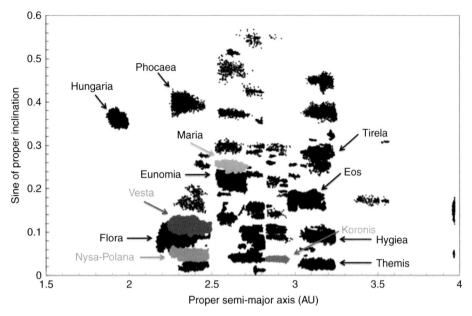

Figure 5.1 Plot of sine of proper inclination versus semi-major axis for family members from the ~120 families defined by Nesvorný et al. (2015). (A black and white version of this figure will appear in some formats. For the color version, please refer to the plate section.)

5.2.2 Vesta Family

The (4) Vesta family is a large inner main-belt family composed primarily of V-type asteroids. Vesta contains over 15 000 identified members (Nesvorný et al., 2015). Vesta's family is unusual compared to other families in that it is the result of a cratering event and not the product of the disruption of the parent body. Binzel and Xu (1993) identified small (~10 km diameter) objects with HED-like spectra in the Vesta family and outside the Vesta family between Vesta and the 3:1 mean-motion and ν_6 meteorite-supplying resonances. These bodies are typically called Vestoids. The distribution of these Vestoids shows that fragments from Vesta extend far from the defined Vesta family. Near-infrared spectra (Burbine et al., 2001; Kelley et al., 2003; Moskovitz et al., 2008a; Mayne et al., 2011; Hardersen et al., 2014) confirmed the spectral similarity of Vestoids to HEDs with interpreted mineralogies consistent with eucrites, howardites, or diogenites.

5.2.3 Flora Family

The (8) Flora family is a large inner main-belt family composed primary of S-complex asteroids. S-complex asteroid (8) Flora ($a = 2.20$ AU) is the largest body in the Flora family. The Flora family has over 13 000 identified family members

Table 5.3 *List of large asteroid families from Nesvorný et al. (2015). The*
semi-major axis ranges are from Nesvorný (2015).

Name	d_{cut} (m/s)	Number of members	Minimum to maximum semi-major axis (AU)
(3) Juno	55	1684	2.5893–2.7050
(4) Vesta	50	15 252	2.2407 2.4897
(8) Flora	60	13 786	2.1624–2.4049
(10) Hygiea	60	4854	3.0290–3.2420
(15) Eunomia	50	5670	2.5257–2.7102
(20) Massalia	55	6424	2.3226–2.4801
(24) Themis	60	4782	3.0190–3.2398
(25) Phocaea	150	1989	2.2474–2.4603
(31) Euphrosyne	120	2035	3.0690–3.2303
(44) Nysa–(142) Polana	50	19 073	2.2412–2.4825
(128) Nemesis	50	1302	2.6944–2.7994
(145) Adeona	50	2236	2.5494–2.7135
(158) Koronis	45	5949	2.8273–2.9638
(163) Erigone	50	1776	2.3143–2.4224
(170) Maria	60	2940	2.5220–2.7204
(221) Eos	45	9789	2.9533–3.1549
(298) Baptistina	45	2500	2.2010–2.3309
(363) Padua	45	1087	2.6776–2.7994
(375) Ursula	70	1466	3.0401–3.2387
(434) Hungaria	100	2965	1.8633–1.9985
(480) Hansa	200	1094	2.5377–2.7445
(490) Veritas	30	1294	3.1561–3.1839
(668) Dora	45	1259	2.7405–2.8121
(702) Alauda	120	1294	3.0368–3.2507
(808) Merxia	55	1215	2.6841–2.8136
(847) Agnia	30	2125	2.7535–2.8190
(1272) Gefion	50	2547	2.7046–2.8159
(1400) Tirela	50	1395	3.0766–3.1810
(1644) Rafita	70	1295	2.5305–2.6518
(1726) Hoffmeister	45	1810	2.7415–2.8198
(2085) Henan	50	1872	2.6183–2.8191
(2732) Witt	45	1816	2.6665–2.8061

with an age of ~1 Ga. (Nesvorný et al., 2015). Gaffey (1984) interpreted rotational spectral variations on Flora as indicated a differentiated body. Mineralogical analyses of Flora family member (951) Gaspra using Galileo spacecraft data (Section 8.4.1) indicated a body that is more olivine-rich than ordinary chondrites (Granahan, 2011). However, the prevailing view today is that the original Flora parent body is not differentiated since Flora family members have spectral similarities and interpreted mineralogies consistent with LL chondrites (Vernazza et al.,

2008). Vernazza et al. (2008) argue that the Flora family is the parent body of the LL chondrites.

5.2.4 Koronis Family

The (158) Koronis family is a large outer main-belt family composed primarily of S-types in a background of C-complex asteroids. The Koronis family has ~6000 identified members. Ivezić et al. (2002) and Parker et al. (2008) found using SDSS colors that the Koronis family have relatively homogeneous colors that are distinct from the background objects. The Koronis family has ~6000 identified members (Nesvorný et al., 2015). The age of the Koronis family is estimated to be ~2 Ga (Marzari et al., 1995).

Using Band I centers and Band Area Ratios determined from Galileo data (Section 8.4.2), Granahan et al. (1995) and Granahan (2002, 2013) found that Ida and its moon Dactyl had interpreted mineralogies similar to LL chondrites. Rivkin et al. (2011b) found a number of small (< 5 km) bodies had broadband visible colors consistent with Q-types. Also using broadband visible colors, Thomas et al. (2012) found that the spectral slopes of Koronis members tends to increase with increasing size, consistent with a transition from Q- to S-types due to "space weathering." The spectra of Koronis family members is consistent with an ordinary chondrite assemblage (Chapman, 1996; Rivkin et al., 2011b; Thomas et al., 2011a). Genge (2008) also found that a large proportion of micro-meteorites (50–400 μm in size) had mineralogies similar to ordinary chondrites. Since the Koronis family should be a significant supplier of Solar System dust, Genge (2008) interpreted the Koronis family as having an ordinary chondrite-like composition.

5.2.5 Eos Family

The (221) Eos family is an outer main-belt family dominated by K-types. The family has over 9000 members (Nesvorný et al., 2015). The Eos family is a large outer main-belt family composed primarily of K-types. Historically, Eos family members have been linked (e.g., Bell, 1988) with CO3 or CV3 chondrites due to the spectral similarity of these family members to these meteorites in the visible and near-infrared. Both these meteorites and asteroids have subdued spectral features due to olivine. Mothé-Diniz et al. (2008) mineralogically characterized members of the Eos family and concluded that their calculated forsterite-rich olivine mineralogies with minor low-calcium pyroxene for these objects is consistent with a partially differentiated ordinary chondrite precursor. While the mechanism for producing high-Mg olivine is unclear (nebular process versus differentiation),

Mothé-Diniz et al. (2008) suggested that Eos family could possibly be made of mantle fragments of a partially differentiated object.

5.2.6 Baptistina Family

The Baptistina family is an inner main-belt family that has 2500 identified members (Nesvorný et al., 2015). Bottke et al. (2007) identified the breakup that led to the creation of the Baptistina family as the possible source body of the Cretaceous–Paleogene (K–Pg) extinction event. A fossil meteorite found at the K–Pg boundary (Kyte, 1998) appears most similar to a carbonaceous chondrite. Chromium isotopic measurements (Shukolyukov and Lugmair, 1998) of sediments at the K–Pg boundary also argue that the impactor was a carbonaceous chondrite. Bottke et al. (2007) estimated that the breakup occurred ~160 Ma years ago. However, near-infrared data of Baptistina and Baptistina family members found that most observed objects had primarily S-complex spectra (Reddy et al., 2009, 2011).

5.2.7 Gefion Family

The (1272) Gefion family is a large middle main-belt family composed primarily of S-types. The Gefion family has over 2500 identified members (Nesvorný et al., 2015). Nesvorný et al. (2009) believes that the original Gefion parent body may be the parent body of L chondrites from its estimated age (~485 Ma), location, and interpreted mineralogy. Korochantseva et al. (2007) found an age of ~470 Ma for the breakup of the L chondrite parent body from ^{39}Ar–^{40}Ar dating of shocked L chondrites and found that this age correlated with the age of fossil L chondrites located in Ordovician sediments and at the Lockne impact crater (e.g., Schmitz et al., 2001, 2011). The relatively short cosmic-ray exposure ages (< 1 Ma) of some L chondrites (e.g., Eugster et al., 2006) is consistent with the Gefion family's location near the 5:2 meteorite-supplying resonance.

5.3 Light Curves

Light curves plot the magnitude (brightness) of an object versus time (Figures 5.2 and 5.3) and allow the period and amplitude to be determined (Warner, 2006). The time is usually given as a phase where several "cycles" of data are merged together into a single plot. To merge data for several nights, the data must be corrected for distance from the Sun and phase angle. The data for each cycle are overlapped until they align. The period is usually given in hours and the amplitude is given in magnitudes.

The magnitude of an asteroid when observed from Earth will vary depending on its shape and albedo. A perfectly spherical asteroid that is homogeneous in albedo

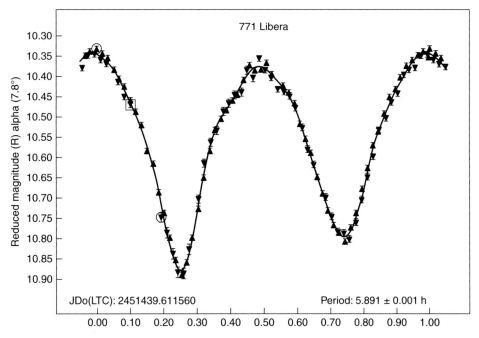

Figure 5.2 Light curve for (771) Libera. The light curve plots reduced magnitude versus phase. Note the two maxima and minima. Credit: Brian D. Warner, Center for Solar System Studies.

will have a light curve that does not vary in brightness with a light curve that is exactly flat. However, since asteroids have irregular shapes, the brightness of the asteroid will vary as it rotates since larger and smaller surface areas will reflect different amounts of light back to Earth.

Light curves have two maxima and two minima since four different sides will be apparent to the observer. The rotation period is the time it takes the asteroid to make one revolution on its axis. The rotation period can be calculated by determining how long it takes the magnitude of the asteroid to vary through its two maxima and two minima and return to its original magnitude (Figures 5.2 and 5.3). Most asteroids have light curves with a repeatable single period (Kaasalainen et al., 2002). However if the period does not repeat, the asteroid could possibly be a tumbler or in a binary system. Tumblers do not rotate around their principal axis, while binary systems have a moon. The damping time is the estimated time needed for a tumbler with a particular period and diameter to become a principal axis rotator, which is the lowest energy state (Pravec et al., 2002). Slower rotators and larger bodies have larger damping times (Figure 5.4).

The amplitude of the light curve is the magnitude difference between the maximum and minimum brightnesses. The amplitude will be a function of the shape

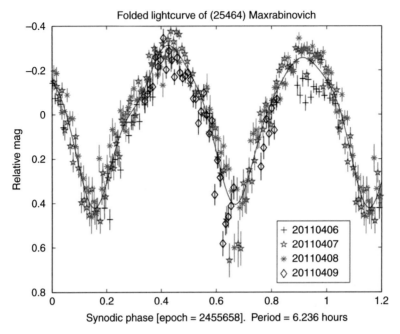

Figure 5.3 Light curve for (25464) Maxrabinovich (Polishook, 2012). The light curve plots relative magnitude versus phase. Note the two maxima and two minima. Credit: David Polishook, Weizmann Institute of Science, and The Minor Planet Bulletin. (A black and white version of this figure will appear in some formats. For the color version, please refer to the plate section.)

and/or albedo variations on the surface. Assuming a triaxial ellipsoid for the body with faces with sizes of $a \geq b \geq c$, the axial (a/b) ratio approximately given by the formula (Tegler, 2014)

$$\frac{a}{b} = 10^{[\Delta m/(2.5 \ magnitudes)]} \qquad (5.27)$$

where Δm is the amplitude in magnitudes. The angle between the rotation axis and the line of sight is 90° (Rabinowitz et al., 2006). The axial ratio will be a minimum value since it also depends on other factors such as phase angle and albedo differences across the surface.

Example 5.5

What is the axial ratio for a body with a light curve amplitude of 0.3? Assume that the angle between the rotation axis and the line of sight is 90°.

The axial ratio is given by the formula

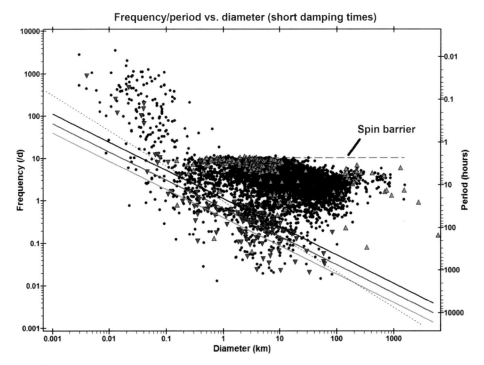

Figure 5.4 Frequency (cycles per day) (or period in hours) versus diameter (km) for asteroids (Warner et al., 2009). Red triangles are binary objects. Green triangles are tumblers. The dashed horizontal line is the spin barrier. The three solid diagonal lines represent the "tumbling barrier" based on damping times of 200 Ma (black line), 1 Ga (green line), and 4.5 Ga (red line) where an object below the line is a good candidate for being a tumbler. The thin dashed diagonal line is based on the collisional lifetime of the objects. Credit: Brian D. Warner, Center for Solar System Studies. (A black and white version of this figure will appear in some formats. For the color version, please refer to the plate section.)

$$\frac{a}{b} = 10^{0.3/2.5} = 1.32. \tag{5.28}$$

Therefore assuming a triaxial ellipsoid, the axial ratio will be 1.32.

5.3.1 Measuring Light Curves

Light curves are usually measured using differential photometry. Differential photometry calculates the magnitude difference between the studied object and stars with known magnitudes in the same image. The asteroid and comparison (standard) stars are observed at the same time in the same region of the sky so all objects are observed at approximately the same air mass. So any change in the magnitude difference between the asteroid and the comparison stars will reflect only changes in the magnitude of the asteroid. Light curves can be measured with or without a filter. Warner (2006) gives a detailed discussion on the procedures for obtaining light curves.

Asteroid light curves are often published in *The Minor Planet Bulletin*. *The Minor Planet Bulletin* is the official publication of the Minor Planets section of the Association of Lunar and Planetary Observers (ALPO) and is published four times a year. The asteroid light curve database compiles all the light curve data in one place on the Internet (Warner et al., 2009).

5.3.2 Spin Barrier

For asteroids greater than ~200 meters in diameter, almost all measured objects have rotation periods smaller than ~2.2 hours (~11 cycles/day) (Figure 5.4). This 2.2 hour period is called the spin barrier or spin limit. The prevailing belief is that asteroids are rubble piles, a conglomeration of fragments that are held together by gravity. If a rubble pile did spin faster, the asteroid would break apart.

A spinning body produces a centripetal force that is directed outwards. If the centripetal force is less than or equal to the gravitational force (which is directed inwards) on a particle on the surface, the particle will not be ejected off the surface while the body rotates. Some relatively simple calculations support this idea. We will assume a spherical body with a cohesionless particle on the equator. The centripetal force (F_c) on a particle on the equator is

$$F_c = \frac{m_2 v^2}{R} \tag{5.29}$$

where m_2 is the mass of a particle on the surface and v is the rotational speed ($2\pi R / T$ where T is the rotation period in seconds and R is the radius of the body in meters). The gravitational force (F_g) is

$$F_g = G\frac{m_1 m_2}{R^2} \tag{5.30}$$

where G is the gravitational constant and m_1 is the mass of the body in kilograms. The mass of the body is related to its density by $m_1 = \rho V$ where V is the volume and equal to $(4/3)\pi R^3$. Equating the centripetal and gravitational forces, doing the necessary substitutions, and solving for the rotational period (T) produces the equation

$$T = \sqrt{\frac{3\pi}{G\rho}} \tag{5.31}$$

This equation shows that for a body where the centripetal force is exactly equal to gravitational force for a particle on the equator of a spherical body, the rotational period is proportional to the inverse of the square of the density of the object. For a density of 2250 kg/m^3 (2.25 g/cm^3), the rotational period will be 7925 seconds (2.2 hours). Increasing the density will decrease this limiting rotational period. Remember that this density is a rough approximation for the density needed for the asteroid not to fly apart. Any asteroid will not be exactly spherical. Particles off the equator will experience a different centripetal force. The gravitational force and centripetal force will change as you go interior into the body.

This spin barrier of ~2.2 hours for asteroids greater than ~200 meters in diameter implies that these objects have densities in the range ~2–3 g/cm^3. This is lower than the typical average grain densities for meteorites (Table 3.8), implying that these asteroids have significant porosities and, therefore, are rubble piles.

The particles in these bodies appear not to be cohesionless. For example, Rozitis et al. (2014) studied (29075) 1950 DA, which is rotating faster (2.1216 hours) than would be allowed for an object of its density (1.7 ± 0.7 g/cm^3). Polishook et al. (2016) found that ~2.3 km S-type (60716) 2000 GD$_{65}$ had a rotation period less than 2 hours (1.9529 ± 0.0002 hours). Some type of force must be holding these particles together on these objects and not allowing the body to fly apart. This force is believed to be attractive van der Waals forces, which are relatively weak forces of attraction between electron-rich regions of one molecule (or grain) and electron-poor regions of another molecule (or grain).

Smaller objects tend to have much faster spin rates, arguing that they are intact, monolithic bodies. The asteroid with the fastest measured rotation rates is 2014 RC with a measured period of 15.8 seconds. The asteroid with the second fastest rotation rate is 2010 JL$_{88}$ with a period of 24.5 seconds.

5.3.3 Spin State

The spin state (or spin vector) of an asteroid will be its direction of rotation (prograde or retrograde) and obliquity (tilt between the asteroid's equator and its orbital plane). The spin state of an asteroid can be determined by determining light curves for an asteroid over many different phase angles (Binzel, 2003).

Slivan (2002) found that four of the ten observed Koronis family members were prograde rotators with similar obliquities and rotation periods. The six retrograde rotators had a much larger range of obliquities and rotation periods with either very fast or very slow rotators. The observed spin states would be typically expected to be randomly oriented after family formation and later collisional evolution (Slivan et al., 2009). This alignment of prograde orbits was dubbed a Slivan state (Vokrouhlický et al., 2003). Vokrouhlický et al. (2003) found that the spin distribution of the prograde-rotating Koronis family members are the result of the effects of YORP torques combined with solar and planetary gravitational torques acting over 2–3 Ga (Slivan et al., 2009). Further studies of Koronis family members did find one prograde-rotating body that did not have spin properties that clustered with the other prograde objects. Slivan states were also found in the Flora family (Kryszczyńska, 2013).

5.4 Radar

Radar (Radio Detection and Ranging) astronomy actively transmits microwaves at an asteroid and measures the reflected signal. Radar observations are different from other types of observing techniques since radiation is transmitted from Earth to the asteroids and it can be used 24 hours a day. Radar can be used determining shapes and spin state, finding moons, doing astrometry, and estimating metal contents. Since the strength of the returned radar signal is proportional to the inverse of the fourth power of the distance to the target, radar can only be used to study bodies in the Solar System.

The Arecibo Observatory (Figure 5.5) is the largest single aperture radio telescope with a diameter of 305 meters (Altshuler, 2002). The Arecibo Observatory is located on the island of Puerto Rico. Arecibo has three radar transmitters. The transmitter that is used for planetary studies is the 1 MW transmitter, which emits a 2.38 GHz (12.6 cm) signal (S-band). The ~300-meter collector is constructed inside a large depression that was produced by a karst sinkhole.

The Goldstone observatory has a 70-meter antenna (Slade et al., 2011). Goldstone is located in the Mojave Desert in California. Goldstone has a 500 kW transmitter, which emits an 8.56 GHz (3.5 cm) signal (X-band).

5.4.1 Delay-Doppler Imaging

Asteroid images are determined using delay-Doppler imaging (Figure 5.6). The radar signal is transmitted toward the asteroids and the time delay (time it takes for the signal to be reflected back to the detector) is measured. Since the radar waves travel at the speed of light, the reflected radar signals can be used to determine the distance to the asteroid. Regions on the asteroid that are closer to the detector will produce radar signals with smaller time delays than those that are farther way. Measuring these time delays produces a two-dimensional image at any particular time. Asteroid moons can also be detected. If the body is rotating, the frequency of the reflected radio waves will change due to the Doppler shift, which allows the rotation speed to be determined.

Figure 5.5 Image of Arecibo Observatory. The telescope is 305 meters in diameter. Credit: National Oceanic and Atmospheric Administration. (A black and white version of this figure will appear in some formats. For the color version, please refer to the plate section.)

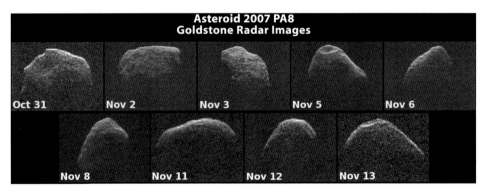

Figure 5.6 A collage showing nine radar images of near-Earth asteroid (214869) 2007 PA$_8$. The images show possible craters and boulders. Note the irregular, asymmetric shape. The asteroid is ~1.6 km long. Credit: NASA/JPL-Caltech.

The formula for the Doppler shift is

$$\frac{\Delta\lambda}{\lambda_0} = \frac{v}{c} \tag{5.32}$$

where $\Delta\lambda$ is the change in wavelength relative to the rest wavelength (λ_0), v is the radial (line-of-sight) velocity, and c is the speed of light.

So if you can measure the change in wavelength, you can measure the velocity relative to the observer. If the part of the body is moving toward you, the radio waves are shifted toward shorter wavelengths (blueshifted) and if part of the body is moving away from you, the radio waves are shifted toward longer wavelengths.

If radar signals are transmitted for a period of time, these two-dimensional images can be combined to produce a three-dimensional image. Since the returned radar signal falls off as the inverse of the distance to the fourth power, Doppler imaging is only used for near-Earth objects.

5.4.2 Surface Roughness and Metal Content

Radar observations of asteroids can also give an indication of the surface roughness and metal content. The transmitted radar signal (Ostro, 1993) is usually circularly polarized and the reflected radar signal can either have the same circular polarization (called SC) or the opposite polarization (called OC). Normal reflection from a plane mirror yields echoes that are almost entirely in opposite polarization, while multiple scattering from a surface can cause echoes that are in the same polarization. The circular polarization ratio (SC/OC) measures the wavelength-scale roughness of the surface (Benner et al., 2008).

The radar albedo for an asteroid is the ratio of a target's OC radar cross section to its projected area (Ostro, 1993). Radar albedos are a complicated function of the near-surface bulk density (which is a function of the solid rock density and the porosity of the surface) and the mineralogy of the object. Increasing the metal content of the surface or decreasing the surface porosity will increase the radar albedo of an object. Radar observations have been done of over 100 asteroids in the main belt and near-Earth population (Magri et al., 1999, 2007).

A number of observed M-type asteroids have been found to have high radar albedos (Shepard et al., 2010, 2011, 2015) that appear to almost certainly indicate metallic iron-rich surfaces. Shepard et al. (2010) calls these bodies Mm-types.

5.5 Minor Planet Moons

Many minor planets have companions that orbit them. If the two bodies are approximately the same size, the two objects are part of a binary system. If one body is much larger, the larger body is called the primary and the smaller one is called the secondary. The secondary is often called a moon or satellite.

Over 270 moons have been discovered around minor planets with some having multiple moons. Approximately 70% of these moons are found around bodies considered asteroids and the remaining moons are found around outer Solar System bodies. Most (~96%) of these systems are binary. Approximately 4% are triple systems. The only known sextuple system is the Pluto system with five moons orbiting Pluto (Charon, Nix, Hydra, Kerberos, Hydra).

Asteroid moons had been proposed to exist for many years (Binzel and van Flandern, 1979; van Flandern et al., 1979). For example, McMahon (1978) used a stellar occultation to identify a moon around (532) Herculina. Occultations occur when an astronomical body passes in front of a star and blocks light from the star from reaching the observer. The brightness of the star is measured. Multiple chords (lines crossing the object during the occultation) will be observed by (usually) multiple observers on different sites on the Earth. The length of the chords will allow the dimensions of the body to be estimated. Further dimming of the star will allow the presence of an atmosphere, rings, and moons to be determined. McMahon (1978) and a number of other observers observed an occultation by Herculina and then a secondary occultation, which was attributed to a moon. This "discovery" was argued to be the confirmed satellite of a minor planet (Dunham, 1978). However, Hubble Space Telescope observations of Herculina did not identify any satellites (Storrs et al., 1999).

The first asteroid moon that was definitively discovered was Dactyl, which was discovered through imaging of (243) Ida by the Galileo spacecraft (Figure 5.7). This moon was named Dactyl after the mythological people that were said to inhabit Mount Ida on the island of Crete. The provisional name for the moon of a minor planet is S for satellite, a slash (/), the year of the discovery, the number of the primary in parentheses, and a number indicating the discovery sequence. For example, the provisional designation for Dactyl was S/1993 (243) 1. The official name for a moon of a minor planet is the number of the primary in parentheses, the name of the primary, a Roman numeral indicating the discovery sequence (e.g., I for the first moon, II for the second), and the name of the moon. The official name for Dactyl is (243) Ida I Dactyl.

The first asteroid to have two moons discovered around it was (87) Sylvia. Since Sylvia was named after Rhea Silvia, the mother of the mythological founders (Romulus and Remus) of Rome, S/2001 (87) 1 was named (87) Sylvia I Romulus and S/2004 (87) 1 was named (87) Sylvia II Remus. Approximately 60 near-Earth asteroids have moons around them.

5.5.1 Density

The density of a body is its mass divided by its volume. The mass of a planetary body can be determined by its gravitational effect on another body. There are a number of different ways to determine these gravitational effects (Britt et al., 2002;

Figure 5.7 Galileo green-filter image of (243) Ida and its moon Dactyl. Ida is 31.4 km in diameter, while Dactyl is 1.4 km in diameter. Credit: NASA.

Carry, 2012). For example, an asteroid can influence the orbit of smaller asteroids during a close encounter. Another way is due to the gravitational influence of a number of asteroids can affect the positions of planets, satellites, and spacecraft. Both of these methods work best for the largest asteroids.

The Doppler shift of radio signals transmitted by a spacecraft can also be measured to determine the mass of the planetary body either through the amount of deflection of the spacecraft's trajectory or the spacecraft's orbit around the body. The orbit of a moon around a minor planet can be used to determine the mass of the primary body.

The Doppler shift method is considered the most precise for determining masses but is limited by the relatively small number of spacecraft that have flown by or orbited minor planets. The second most precise technique is the orbit of the moon and tends to be the most productive with the large number of known binary systems. To determine the mass of the asteroid during the flyby, the perturbed velocity of the asteroid is estimated from the additional Doppler shift of the transmitted radio signal in comparison with the expected Doppler shift of an unperturbed trajectory (Pätzold et al., 2011).

The presence of a moon (or a spacecraft in orbit) allows the mass and density of the minor planet to be estimated if the orbiting body is much less massive than the minor planet. If you again equate the gravitational and centripetal force and solve for the mass (M) of the primary body, the formula for the mass becomes

$$M = \frac{v^2 r}{G} \tag{5.33}$$

where v is the velocity of the moon and r is the orbital distance of the center of the moon to the center of mass of the system. Since the mass of the moon is much smaller than the primary, the center of mass of the system is approximately the center of the primary. Since the rotational speed is equal to $2\pi r/T$ where T is the orbital period of the moon, the formula to calculate the mass of the minor planet becomes

$$M = \frac{4\pi^2 r^3}{GT^2} \tag{5.34}$$

This is a version of Newton's reformulation of Kepler's third law of planetary motion that works for a body orbiting an object with any mass. The period squared of the orbit can be seen to be proportional to the semi-major axis of the distance cubed.

The volume (V) of a sphere can then be computed by the formula

$$V = \frac{4}{3}\pi R^3 \tag{5.35}$$

where R is the radius. Most asteroids are assumed to be spheres. The density will be calculated by dividing the mass by the volume.

Example 5.6

If you assume the moon of a minor planet has an orbital period of 31 hours (T), is located 95 km (r) from the center of mass of the body, and the primary body is spherical with a radius of 15.7 km, what is the density of the primary object?

The orbital period needs to be determined in seconds, which will be 31 hours × 60 minutes/hr × 60 minutes/s. The orbital period will then be 111 600 seconds. The orbital distance needs to be in meters and, therefore, will be 95 000 m.

The formula for the mass then becomes

$$M = \frac{4\pi^2 (95000 \text{ m})^3}{(6.67 \times 10^{-11} \text{ m}^3 \text{ kg}^{-1} \text{ s}^{-2})(111600 \text{ s})^2}. \tag{5.36}$$

The mass of the primary will then be 4.1×10^{16} kg.

The volume of the primary will be $(4/3)\pi R^3$. Since the radius is 15 700 m, the volume will then be 1.62×10^{13} m^3. The density will be the mass divided by the volume and will be ~2500 kg/m^3.

Carry (2012) compiled the densities of ~300 small bodies (asteroids, comets, and trans-Neptunian objects). As expected, S-complex bodies (~2700 kg/m^3) were denser than C-complex bodies (~1600 kg/m^3). X-complex bodies have a wide range of inferred densities. Comets and trans-Neptunian objects have extremely low densities. Using the most precise data, comets were calculated to have densities of ~500 kg/m^3 and trans-Neptunian objects have densities of ~1000 kg/m^3.

5.6 Distribution of Taxonomic Classes

It has long been known that the abundances of different taxonomic classes varied with heliocentric distance. With ~100 classified objects, Chapman et al. (1975) found that objects that were classified as C-asteroids become much more abundant in the outer belt.

Plotting the distribution of the major taxonomic types (E, R, S, M, F, C, P, D) identified at the time, Gradie and Tedesco (1982) found that the distribution varied systematically with heliocentric distance. However to plot the "true" distribution instead of the apparent distribution, observational biases must be corrected for. These biases include the incompleteness of the observed asteroid population compared to the actual population, the tendency to observe bodies that are nearer, larger, and higher albedo because these bodies are brighter and thus easier to observe, and the observations of multiple asteroid family members.

Gradie and Tedesco (1982) found that E-types were preferentially found in the Hungaria region, the S-complex (called S-types at the time) distribution peaks at ~2.2 AU, the C-complex (called C-types) distribution peaks at ~3 AU, P-types are preferentially found in the Hilda region at ~4 AU, and D-types are preferentially found in the Jupiter Trojan region at 5.2 AU. Using a much larger number of classified asteroids with diameters equal to or greater than 20 km, Bus and Binzel (2002b) noted that their taxonomic distribution matches pretty well the distribution of Gradie and Tedesco (1982).

Bell (1986) and Bell et al. (1989) inferred degrees of metamorphic heating for each of the Tholen (1984) classes plus the K-class and placed the classes into "primitive" (D, P, C, K, Q), "metamorphic" (T, B, G, F), and "igneous" (V, R, S, A, M, E) superclasses. The heat source was assumed to be due to [26]Al and/or solar wind induction heating and had to vary in intensity with distance from the Sun. Bell et al. (1989) found that the igneous superclass objects dominated the main asteroid belt interior to ~2.7 AU, the primitive superclass objects dominate the belt past ~3.4 AU, and the metamorphic superclass objects are relatively rare and peak in the belt from ~3.0 AU. This interpretation is somewhat controversial today since a high percentage of S-complex bodies is now thought to have mineralogies similar to ordinary chondrites (e.g., Vernazza et al., 2014).

Mothé-Diniz et al. (2003) looked at the bias-corrected taxonomic distribution for ~2000 main-belt objects classified in the visible from ~2.1 to ~3.3 AU. The bias correction corrects for observational biases such as tending to observe the brightest asteroids, which tend to have the highest albedos, largest diameters, and/or closest to Earth. Mothé-Diniz et al. (2003) used the typical boundaries for the inner, middle, and outer belt to look at the distribution of X-, C, and X-complex objects. As expected, Mothé-Diniz et al. (2003) found that that S-complex objects are the most abundant asteroid class in the inner and middle belt, while C-complex objects are the most abundant in the outer belt, especially past 3.05 AU. Between 2.1 and 2.5 AU, Alvarez-Candal et al. (2006) found that the concentration of S-complex objects decreases with increasing heliocentric distance and the concentration of C-complex objects increases. However, Michtchenko et al. (2010) found that S-complex bodies are evenly distributed in this inner main-belt region. X-complex objects are the least abundant complex in the inner and middle of the belt (Mothé-Diniz et al., 2003) but become more abundant in the outer belt. Using SDSS colors, the taxonomic distributions of Carvano et al. (2010) were pretty consistent with the distributions seen by Mothé-Diniz et al. (2003).

On average, the spectral slopes of low-albedo asteroids in the visible and near-infrared tend to get redder with increasing distance from the Sun (e.g., Gradie and Tedesco, 1982; Vilas and Smith, 1985; Jewitt and Luu, 1990; Emery and Brown, 2003). From SDSS colors, the Cybele region (~3.3–3.7 AU) [named after (65) Cybele] tends to contain more X-types (P-types if these objects are assumed to have low albedos) than D-types (Gil-Hutton and Licandro, 2010). Using SDSS colors, Gil-Hutton and Brunini (2008) found approximately equal numbers of D- and X-types in the Hilda region. The differences in spectral slopes (D-, P-, C-types) between these outer region objects have been argued to be due to space weathering (e.g., Gil-Hutton and Brunini, 2008) and/or compositional differences (e.g., Emery et al., 2011).

DeMeo and Carry (2013, 2014) updated the distribution of taxonomic classes by looking at the distribution by mass of different taxonomic types versus heliocentric distance (Figure 5.8). Mass instead of the number of objects was used to produce the distribution since it should better reproduce the distribution of different types of material in the belt. Previous studies would weight a 500 km diameter and a 20 km equally in their study.

DeMeo and Carry (2013, 2014) used SDSS data for ~60 000 bodies, which they bias-corrected according to size, distance, and albedo. Mass was calculated by assuming a density (Carry, 2012) for each taxonomic class and then determining its volume by using its absolute magnitude to calculate its diameter. Since SDSS

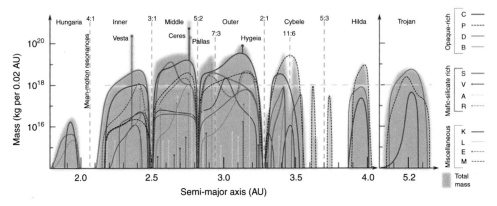

Figure 5.8 Distribution of taxonomic classes by mass (kg per 0.2 AU) versus semi-major axis (AU) (DeMeo and Carry, 2014). The total mass is in gray. Different regions of the main asteroid belt and beyond the asteroid belt plus mean-motion resonances are labeled. The horizontal line at 10^{18} kg is the limit of the work on determining taxonomic distributions in the 1980s. Credit: Francesca DeMeo, Massachusetts Institute of Technology, and Nature Publishing Group. (A black and white version of this figure will appear in some formats. For the color version, please refer to the plate section.)

could not measure the brightest asteroids (H < 12), the taxonomic classifications from visible spectroscopic surveys for these ~1500 asteroids were used.

DeMeo and Carry (2013, 2014) found E-types dominate the Hungaria region, S-complex objects dominate the inner belt (excluding Vesta), S- and C-complex bodies are in equal proportion in the middle belt (excluding Ceres and Pallas), the outer belt is primarily C-complex bodies, the P-types dominate the Cybele and Hilda regions, and the Jupiter Trojans are primarily P-types. These results are very similar to what was found by Gradie and Tedesco (1982) with a much smaller sample.

5.7 Planetary Migration

Historically, the planets and minor planets were thought to have formed near their present-day orbits. However, Goldreich and Tremaine (1980) found that Jupiter's semi-major axis would change over time as it gravitationally interacts with the protoplanetary disk. During their modeling of the final stages of accretion of Uranus and Neptune, Fernández and Ip (1984) found that Saturn, Uranus, and Neptune tend to gain angular momentum and move outwards as they scatter planetesimals to the inner Solar System, while Jupiter (the main ejector due to its large mass) loses angular momentum and migrates toward the Sun. This type of migration was confirmed by the discovery of numerous extrasolar planets with masses and sizes similar to Jupiter but extremely close to their parent stars argues pretty conclusively that the planets migrated there. These Jupiter-sized planets must have formed past the frost line where they could accumulate enough gas and ice to grow to these sizes. These bodies are called "hot Jupiters." "Hot Jupiters" are the easiest extrasolar planets to discover using radial velocity and transit methods. Their large sizes and small semi-major axes produce the largest Doppler shifts on their parent star's spectral lines using the radial velocity method for detecting planets and the most frequent "dips" in the star's light curve using the transit method compared to smaller planets at larger semi-major axes. However, Wright et al. (2012) estimates that only ~1% of solar-type stars have "hot Jupiters."

The Nice model argues that the many characteristics of the Solar System are products of the migration of the giant planets. This model was first proposed in a series of three interrelated papers (Tsiganis et al., 2005; Morbidelli et al., 2005; Gomes et al., 2005). This model was later named for the location (Nice, France) where it was developed. These migrations would have caused the Late Heavy Bombardment, which would have destabilized the planetesimal disk (Gomes et al., 2005). The Late Heavy Bombardment (LHB) occurred between ~3.8 Ga and 4.1 Ga when there was a huge spike in the impact flux on the Moon (and presumably the whole Solar System). The model also predicts that Jupiter would have captured Trojan asteroids during the planetary migrations (Morbidelli et al., 2005). The Nice model was later revised (Levinson et al., 2011) with more realistic initial conditions and is called the Nice 2 (originally proposed as Nice II) model.

Walsh et al. (2011) proposed the Grand Tack Hypothesis where Jupiter migrated to ~1.5 AU and truncated the planetesimal disk at ~1 AU. Saturn is also migrating too. The "Tack" refers to the heliocentric distance where Jupiter stops migrating inwards. The Grand Tack was proposed to solve the "Mars problem" where computer simulations in forming planets at Mars-like distance from the Sun that are too massive compared to Mars (e.g., Raymond et al., 2009). The Grand Tack occurs in the first ~5 Ma of Solar System history, while the Nice model occur 500 Ma later. During their migration into the inner Solar System, Jupiter and Saturn initially scatter inner Solar System planetesimals (assumed to be S-complex bodies) away from the Sun (Figure 5.9). As Jupiter and Saturn then migrate away from the Sun, the scattered S-complex bodies are scattered again into the inner asteroid belt region while the previously undisturbed C-complex bodies are scattered into the outer asteroid belt region. The Grand Tack would result in the distribution of taxonomic types currently found in the asteroid belt where S-complex bodies are preferentially found in the inner main belt while C-complex bodies are primarily found in the outer main belt.

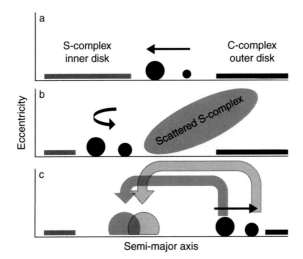

Figure 5.9 Cartoon of the Grand Tack during the first 5 Ma of Solar System history. Eccentricity versus semi-major axis is plotted. (a) Initially S-complex bodies form interior to Jupiter (large circle) and Saturn (small circle) and C-complex bodies form exterior to these giant planets. (b) As Jupiter and Saturn migrate toward the Sun, S-complex bodies are scattered outward from the Sun. (c) As Jupiter and Saturn migrate away from the Sun, the scattered S-complex bodies are scattered again toward the dynamically stable asteroid belt region and the previously undisturbed C-complex bodies are scattered into the asteroid belt region. The S-complex bodies will be preferentially found in the inner main belt while the C-complex bodies will be preferentially found in the outer main belt. Credit: Kevin Walsh, Southwest Research Institute. (A black and white version of this figure will appear in some formats. For the color version, please refer to the plate section.)

Questions

1) Two asteroids are located at 2.2 AU. One body is the parent body of the aubrites while the other is the parent body of the CM chondrites? Would you expect the aubrite parent body or the CM parent body to have the hotter average surface temperature? Why?

2) What is the thermal inertia of an asteroid with a thermal conductivity of 0.7 W/(m K), a bulk density of 1500 kg/m^3, and a specific heat capacity of 100 J/(kg K)? Would you expect the regolith to be fine-grained?

3) What is the subsolar temperature for an asteroid if the visual geometric albedo is 0.05, slope parameter is 0.15, beaming factor is 0.756, emissivity is 0.9, and the distance from the Sun is 1.2 AU?

4) What is the escape velocity from the 939-km diameter Ceres, which has a mass of 9.40 × 10^{20} kg (Russell et al., 2015)?

5) What is the distance metric between (4) Vesta and (1929) Kollaa? The proper elements for Vesta and Kollaa and the mean motion of Vesta are given in Table 5.4.

Table 5.4 *Proper elements for (4) Vesta and (1929) Kollaa and the mean motion for (4) Vesta.*

Name	a_p (AU)	e_p	$\sin i_p$	n (degrees/year)
(4) Vesta	2.362	0.0988	0.111	99.2
(1929) Kollaa	2.363	0.1141	0.123	

6) What is the axial ratio for a body with a light curve amplitude of 0.5? Assume that the angle between the rotation axis and the line of sight is 90°.
7) What is the spin barrier?
8) Estimate the mass of a 100-km diameter S-type asteroid.
9) Why is identifying a moon in orbit around a minor planet important for calculating its density?
10) If you assume the moon of a minor planet has an orbital period of 16.5 hours (T), is located 170 km (r) from the center of mass of the body, and the primary body is spherical with a radius of 44 km, what is the density of the primary object?

6

Comets and Outer Solar System Bodies

6.1 Comets

Comets are different than asteroids since comets are often bright enough to be visible in the night sky with the naked eye. Some comets have even been bright enough to be visible during the day. When observing a comet from Earth, what the observer sees is the coma (atmosphere) and the tail of the comet (Figure 6.1).

The comae and tails of comets are due to the sublimation of volatile material from the nucleus as it approaches the Sun. The sublimation of water ice tends to become significant at heliocentric distances of 4–6 AU (Meech and Svoreň, 2004). Frederick Whipple (1906–2004) proposed in a landmark paper (Whipple, 1950) that comets were "icy conglomerates" which was referred to by others as meaning "dirty snowballs." As the temperature rises on the near-surface of the comet, sublimation of volatiles increases with gas and dust released from the surface. Comets tend to have a distinct gas tail and a dust tail. The gas tail points away from the Sun while the dust tail follows the comet's orbit (Figure 6.2). The size of the tail and coma will vary as a comet gets closer and then farther from the Sun in its orbit.

Comets are generally considered the most pristine bodies remaining in the Solar System (Despois et al., 2005). The chemistry of the comet is considered the best remaining analog to the chemistry of the protosolar molecular cloud.

6.1.1 Ancient Observations of Comets

Comets have been known to exist for thousands of years. Approximately 4000 comets have been uniquely identified over recorded history. Depictions of comets throughout history all show their bright coma (atmosphere) and tail. Comets were usually thought of as bad omens when apparent in the sky since these objects would appear without any warning. Some cultures thought that comets looked like a fiery sword. A comet observed in AD 60 was thought to portend the end of

Figure 6.1 Image of Halley's comet taken in 1910. Credit: Edward Emerson Barnard (1857–1923), Yerkes Observatory, and New York Times.

Figure 6.2 Image of comet Hale–Bopp taken in April 1997. The blue gas tail is on the left of the image and the white dust tail is on the right. Credit: E. Kolmhofer and H. Raab, Johannes-Kepler-Observatory. (A black and white version of this figure will appear in some formats. For the color version, please refer to the plate section.)

Figure 6.3 Depiction of Halley's comet on the Bayeux Tapestry. Credit: Myrabella.

the rule of Nero (37–68) so Nero massacred his challengers (Schechner, 1999). Moctezuma II (also known as Montezuma II) (c. 1466–1520) saw a comet in 1519 that foreshadowed the end of the Aztec empire (Newcott, 1997).

The Bayeux Tapestry is a large embroidered cloth (~70 m × ~0.5 m) from the eleventh century which has one of the earliest surviving depictions of a comet (Figure 6.3). The Bayeux Tapestry depicts the eleventh-century Norman (from the Normandy region of France) conquest of England by William the Conqueror (c. 1028–1087). The tapestry depicts the apparition of Halley's comet in 1066. The appearance of the comet was considered a bad omen for King Harold II of England (c. 1022–1066) who died at the Battle of Hastings, the decisive battle in the Norman Conquest.

6.1.2 Naming

When a comet is discovered, it is first given a designation to indicate its orbit and a designation to indicate when it was discovered. The designation "P/" indicates a periodic comet. Periodic comets (also known as short-period comets) have orbital periods less than 200 years or confirmed observations at more than one perihelion passage. When a comet has a second perihelion passage, a number is placed before the "P/" to indicate the number of comets that have been observed to have a second passage through the Solar System. For example, "27P/" refers to the 27th comet that has been observed to have a second perihelion passage.

Comets have orbits that follow Kepler's laws except that non-gravitational forces due to outgassing can affect the comet's orbit. Jupiter-family comets have orbital periods less than 20 years and have gravitationally interacted with Jupiter (Lowry et al., 2008). They usually have T_J values between ~2 and ~3. Jupiter-family comets are given the "/P" designation, but are not given any other special designation. Halley-type comets (or intermediate-period comets) have orbital periods between 20 and 200 years. Comets that come very close to the Sun are often called sungrazing comets.

"C/" indicates a non-periodic comet. Non-periodic comets (also known as long-period comets) have orbital periods of 200 years or greater. "X/" indicates a comet where a meaningful orbit could not be calculated. A few of the designations can be changed for the comet after later observations. "D/" indicates a comet that no longer exists (e.g., has broken up) or has disappeared. "A/" indicates a body that was originally identified as a comet, but is actually a minor planet.

The comet designation is similar to the preliminary designation of asteroids with the year of the discovery, a letter to indicate the half-month, and a number to indicate the order of discovery in the half-month.

A comet is then given a name to indicate its discoverer or discoverers, which is what the comet is typically called. Besides people, comets are also named after the observing program or spacecraft that discovered it. A discoverer is defined as the person or persons or observing program or spacecraft that first detects the comet (visually, photograph, or electronic image). The discoverer or discoverers are responsible for obtaining and communicating to the Central Bureau for Astronomical Telegrams (CBAT) accurate information on the comet's positions and physical appearance. When it is a person or persons, only their last names are used. Currently, comets are usually named after one or two people. For two people, a dash is used between the names. Up until 1995, a number was used to indicate if the person, team, or spacecraft has discovered multiple comets. Comets discovered hundreds of years ago were often named after the person who computed its orbit. Table 6.1 lists a number of famous comets and their formal designations.

Comets that are extremely bright are often called "Great Comets" or "The Great Comet of ... ," followed by a year. Comets that are predicted to be extremely bright are often called the "Comet of the Century."

Table 6.1 *List of unofficial and official comet names for a number of well-known comets*

Unofficial comet name	Formal designations	
Halley	1P/1682 Q1	1P/Halley
Encke	2P/1818 W1	2P/Encke
Swift–Tuttle	109P/1862 O1	109P/Swift–Tuttle
Borelly	19P/1904 Y2	19P/Borrelly
Tempel 1	9P/1867 G1	9P/Tempel
Churyumov–Gerasimenko	67P/1969 R1	67P/Churyumov–Gerasimenko
Wild 2	81P/1978 A2	81P/Wild
Shoemaker–Levy 9	D/1993 F2	D/1993 F2 (Shoemaker–Levy 9)
Hale–Bopp	C/1995 O1	C/1995 O1 (Hale–Bopp)
Hyakutake	C/1996 B2	C/1996 B2 (Hyakutake)
ISON	C/2012 S1	C/2012 S1 (ISON)

6.1.3 Compositions

The composition of comets is determined through spectroscopy. Spectra of comets consist of a reflected solar spectrum from the dust of the coma (or tail) and emission lines of cometary molecules.

How quickly different molecular species sublimate off the surface is called the production rate. The production rate is the number of molecules (or atoms) released by a comet off the whole nucleus surface per unit time (Sekanina, 1991). The number of molecules (N_i) in the coma for each molecular species i will be equal to the production rate (Q_i) for each molecular species times the mean lifetime (τ_i) of the molecular species (Swamy, 2010). The formula can be written as

$$N_i = Q_i \tau_i. \tag{6.1}$$

The volatiles in comets are predominately composed of water and carbon monoxide. Approximately 60 different volatile molecular species have been identified in comets to date. Some of the most abundant molecules are given in Table 6.2.

6.1.4 Halley's Comet

The most famous comet in human history is Halley's comet. In 1705, English astronomer Edmund (or Edmond) Halley (1656–1742) (Hughes and Green, 2007) published "A Synopsis of the Astronomy of Comets" where he argued that the comets that appeared in 1531, 1607, and 1682 were actually the same comet. Halley determined that this comet had an elliptical orbit and predicted it would return in 1758. The comet did reappear on Christmas night in 1758. German astronomer

Table 6.2 *List of abundant volatile molecules identified in comet Hale–Bopp (Despois et al., 2005)*

Molecule	Formula	Hale–Bopp abundance $(H_2O = 100)$
water	H_2O	100
carbon monoxide	CO	23
carbon dioxide	CO_2	6
methanol	CH_3OH	2.4
methane	CH_4	1.5
hydrogen sulfide	H_2S	1.5
formaldehyde	H_2CO	1.1
ammonia	NH_3	0.7
ethane	C_2H_6	0.6
dimethyl ether	CH_3OCH_3	<0.45
carbonyl sulfide	OCS	0.4
sulfur monoxide	SO	0.3
ethylene glycol	CH_2OHCH_2OH	0.25
hydrogen cyanide	HCN	0.25
hydroxylamine	NH_2OH	<0.25
nitrous oxide	N_2O	<0.23
acetylene	C_2H_2	0.2
sulfur dioxide	SO_2	0.2

Johann Palitzsch (1723–1788) made the observation. This comet was then named Halley's comet in honor of Edmund Halley. It was later discovered through ancient Chinese texts that Chinese astronomy had a nearly unbroken record of observations of Halley's comet since 240 BC (with the exception of 164 BC) (Newburn and Yeomans, 1982). Stephenson et al. (1985) have found possible references to Halley's comet in ancient Babylonian clay tablets.

Halley's comet currently has a period of 75.3 years. Halley's comet is the only known comet that can potentially be seen twice during your lifetime with your naked eyes. Mark Twain (1835–1910) was born right after Halley's comet appeared in 1835 and died right after it reappeared in 1910 (Figure 6.1). Halley's comet last appeared in 1986 and should appear again in 2061.

During its apparition in 1986, Halley's comet was the first comet to have its nucleus observed by spacecraft. Soviet spacecraft Vega-1 and Vega-2 and the European Space Agency (ESA) spacecraft Giotto all imaged the nucleus. Vega-1 and Vega-2 first both deployed landers to Venus and then flew by Halley's comet. Vega-1 flew by the nucleus at a distance of 8890 km while Vega-2 flew by the nucleus at a distance of 8030 km. Giotto flew by comet Halley in 1986 and then Comet Grigg–Skjellerup in 1990. Giotto passed within 605 km of the nucleus of

Halley's comet. Two other spacecraft that were part of the Halley armada were the Japanese Institute of Space and Astronautical Science (ISAS) spacecraft Sakigake (Japanese for "pioneer") and Suisei (Japanese for "comet"). Sakigake observed solar wind magnetic field and plasma activities near Halley's comet, while Suisei took UV images of the comet.

Images of the interior of the comet showed that it had an extremely dark nucleus with a bright jet of gas and dust emanating from one region of the nucleus. The nucleus of Halley's comet was found to have dimensions of $7.2 \times 7.22 \times 15.3$ km^3 (Merényi et al., 1990). The nucleus had a visual geometric albedo of ~0.04 and a Bond albedo as low as ~0.02 (Whipple, 1989).

For Earth-based observations, the International Halley Watch (IHW) was organized to collect and archive observations of the comet. During the 1986 appearance, the tail of the comet was found to be at least 50 million km long. The size of its coma is on the order of ~100 000–200 000 km in diameter as determined from dust counters (Oberc, 1996).

6.1.5 Comet Swift–Tuttle

Comet Swift–Tuttle was an extremely bright comet that was discovered independently in 1862 by Lewis Swift (1820–1913) and Horace Parnell Tuttle (1837–1923). Swift–Tuttle is the parent body of the Perseids meteor shower. Marsden (1973) redetermined the orbit of Swift–Tuttle and linked Swift–Tuttle with comet Kegler, which was discovered in 1737 by the Jesuit missionary Ignatius Kegler (c.1680–1746) (Jenniskens, 2006). Marsden (1973) predicted that it could return in 1992. As predicted, comet Swift–Tuttle was rediscovered in 1992 and was determined to have an orbital period of 133.3 years.

Chinese records indicate that comet Swift–Tuttle was initially seen in 69 BC and AD 188 (Yau et al., 1994). The comet was not bright enough to be seen with the naked eye during other returns. The comet will reappear in 2126.

6.1.6 Comet Shoemaker–Levy 9

Shoemaker–Levy 9 was the ninth comet discovered by Carolyn and Eugene Shoemaker (1928–1997) and David Levy in March 1993 using the 0.46-m Schmidt telescope at Palomar Observatory. The comet was unusual since it appeared as a linear train with a number of condensations (Marsden, 1993a). These ~20 sub-nuclei (Figure 6.4) were later said to be "strung out like pearls on a string" by Jane Luu and David Jewitt (Marsden, 1993b). Shoemaker–Levy 9 was found to be in orbit around Jupiter and appears to have broken apart during a close encounter with Jupiter in 1992 (Marsden, 1993c). Shoemaker–Levy 9 passed within the Jupiter's Roche limit, which

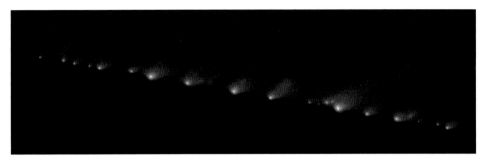

Figure 6.4 A Hubble Space Telescope image of comet Shoemaker–Levy 9, taken in May 1994 with the Wide Field Planetary Camera 2 (WFPC2). Credit: NASA. (A black and white version of this figure will appear in some formats. For the color version, please refer to the plate section.)

is the distance from a celestial body where an object held together by its own gravity will break apart, of Jupiter. The comet was then predicted in May 1993 to impact Jupiter in July 1994 (Marsden, 1993d, 1993e). This was the first celestial object to have been successfully predicted to strike a planetary body. The fragments were named in impact order with A first and omitting I, O, X, Y, and Z (Harrington et al., 2004). Fragments P and Q split after being named and became P2, P1, Q2, and Q1 in impact order. Fragments F, J, P2, P1, and T disappeared before impact.

At least 16 fragments struck Jupiter from July 16 to July 22, 1994 (Harrington et al., 2004). The effects of these impacts were observed by numerous large and small telescopes on Earth and in space (e.g., Hubble Space Telescope, Galileo). Large dark spots due to the impacts were visible on Jupiter's surface. The largest spot (due to fragment G) was over 12 000 km in diameter.

6.1.7 Comet Hale–Bopp

One of the brightest comets in recent history was Hale–Bopp (Figure 6.2). Alan Hale and Thomas Bopp discovered Hale–Bopp independently in 1995 in the constellation Sagittarius. Hale discovered the comet while observing using a telescope in his driveway, while Bopp was looking through a telescope in the desert outside Phoenix, Arizona (Newcott, 1997). Hale had a PhD in astronomy and had previously observed ~200 comets, while Bopp was an amateur astronomer. Hale immediately sent an email to the Central Bureau for Astronomical Telegrams, while Bopp actually sent a telegram a few hours after his discovery.

Hale–Bopp has a period of 2534 years. Analysis of Hubble Space Telescope images found that Hale–Bopp has an effective diameter of 27–42 km (Weaver et al., 1997), which is much larger than Halley's comet.

Hale–Bopp is one of the best-studied comets due to its brightness (magnitude of 10.5 at 7.15 AU) and its discovery far from the Sun (McFadden, 1999). More than 50 observed molecules, radicals, and ions were discovered in Hale–Bopp (McFadden, 1999). A number of new molecules (SO, SO_2, HC_3N, NH_2CHO, HCOOH, $HCOOCH_3$) were identified in Hale–Bopp (Bockelée-Morvan et al., 2000).

6.1.8 Comet Wild 2

Wild 2 was the second comet discovered by Paul Wild (1925–2014). Wild 2 has an orbital period of 6.4 years. Wild 2 was the target of the Stardust mission, which was a NASA mission launched on February 7, 1999 to collect dust samples from the coma of a comet and return the samples back to Earth. The principal investigator (PI) of the mission was Donald Brownlee.

Comet Wild 2 is a Jupiter-family comet that passed close by Jupiter in 1974, which changed its orbital period from ~40 years to 6.17 years and its perihelion distance from 4.9 AU to 1.49 AU. The Wild 2 nucleus had a ~5 km diameter and was relatively spherical. Wild 2 was covered with circular depressions that appeared to be impact craters (Figure 6.5) (Brownlee et al., 2004). These crater-like features were grouped into two types: pit halo and flat floor. "Pit-halo" features have a rounded central pits surrounded by an irregular region of partially excavated material while "flat-floor" features lack these halo regions and have extremely steep cliffs. The visual geometric albedo of the nucleus was ~0.03.

Cometary grains were collected from the 6.1 km/s impacts of grains into a low-density silica aerogel and an aluminum foil (Brownlee, 2012). The collected material was dominated by material that was similar to the common high-temperature components in chondrites such as mineral, CAI, and chondrule fragments, which initially formed at temperatures of ~1000–2000 K (Brownlee et al., 2012). These high-temperature fragments must have formed relatively close to the Sun. Almost all the grains collected by Stardust had isotopic compositions (hydrogen, carbon, nitrogen, oxygen) consistent with a Solar System origin (McKeegan et al., 2006). Only a relatively few isotopically anomalous presolar grains have been identified among the Stardust samples (e.g., McKeegan et al., 2006).

6.1.9 Comet Borrelly

The nucleus of comet Borrelly was imaged at a distance of 3500 km by Deep Space 1 when it was 1.36 AU from the Sun (Boice et al., 2002). The nucleus of Borrelly is extremely elongated with dimensions of $8 \times 4 \times 4 \text{ km}^3$ (Lamy et al., 1998) (Figure 6.6). The surface of Borrelly is devoid of unambiguous impact craters, arguing that its surface is too active to retain craters (Britt et al., 2004; Basilevsky and Keller, 2006).

Figure 6.5 Image of the comet Wild 2 nucleus. The nucleus is ~5 km in diameter. Note the large number of crater-like features on the surface. Credit: NASA.

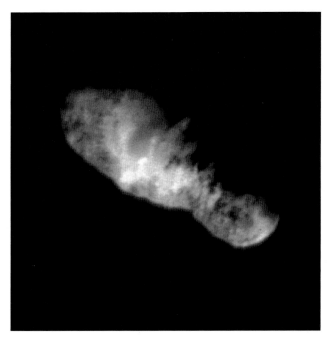

Figure 6.6 Image of the comet Borrelly nucleus. The nucleus is ~5 km in diameter. Note the lack of craters on the surface. Credit: NASA.

6.1.10 Comet Tempel 1

Comet Tempel 1 was the first comet discovered by Wilhelm Tempel (1821–1889) who observed it in 1867. Currently, Tempel 1 has an orbital period of 5.5 years. Tempel 1 was the target of the Deep Impact mission, which released an impactor that struck the comet (A'Hearn et al., 2005). The spacecraft then flew by the comet to image the effects of the impact. The spacecraft had two cameras: the High Resolution Imager (HRI) and the Medium Resolution Imager (MRI). The impactor also had a camera called the Impactor Targeting Sensor (ITS). The impactor was composed entirely of copper, since copper was not expected to be present in the comet and, therefore, any spectral signatures due to copper could be ignored.

The impact produced a flash on the comet and an impact plume (A'Hearn et al., 2005). Emission features in the plume in included H_2O, HCN, CO_2, and a feature due to the C–H stretching mode (indicating organics). The impact also excavated a number of very fine particles. The strength of the excavated material was smaller than 65 Pa. The estimate of the bulk density of the Tempel 1 nucleus is ~600 kg/ m^3. The Deep Impact mission was then extended and called the EPOXI (Extrasolar Planet Observation and Deep Impact Extended Investigation) mission where the spacecraft observed extrasolar planets and flew by the Moon and 103P/Hartley (comet Hartley 2).

The Stardust mission was also extended so it could flyby comet Tempel 1 and was renamed Stardust-NExT. Stardust-NExT identified a 49 ± 12 meter diameter crater with an ejecta blanket that is 85–120 m in diameter (Richardson and Melosh, 2013). To produce an ejecta blanket of this size, the material properties of the surface were estimated to be similar to lightly packed, "dry" mountain snow.

6.1.11 Comet Churyumov–Gerasimenko

Klim Ivanovich Churyumov (1937–2016) and Svetlana Ivanovna Gerasimenko discovered comet Churyumov–Gerasimenko in 1969. Churyumov–Gerasimenko has a period of 6.4 years. Comet 67P/Churyumov–Gerasimenko was the target of the Rosetta spacecraft, which was launched by the European Space Agency in March 2004 to orbit the comet and launch a lander (Philae) to the surface. Philae landed on the surface; however, its harpoons did not deploy and it bounced a number of times on the surface. Philae worked for three days before its battery power ran out. Philae's landing site did not allow its solar panels to receive enough sunlight to power its instruments. The instrument on Philae that was used to determine the composition of the comet was COSAC (Cometary Sampling and Composition Experiment), which contained a gas chromatograph (GC) and a time-of-flight mass spectrometer (TOF-MS). Also on Philae was the Sample, Drilling, and Distribution system (SD2) that could deliver samples to COSAC. Since Rosetta will allow for a much better understanding of comets, the mission was named after the Rosetta Stone. The Rosetta Stone was used to decipher Egyptian hieroglyphs since the ancient stone contained the same text written in three different languages.

Figure 6.7 Mosaic of four images taken by Rosetta's navigation camera (NAVCAM) at a distance of 28.6 km from the center of comet Churyumov–Gerasimenko. The large lobe has dimensions of 4.1 × 3.3 × 1.8 km^3 while the small lobe has dimensions of 2.6 × 2.3 × 1.8 km^3. Credit: ESA/Rosetta/NAVCAM, CC BY-SA IGO 3.0.

Images of comet Churyumov–Gerasimenko showed its nucleus to be composed of two lobes connected by a short neck (Figure 6.7) (Sierks et al., 2015). The two lobes could either represent a contact binary or a single body that has lost considerable material in one region (the neck) of the nucleus. Activity of the nucleus at heliocentric distances greater than 3 AU are predominately from the neck. The mass of the body was calculated using spacecraft velocity perturbations. The average bulk density was calculated to be 0.533 ± 0.006 kg/m^3, implying a porosity of 72–74% (Pätzold et al., 2016).

Rosetta observed a wide variety of morphologies on the nucleus (Thomas et al., 2015). Five different types of terrains were observed: dust-covered terrains, brittle materials with pits and circular structures, large-scale depressions, smooth terrains, and exposed consolidated surfaces. Considerable fracturing was seen on the surface.

After Philae landed on the surface, COSAC took a mass spectrometer spectrum in sniffing mode (no active sampling) and identified 16 molecules on the surface (Goesmann et al., 2015). Due to the relatively low resolution of the mass spectrometer, single mass peaks could not be subdivided into different molecular species. To determine the molecules that were present, all conceivable molecules and their fragmentation patterns were evaluated with the measured spectrum and those with fragmentation patterns that were not consistent with the spectrum were eliminated. Water was the abundant molecule (~80%) that was detected. Four organic molecules that had not previously been detected on comets were identified. ROSINA (Rosetta Orbiter Spectrometer for Ion and Neutral Analysis) made the first in situ detection of N_2 (molecular nitrogen) in a comet (Rubin et al., 2015). The low N_2/CO_2 ratio compared to the solar nebula value implies that the cometary grains formed at low temperatures (~30 K).

Rosetta orbited the comet for over two years. Due to dwindling solar power, the Rosetta spacecraft ended its mission by "softly" crashing into the comet's surface on September 30, 2016.

6.1.12 Oort Cloud

The Oort cloud is a theoretical spherical cloud of icy bodies that surrounds the Solar System from distances of a few thousand AU to 50 000–100 000 AU. The Oort cloud is named after Jan Oort (1900–1992) who proposed that long-period comets originated from an orbiting cloud at the far reaches of the Solar System (Oort, 1950). With the Kuiper belt and main asteroid belt, the Oort cloud is one of three known reservoirs of comets (Dones et al., 2015).

These bodies have stable orbits until they are gravitationally perturbed. For example, passing stars can perturb these Oort cloud objects (e.g., Mamajek et al., 2015). Some of these bodies are perturbed toward the inner Solar System where they will be identified as comets.

6.2 Asteroid–Comet Relationships

Some comets were later identified as asteroids when they lost their tails in subsequent images, while some asteroids were later identified as comets when tails appeared in later images. The first comet to be later identified as an asteroid was comet 107P/Wilson–Harrington, which is also known as (4015) Wilson–Harrington. Wilson–Harrington has the longest name of any asteroid and exceeds the present-day 16 character limit for asteroid names. Wilson-Harrington was discovered in 1949 and initially identified as a comet before it was lost. In 1979, an Apollo asteroid was discovered by Eleanor "Glo" Helin (1932–2009) at Palomar Observatory and was given the designation of (4015) 1979 VA. In the process of identifying prediscovery (precovery) images of asteroids in the Palomar Sky survey, Ted Bowell (Marsden, 1992) identified 1979 VA in a 1949 plate and found that it had a faint tail and that its orbit matched comet Wilson-Harrington. Centaur (2060) Chiron also shows cometary activity and is also designated as 95P/Chiron.

Main-belt comets are objects in the asteroid belt that show cometary activity. The first known main-belt comet was (7968) Elst–Pizarro (133P/Elst–Pizarro)

Table 6.3 *A list of numbered minor planets that are also listed as comets*

Minor planet name	Comet name	Semi-major axis (AU)
(2060) Chiron	95P/Chiron	13.64
(4015) Wilson–Harrington	107P/Wilson–Harrington	2.64
(7968) Elst–Pizarro	133P/Elst–Pizarro	3.16
(60558) Echeclus	174P/Echeclus	10.68
(118401) LINEAR	176P/LINEAR	3.22

(Hsieh et al., 2004). When this object was first discovered in 1979, it appeared like any other asteroid in images and it also had a typical asteroidal orbit. However, images of this body by Eric Elst and Guido Pizarro in 1996 found it to have a tail (Marsden, 1996). Asteroid (596) Scheila was observed by Steve Larson to display cometary activity (Green, 2010), but has not been given an official comet designation. A list of numbered minor planets that have also been identified as comets are given in Table 6.3. Also, a number of comets (e.g., 238P/Read) have orbits within the asteroid belt.

Jewitt (2012) argues that a number of processes can cause an asteroid to have cometary activity. The repetitive nature of the activity on a number of bodies argues that their cometary features are due to the sublimation of water ice. The cometary nature of Scheila appears to be due to an impact. Activity on Phaethon may be due to cracking or dehydration at extreme perihelion temperatures.

Fernández et al. (2005) found that almost all NEAs and damocloids with T_J values less than 2.6 have comet nucleus-like geometric albedos (< 0.075) in the R-band (~ 0.658 μm). This result is consistent with these bodies being extinct comets.

6.3 Trojans

Trojans are bodies that share an orbit with another larger body but are found at either of two points of stability (called Lagrangian points) that are either $\sim 60°$ ahead (L_4 point) or behind (L_5 point) the larger body's orbit. Trojans are in a co-orbital configuration with a planet, meaning that they are in a 1:1 mean-motion resonance with the planet. Trojans have been detected near Venus, Earth, Mars, Jupiter, Uranus, and Neptune; however, most Trojans are found near giant planets. Only one Earth Trojan (2010 TK$_7$) has been identified (Connors et al., 2011), while four Mars Trojans have been identified. The orbit of 2010 TK$_7$ has an orbit that is stable for at least 10000 years. Inclusion in the Minor Planet Center list of Mars Trojans requires that they be confirmed as Trojans by numerical integrations for at least 10^5 yr (Christou, 2013).

For two large bodies (Sun and a planet) in orbit around each other, there are five configurations where a much smaller body can remain in a theoretically stable position relative to the other two bodies due to gravity. They are known as the L_1, L_2, L_3, L_4, and L_5 points. The L_1, L_2, and L_3 points were discovered by Leonhard Euler (1707–1783), while the L4 and L5 points were discovered by Joseph-Louis Lagrange (1736–1813).

The L_1, L_2, and L_3 points occur on a straight line between the Sun and the planet. The L_1 point falls between the Sun and the planet, the L_2 point falls outside the planet's orbit, and the L_3 point falls on the other side of the Sun. Objects at the L_1, L_2, and L_3 points are not truly in stable orbits since any gravitational perturbation can move a body out of these positions.

6.3.1 Jupiter Trojans

Over ~6000 Jupiter Trojans are currently known; however, estimates of the total number of Jupiter Trojans with diameters greater than a kilometer is ~10^6 (Yoshida and Nakamura, 2005). The first known Trojan [(588) Achilles] was discovered near Jupiter in 1906. From the Nice model, Jupiter Trojans are commonly believed to be captured bodies from the primordial Kuiper belt (Emery et al., 2015).

Jupiter Trojans are extremely dark with an average visual geometric albedo of 0.07 ± 0.03 from WISE measurements (Grav et al., 2011). Jupiter Trojans are dominated by the extremely red D-types, implying organic-rich mineralogies (Gradie and Veverka, 1980). However, Emery and Brown (2003, 2004) argue that the red spectral slope is due to space-weathered anhydrous silicates and not organic material (Section 4.9.5).

6.3.2 Uranus Trojans

Only one Uranus Trojan (2011 QF_{99}) has been identified (Alexandersen et al., 2013). Alexandersen et al. (2013) finds that this object oscillates around Uranus' L4 point for greater than 70 000 years and remains co-orbital for ~1 million years before becoming a centaur.

6.3.3 Neptune Trojans

The first discovered Neptune Trojan was 2001 QR_{322}. Eleven Neptune Trojans have been subsequently discovered. Sheppard and Trujillo (2010) estimate that there are approximately 400 Neptune Trojans with radii larger than 50 km.

6.4 Trans-Neptunian Objects

Trans-Neptunian objects all have semi-major axes greater than Neptune's and include Kuiper belt objects, scattered disk objects (SDOs), and Oort cloud objects. Currently, over 1600 trans-Neptunian objects have been discovered. A list of notable trans-Neptunian objects are given in Table 6.4.

Table 6.4 *Notable trans-Neptunian objects*

Name	Type	Semi-major axis (AU)	Diameter (km)	Visual geometric albedo
(15760) 1992 QB$_1$	cubewano	43.80	–	–
(20000) Varuna	cubewano	43.12	500 ± 100	$0.16^{+0.10}_{-0.08}$
(28978) Ixion	plutino	39.50	650^{+260}_{-220}	$0.12^{+0.14}_{-0.06}$
(50000) Quaoar	cubewano	43.37	1110 ± 5	0.109 ± 0.007
(90377) Sedna	scattered disk	507	<1600	>0.16
(90482) Orcus	plutino	39.47	$946.3^{+74.1}_{-72.3}$	$0.1972^{+0.0340}_{-0.0276}$
(134340) Pluto	plutino	39.53	2374 ± 8	0.49–0.66
(136108) Haumea	–	43.31 ($1920 \times 1540 \times 990$ km^3)	$0.804^{+0.062}_{-0.095}$	
(136199) Eris	scattered disk	67.69	2326 ± 12	$0.96^{+0.09}_{-0.04}$
(136472) Makemake	cubewano	45.76	1428^{+48}_{-24}	$0.81^{+0.03}_{-0.05}$
2012 VP$_{113}$	scattered disk	262	–	–

The semi-major axes are from the Minor Planet Center database. The diameter and albedo of Varuna, Ixion, Sedna, and Orcus are from Stansberry et al. (2008). The diameter and albedo for Quaoar are from Braga-Ribas et al. (2013). The diameter of Haumea is from Lockwood et al. (2014) and its albedo is from Fornasier et al. (2013). The diameter of Pluto is from Stern et al. (2015) and its range in visual geometric albedo is from Lellouch (2011). The diameter and albedo for Eris are from Sicardy et al. (2011). The diameter and albedo of Makemake are from Brown (2013). The diameter of Makemake is the average of the equatorial and polar diameter and the uncertainties are the upper value of the errors for the diameters.

Kuiper belt objects have semi-major axes that vary from ~30 AU (orbit of Neptune) to ~50 AU from the Sun. The outer boundary is defined approximately by the 1:2 mean-motion resonance (47.8 AU) with Neptune. The Kuiper belt is named after Gerard Kuiper (1905–1973) who proposed that a belt of comet-like objects once existed outside Neptune's orbit, but a significant population of these objects do not exist today (Kuiper, 1951). This region is also called the Edgeworth–Kuiper belt to also honor Kenneth Edgeworth (1880–1972). Edgeworth had proposed in 1943 a large number of small bodies in the outer reaches of the Solar System that sometimes travel to the inner Solar System to be seen as comets (Edgeworth, 1943). However, the work of Edgeworth was not referenced by Kuiper in his 1951 paper. Frederick Leonard (1896–1960) proposed earlier that there might be more objects like Pluto beyond Neptune (Leonard, 1930). Jewitt (2010) argues that the person who should be given credit for most accurately predicting this belt of objects is Julio Fernández. Fernández (1980) predicted the presence of a belt of comets between 35 and 50 AU that would supply short-period comets to the inner Solar System as Neptune gravitationally perturbs them.

In the late 1980s, David Jewitt and Jane Luu initiated a discovery survey to search for Kuiper belt objects using first the 1.3-m telescope at Kitt Peak Observatory in Arizona and then the University of Hawaii 2.2-m telescope on Mauna Kea (Jewitt, 1999, 2010). For almost five years, they were unsuccessful (e.g., Luu and Jewitt, 1988). The discovery method was blinking the CCD images and looking for slow moving objects. The difficulty in discovering trans-Neptunian objects is their faintness and slow movement in the sky. Remember that the brightness of distant minor planets varies as the inverse of the fourth power of the distance from the Sun squared. Possible reasons for the lack of discoveries included that these bodies did not actually exist due to gravitational interactions with outer planets or they were just too faint to be easily seen due to their faintness. However, the first Kuiper belt object that was discovered by Jewitt and Luu (1993) at the University of Hawaii 2.2-m telescope was (15760) 1992 QB$_1$. This body had a +22.8 R-band magnitude. The body was estimated to be far beyond Neptune. Jewitt and Luu nicknamed the object "Smiley" after the British spymaster George Smiley in the John Le Carré novels (Leverington, 2007). The diameter of (15760) 1992 QB$_1$ was estimated to be ~250 km if a comet nucleus-like albedo was assumed. Six months later, Jewitt and Luu discovered a second Kuiper belt object, (181708) 1993 FW. This body was nicknamed "Karla" after Smiley's Russian nemesis in the Le Carré novels. For their work in "discovering" the Kuiper belt, Jewitt and Luu were awarded the 2012 Shaw Prize in Astronomy and also the 2012 Kavli Prize in Astrophysics with Michael Brown who is a prolific discoverer of trans-Neptunian objects.

Kuiper belt objects are subdivided according to whether they are in or not in a mean-motion resonance with Neptune. These mean-motion resonances include the 2:3, 3:5, 4:7, and 1:2. For example, a body in a 2:3 resonance would travel twice around its orbit in the same time that Neptune travels three times in its orbit. Plutinos are Kuiper belt objects in a 2:3 resonance ($a = 39.4$ AU) with Neptune and are named after Pluto since these bodies have orbits similar to Pluto. Twotinos are objects that are in a 1:2 resonance ($a = 47.8$ AU) with Neptune. Many researchers have considered this 1:2 resonance with Neptune as the outer edge of the Kuiper belt.

Objects not in resonance are called classical Kuiper belt objects or cubewanos after 1992 QB$_1$ ("QB1os"). "Cold" classical Kuiper belt objects have low inclinations and nearly circular orbits, while "hot" classical Kuiper belt objects have more inclined and more eccentric orbits.

Scattered disk objects have semi-major axes greater than ~50 AU and perihelia greater than 30 AU (Gomes et al., 2008). These bodies appear to have been gravitationally scattered by the giant planets. Clusters in orbital properties for scattered disk

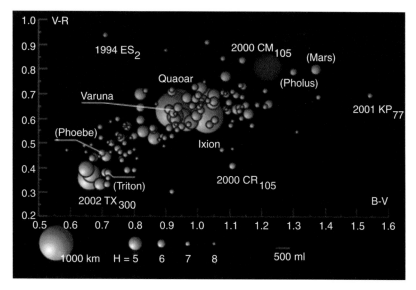

Figure 6.8 Color–color plot of trans-Neptunian objects for bodies discovered as of 2006. Larger objects are plotted with larger circles. For reference, Mars, Phoebe (moon of Saturn), Pholus (centaur), and Triton (moon of Neptune) are also plotted. Mars and Triton are not plotted using the same size scale as the trans-Neptunian objects. Credit: Drbogdan. (A black and white version of this figure will appear in some formats. For the color version, please refer to the plate section.)

objects have been argued to be possibly due to the presence of a distant planet in the outer Solar System (Trujillo and Sheppard, 2014; Batygin and Brown, 2016). These objects tend to have argument of perihelia (ω) that cluster around 0, which means that these bodies' perihelia lie in the ecliptic and move south to north when crossing the ecliptic. These bodies also cluster in terms of perihelion distances and orbital planes.

6.4.1 Compositions

Trans-Neptunian objects tend to be divided into two color groups: "red" and "blue-gray." Color–color diagrams plot the V–R color versus the B–V color (Figure 6.8). Objects with colors greater than those of the Sun are called "red" while those with colors similar or smaller than the Sun are called "blue-gray" (Doressoundiram et al., 2008). The V–R of the Sun is 0.356 and the B–V of the Sun is 0.653.

The red spectral slope of trans-Neptunian objects is thought to be due to the presence of tholins. Tholins are complex organic compounds that are produced by the irradiation of simple organic compounds (e.g., Sagan and Khare, 1979). The term "tholin" comes from the Greek word "tholus," which means muddy. Tholins tend to have a brownish-red appearance. The exposure of organics to space will alter their spectral properties through irradiation by cosmic rays. Ion irradiation of simple organics (methanol, CH_3OH; methane, CH_4; benzene, C_6H_6) (Brunetto et al., 2006) produced strong reddening and darkening. However, ion irradiation (Moroz et al., 2003, 2004) of complex hydrocarbons showed a decrease in the spectral slope with increasing irradiation.

Tegler and Romanishin (2000) found that Kuiper belt objects with perihelia greater than 40 AU tend to have extremely red colors. Cubewanos tend to be redder than other populations of TNOs (Doressoundiram et al., 2008). There is still considerable disagreement whether the color diversity found among TNOs is due to compositional diversity or some type of evolutionary process (e.g., collisions, surface irradiation) (Doressoundiram et al., 2008).

6.4.2 (134340) Pluto

The most famous Kuiper belt object is (134340) Pluto. Discrepancies (deviations in their predicted astrometric positions) in the orbits of Uranus and also Neptune were believed to be possibly due to the presence of a ninth planet, which led Percival Lowell (1855–1916) during the early twentieth century to lead a project at Lowell Observatory to search for a possible ninth planet (Tombaugh and Moore, 1980). Discrepancies in the orbit of Uranus had previously led to search for an eighth planet, which led to the discovery of Neptune. However, the mass of Neptune was not large enough to account for Uranus' orbital discrepancies. So a ninth planet was thought to exist.

Pluto was discovered in February 18, 1930 by Clyde Tombaugh (1906–1997) at Lowell Observatory. He started searching for Pluto in 1929 using a 13-inch astrograph (Figure 6.9) (Gielas, 2000). Astrograph telescopes have wide field of views. The telescope is a refractor with a Cooke triplet lens (three pieces of glass with different indexes of refraction).

Blinking two plates that were taken six days apart of the part of the constellation Gemini during the previous month, Tombaugh identified an object that he knew immediately was trans-Neptunian due to its slow movement between images. Tombaugh then checked a third plate and confirmed the presence of the object.

Its name was first proposed by Venetia Burney (1918–2009) who was an 11-year-old English schoolgirl. Venetia suggested the name in honor of the Roman god of the underworld to her grandfather who passed the name to a professor of astronomy at the University of Oxford who sent a telegram to Lowell Observatory. Helping the case for the people at Lowell Observatory to support this name was that the first two letters of Pluto were also the initials of Percival Lowell who founded Lowell Observatory.

Figure 6.9 The 13-inch astrograph used by Clyde Tombaugh at Lowell Observatory to discover Pluto. Credit: Pretzelpaws.

As with the other planets, Pluto's name was fitting. Mercury, named after the Roman god of messengers, travels fastest around the Sun. Venus, named after the Roman god of love, is the brightest planet in the sky. Mars, named after the Roman god of war, is the color of blood (the aftermath of war). Jupiter, named after the Roman king of the gods, is the largest planet. The Roman god Saturn is the father of Jupiter and the Greek god Uranus is the father of Cronus (the Greek name for Saturn). Neptune, the Roman god of the sea, has a bright blue ocean-like color. Pluto, named after the Roman god of the underworld, was at the time of discovery the farthest planet known from the Sun.

Pluto also has five known moons (Charon, Nix, Hydra, Kerberos, Styx). James Christy discovered Charon in 1978. While observing photographic plates of Pluto, Christy noticed that Pluto had an irregular shape with a bulge on one side (Figure 6.10) that would appear and disappear. Christy rightly concluded that this was a moon of Pluto. Charon is the ferryman who brought the dead to the

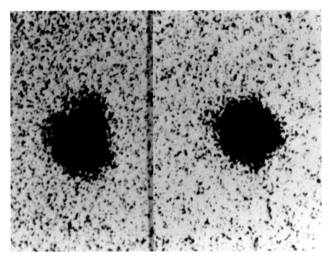

Figure 6.10 Images of (134340) Pluto taken by James Christy that led to the discovery of Charon. Charon can be seen as a bulge in the left image that disappears in the right image. Credit: James Christy, United States Naval Observatory.

underworld. The name was also chosen to commemorate Christy's wife Charlene, who had the nickname "Char." Canup (2005) argues that the Pluto–Charon system is the result of a low-velocity (< 0.9 km/s) oblique collision by an impactor containing 30–50% of the Pluto–Charon system's mass.

The other moons of Pluto were discovered much later due to their small sizes. To characterize the Pluto system before the New Horizons encounter, the Hubble Space Telescope observed Pluto at various times. These studies were done to try to ensure that New Horizons could fly safely through the Pluto system. The Hubble Space Telescope discovered Nix and Hydra in 2005, Kerberos in 2011, and Styx in 2012. These moons are primarily named for mythological characters or places associated with the god Pluto. The exception is Hydra, which is named for the nine-headed serpent that batted Hercules. The name Hydra was chosen to honor Pluto's long status as the ninth planet. Styx is the river that Charon used to ferry the dead to the underworld. Nyx is the goddess of night who is the mother of Charon. Nix is spelled with an "i" so as not to duplicate the name of the asteroid (3908) Nyx. The first letters ("N" and "H") of Nix and Hydra also commemorate the New Horizons mission, and the letter "H" also commemorates the Hubble Space Telescope. Kerberos is the three-headed dog that guards the entrance of the underworld. These moons are predicted to have formed from the disk of material produced in the aftermath of the Charon-forming collision (Stern et al., 2006).

6.4.3 New Horizons

6.4.3.1 Before Encounter

NASA's New Horizons flew by dwarf planet (134340) Pluto and its five moons. In 2015 and 2016, distant observations were made of Kuiper Belt object (15810) 1994 JR_1. In January 2019, New Horizons will also fly by Kuiper belt object 2014 MU_{69}. The PI of the mission is Alan Stern. Instruments on New Horizons include the camera LORRI (Long Range Reconnaissance Imager), an ultraviolet imaging spectrometer called ALICE, a visible CCD imager and near-infared spectrometer that is part of the Ralph telescope, the Venetia Burney Student Dust Counter (VBSDC), Radio Science Experiment (REX), a solar wind analyzer (Solar Wind Around Pluto or SWAP), and the Pluto Energetic Particle Spectrometer Science Investigation (PEPSSI). Ralph is named after Ralph Cramden of *The Honeymooners* whose wife was Alice (like the ultraviolet instrument) (Reuter et al., 2008).

Before the encounter, very little was known about the geology of Pluto and Charon (Moore et al., 2015). Pluto's reflectance spectrum has absorption bands consistent with methane and nitrogen ice on it surface, while Charon's reflectance spectrum has absorption bands consistent with crystalline water ice and ammonia (Sawyer et al., 1987; Binzel, 1988; Brown and Calvin, 2000; Merlin et al., 2010; Merlin, 2015). Both bodies have heterogeneous compositions across their surfaces. Hubble light curves and images of Pluto and Charon also show albedo and color variegations across their surfaces (Buie et al., 2010a, 2010b). However, maps taken by Hubble of Pluto and Charon only have spatial scales of a few hundred kilometers, which does not allow any substantial geological interpretations to be made of their surfaces (Buie et al., 2010b). The resolution of some of the best New Horizon images for Pluto and Charon vary from hundreds of meters to ~30 kilometers per pixel (Stern et al., 2015), which does allow geological interpretations to be made. Before the encounter, Pluto's density was calculated to be 1890 ± 60 kg/m^3 and Charon's density was calculated to be 1720 ± 20 kg/m^3 (Brozović et al., 2015).

Pluto and Charon should have craters on their surfaces (Moore et al., 2015). The icy surfaces of Pluto and Charon would also be expected to have undergone cryovolcanism. Tectonic features on Pluto would be due to the reduction in radiogenic heating over time, which would have led to the freezing of any liquid ocean. Since water expands when it freezes, Pluto would be expected to have extensional tectonic features such as rifts and grabens. Pluto's thin atmosphere, measured through occultations of stars (e.g., Hubbard et al., 1988), was known to be primarily nitrogen (e.g., Lellouch et al., 2009), while Charon has no discernible atmosphere (e.g., Gulbis et al., 2006).

Figure 6.11 This image was produced from four images of (134340) Pluto from New Horizons' Long Range Reconnaissance Imager (LORRI) that were combined with color data from the Ralph/Multispectral Visual Imaging Camera (MVIC). Pluto has a diameter of 2374 km. Note the relatively craterless heart-shaped region called Tombaugh Regio. Credit: NASA/JHUAPL/SwRI. (A black and white version of this figure will appear in some formats. For the color version, please refer to the plate section.)

6.4.3.2 *After Encounter*

New Horizon's closest approach to Pluto occurred on July 14, 2015 (Stern et al., 2015). Pluto was found to have a diverse range of landforms across its surface (Figure 6.11). Pluto's surface has a very distinctive large high-albedo, relatively craterless region that is shaped like a heart (called Tombaugh Regio after the discoverer of Pluto) and is ~1590 km across. Features on Pluto are named after mythological people associated with the underworld and people (writers, scientists, engineers) associated with Pluto and the Kuiper belt. The western lobe of Tombaugh Regio is composed of volatile ices (N_2, CO, CH_4) and has no apparent craters, implying crater retention ages of 10 Ma or less (Moore et al., 2016). However other regions on Pluto appear heavily cratered and have surface ages of ~4 Ga.

The density of Pluto was found to be 1860 ± 13 kg/m^3 (Stern et al., 2015), very similar to the pre-encounter estimate by Brozović et al. (2015). N_2, CO, CH_4, and H_2O were all detected on Pluto's surface (Grundy et al., 2016). Pluto's upper atmosphere was determined to be extremely cold (Gladstone et al., 2016), which

Figure 6.12 This image of Charon combines blue, red and infrared images taken by the New Horizon's Ralph/Multispectral Visual Imaging Camera (MVIC). Charon as a diameter of 1212 km. The colors are processed to best highlight the variation of surface properties across Charon. Note the large canyon across Charon's surface and the reddish color apparently due to tholins around the north pole. Credit: NASA/JHUAPL/SwRI. (A black and white version of this figure will appear in some formats. For the color version, please refer to the plate section.)

causes the escape rate of nitrogen from the surface to be extremely low and would allow Pluto to retain nitrogen on its surface and not have it sublimate off into space. No atmosphere was detected around Charon.

New Horizons also imaged Pluto's moons (Stern et al., 2015; Moore et al., 2016). Charon has a large canyon (or fracture system) that stretches across its observed surface (Figure 6.12). Charon has a density of 1702 ± 21 kg/m³ (Stern et al., 2015), very similar to the pre-encounter estimate by Brozović et al. (2015). Charon is also heavily cratered, also implying surface ages of ~4 Ga (Moore et al., 2016). Features on Charon are named after fictional and mythological destinations, milestones, vessels, and explorers of both space and any other destination. Spectra of Charon are dominated by H_2O ice with weak ammonia absorptions (Grundy et al., 2016). The reddish color of the north pole appears to be due to tholins.

New Horizons also imaged Nix, Hydra, Kerberos, and Styx. Nix (40 km in diameter) and Hydra (45 km) are much larger than Kerberos (10 km) and Styx

Figure 6.13 Panchromatic (single band) image of Nix taken by the Multispectral Visible Imaging Camera (MVIC) aboard New Horizons. Nix has dimensions of $50 \times 35 \times 33$ km^3. Credit: NASA/JHUAPL/SwRI.

(7 km) (Stern et al., 2015). The images of Nix (Figure 6.13) were much higher resolution than the images of Hydra, Kerberos, and Styx due to the closer approach to Nix (Weaver et al., 2016). Images of Nix and Hydra show cratered surfaces. Their visual geometric albedos are extremely high, suggesting water ice compositions. The albedo of Nix is 0.56 ± 0.05, the albedo of Hydra is 0.83 ± 0.08, the albedo of Kerberos is 0.56 ± 0.05, and the albedo of Styx is 0.65 ± 0.07. The high albedo of these moons compared to many Kuiper belt objects (Table 6.4) implies (Weaver et al., 2016) that the moons formed in the aftermath of the Charon-forming collision since Charon also has a water ice-rich surface (Stern et al., 2006). No new moons of Pluto were detected with diameters larger than ~1.7 km (Weaver et al., 2016).

Crater counts on the surfaces of Nix and Hydra surfaces of Nix are consistent with ages of at least 4 Ga (Weaver et al., 2016). Features on Nix are named for deities of the night. Features on Hydra are named after legendary serpents and dragons. Features on Kerberos are named after dogs from literature, mythology, and history. Features on Styx are named after river gods.

6.4.4 (79360) Sila–Nunam

Classical Kuiper belt object (79360) Sila–Nunam is a binary system (Stephens and Noll, 2006) with two bodies of approximately equal size. This system was discovered by Jane Luu, David Jewitt, Chad Trujillo, and Jun Chen in 1997 and was originally thought to be one body. This object is named after two Inuit gods. Sila is the god of the sky while Nunam is the god of the Earth.

6.4.5 (20000) Varuna

Classical Kuiper belt object (20000) Varuna was discovered by Robert MacMillan in November 2000. Varuna is named after the Vedic (ancient Hinduism) god who was the maker and upholder of heaven and Earth. Jewitt et al. (2001) calculated the diameter of Varuna as 900^{+129}_{-145} km and found it to have a red geometric albedo of $0.070^{+0.030}_{-0.017}$.

6.4.6 (50000) Quaoar

Classical Kuiper belt object (50000) Quaoar was discovered by Chad Trujillo and Michael Brown in 2002. At the time, Quaoar was originally thought to be the largest body discovered in the Solar System since Pluto in 1930; however, subsequent estimates of its diameter find that it is actually smaller than Charon (which was discovered in 1977). The minor planet number of 50000 was assigned to this object to honor the discovery of such a large Kuiper belt object. Quaoar is named after the creation god of the Tongva Native Americans who lived in Southern California. Quaoar has one identified moon, Weywot, which was discovered in 2007. Weywot is the son of Quaoar. Quaoar has reflectance spectra with absorption bands that are consistent with a surface mixture of crystalline water ice, methane, and ethane with the possible presence of ammonia hydrate (Jewitt and Luu, 2004; Barucci et al., 2015).

6.4.7 (90377) Sedna and 2012 VP$_{113}$

Trans-Neptunian object (90377) Sedna was discovered by Michael Brown, Chad Trujillo, and David Rabinowitz in 2003 (Brown et al., 2004). The object was originally nicknamed the "Flying Dutchman" (or "Dutch") (Brown, 2010) and was given the provisional designation (2003 VB$_{12}$). Sedna has a semi-major axis of ~507 AU, a perihelion of ~76 AU, and an aphelion of ~938 AU. It has an orbital period of ~11 400 years. At the time of discovery, it was ~100 AU from the Sun, making it the most distant object discovered at the time. Brown et al. (2004) believe that Sedna is a member of the inner Oort cloud. From Spitzer observations, Sedna has a diameter less than 1600 km and a visual geometric albedo greater than 0.16 (Stansberry et al., 2008).

Sedna was discovered using automated software that looked for movement of objects in the images (Brown, 2010). The software would identify ~100 potential objects each night that could be minor planets, which would then be examined by eye to look for actual objects.

Sedna was named after the Inuit goddess of the sea who is thought to live at the bottom of the Arctic Ocean. This name was chosen to honor Sedna as being the coldest and most distance place known in the Solar System at the time.

A second body (2012 VP_{113}) was discovered with a Sedna-like orbit (Trujillo and Sheppard, 2014). It has a semi-major axis of ~262 AU, a perihelion of ~80 AU, and an aphelion of ~444 AU. This body has an orbital period of ~4268 years. Assuming an albedo of 0.15, 2012 VP_{113} would have a diameter of ~450 km (Trujillo and Sheppard, 2014). These two bodies are called Sednoids. Sednoids are defined as scattered disk objects with semi-major axes greater than 150 AU and a perihelion greater than 50 AU.

Sedna has one of the highest absolute magnitudes (+1.5) of Kuiper belt objects measured to date, implying a very high albedo (Trujillo et al., 2005). Reflectance spectra of Sedna (Barucci et al., 2005, 2010) are extremely red between ~0.4 and ~1.8 μm.

6.4.8 (90482) Orcus

Another plutino is (90482) Orcus. Orcus was discovered by Michael Brown, Chad Trujillo, and David Rabinowitz in 2004. Orcus is named after the Etruscan (ancient Italian civilization) god of the underworld. The absolute magnitude of Orcus of +2.4 is one of the highest of any known trans-Neptunian object.

Orcus has a moon named Vanth. Vanth is named after the winged Etruscan god who guides the souls of the dead to the underworld. The name Vanth was chosen after a naming contest run by its discover Michael Brown on his blog. The absolute magnitude of Vanth is +4.9 and is relatively high compared to most of other Kuiper belt moons. The presence of a moon allowed the density of Orcus to be estimated and was found to be approximately 1500 ± 300 kg/m³ (Brown et al., 2010). However because the mass of Vanth is not negligible compared to Orcus, the calculated density is a function of a number of assumptions (e.g., albedo ratio of the two bodies, densities of the two bodies) that are not important when the moon is much smaller than the primary.

6.4.9 (136108) Haumea

There was considerable controversy concerning who actually discovered dwarf planet (136108) Haumea (Brown, 2010). Haumea was discovered in December 2004 from images taken in May 2004 by Michael Brown, Chad Trujillo, and David Rabinowitz at Palomar Observatory. The object was nicknamed "Santa" by the team since it was discovered right after Christmas. The team were not in a hurry to announce their discovery since a moon (nicknamed "Rudolph") around "Santa" discovered in January 2005 allowed the mass of the primary to be calculated and it was found to be ~1/3 the mass of Pluto, which was similar in size to previously discovered trans-Neptunian objects. A second moon (nicknamed "Blitzen") around "Santa" was discovered in June 2005. For a number of abstracts on the body for the Division of Planetary Sciences (DPS) meeting of the American Astronomical Society (AAS) in September 2005, the team used the code "K40506A" for the object (e.g., Rabinowitz et al., 2005). "K" stood for Kuiper belt, "40506" for the date of discovery on May 6, 2004, and "A" for the first object discovered that day (Brown, 2010).

However, before the Brown and his team could announce their discovery, José Luis Ortiz Moreno had emailed the Minor Planet Center on July 27, 2005 that their team had discovered this object from three images taken in March 2003. Haumea was given the provisional designation 2003 EL$_{61}$. After announcing the discovery, the next day the object was identified in precovery digitized images taken in 1955 at Palomar Observatory and a new observation of the night sky in on July 28, 2005. What was not known to Brown and his team was that searching K40506A on the Internet allowed anyone to see that this object had been observed using a 1.3-meter telescope (Small and Moderate Aperture Research Telescope System or SMARTS) in Chile in May 2005. The software the telescope's camera used kept track of what objects were being observed and where the telescope was pointing. Anyone could see where the telescope was pointing on successive nights, which allowed anybody to determine the position of this body in the sky. Interestingly, a computer at the institution of the Ortiz team had accessed these logs on July 26, 2005. This access to these Internet logs was never mentioned by the Ortiz team until it was later noticed by Brown's team.

Both Ortiz and Brown submitted names to the CSBN of the IAU. Ortiz proposed Ataecina, the Iberian goddess of the underworld, even though the body was not a plutino. Brown proposed Haumea. Haumea is the Hawaiian goddess of fertility and childbirth and fits the creation god theme for trans-Neptunian objects. Both of Haumea's moons were discovered at the Keck Telescope in Hawaii (Brown et al., 2006) and the name "Haumea" also commemorates the discovery of Haumea's moons at the observatory in Hawaii. Brown also proposed that Haumea's moons be named Hi'iaka and Namaka. They are officially known as (136108) Haumea I Hi'iaka and (136108) Haumea II Namaka. Hi'iaka and Namaka are both daughters of Haumea. Hi'iaka is the goddess of the Big Island of Hawaii, while Namaka is the goddess of the sea. Hi'iaka (~320 km) is larger than Namaka (~160 km) (Ragozzine and Brown, 2009).

Brown's names were accepted, which validated the claim that Brown's team "really" discovered Haumea. Due to all this controversy, the citation for the name only lists the place of discovery (Sierra Nevada Observatory in Spain) and not the discoverer.

The discovery of Haumea's moons allowed the mass (Ragozzine and Brown, 2009) and density of Haumea to be estimated. Its calculated density is ~2600 kg/m^3 (Probst et al., 2015). Haumea has the fastest rotation rate (3.9 hours) of any large (diameter greater than 100 km) body in our Solar System (Rabinowitz et al., 2006). Haumea also has relatively large light curve amplitude of 0.28 ± 0.04 magnitudes (Rabinowitz et al., 2006), implying a very elliptical body (Lellouch et al., 2010; Lockwood et al., 2014). Its very elliptical shape (1920 × 1540 × 990 km^3) (Lockwood et al., 2014) appears to be due to its extremely fast rotation. Haumea is consided a dwarf planet since it is in hydrostatic equilibrium and its elliptical shape is due to its fast spinning. Haumea is also in a 7:12 mean-motion resonance with Neptune (Brown et al., 2007).

Reflectance spectra of Haumea are relatively featureless in the visible (~0.5–1.0 μm) (Tegler et al., 2007); however in the near-infrared (~1.0–2.4 μm), Haumea has strong absorption features characteristic of water ice at ~1.5 and ~2 μm (Figure 6.14) (Trujillo et al., 2007; Barkume et al., 2008). A reflectance spectrum of Hi'iaka was also found to have absorption features at ~1.5 and ~2 μm (Barkume et al., 2006), implying that the moon is an impact fragment of Haumea.

Haumea is part of the only identified asteroid family among trans-Neptunian objects (Brown et al., 2007). Members of the asteroid family all have reflectance spectra with a strong ~2 μm band due to water ice and similar proper elements. With water ice on the surface (density of ~1000 kg/m^3), Haumea must have a large silicate core (Probst et al., 2015). This family is argued by Brown et al. (2007) to be due to a giant impact that stripped away most of Haumea's icy mantle, created its moons, and caused Haumea to have a rapid rotation.

6.4.10 (136199) Eris

The most famous scattered disk object is arguably dwarf planet (136199) Eris. This body was discovered in January 2005 from images taken in 2003 by Michael Brown, Chad Trujillo, and David Rabinowitz and was later given the provisional designation 2003 UB_{313}. The team nicknamed this object "Xena" after the character

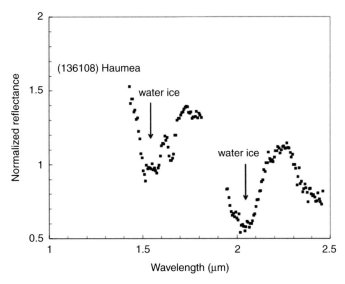

Figure 6.14 Reflectance spectrum of (136108) Haumea from Barkume et al. (2008). Note the strong water ice absorptions at ~1.5 and ~2.0 μm.

in the televisions show *Xena: Warrior Princess* and also used the code "K31021C" for the object.

An orbit was computed for "K31021C" and it was found to be ~120 AU from the Sun when it was discovered (Brown, 2010). "K31021C" had a visual magnitude of +18.8. It was quickly realized that if Pluto was at ~120 AU, Pluto would have a fainter magnitude of +19.7. Since this object would be brighter than Pluto at the same distance from the Sun, this body may be larger than Pluto and could potentially be considered a planet like Pluto. Later using an occultation, Sicardy et al. (2011) estimated the diameter of Eris to be 2326 ± 12 km, which was consistent with Pluto's estimated diameter at the time.

The announcement of the discovery of this object was hurried because of the Haumea controversy. Since "K31021C" was also observed using the SMARTS telescope, its name would be searchable on the Internet if you knew what you were looking for. Its position in the sky could then be determined. To make sure this discovery was not "stolen," the discovery of "K31021C" was announced on July 29, 2005.

Eventually, Eris was named after the Greek goddess of chaos, strife, and discord. The name was chosen partially to reflect the arguments by scientists about whether Eris and Pluto should be called planets.

Eris has one known moon, Dysnomia, which was discovered in September 2005. The moon was nicknamed "Gabrielle" by the team since "Gabrielle" was the sidekick of Xena. Michael Brown and the engineering team of the W. M. Keck Observatory discovered the moon. This moon is named after Eris' daughter.

Since it is a satellite, it is officially known as (136199) Eris I Dysnomia. The discovery of this moon allowed the mass of Eris to be determined very precisely, which also allowed its density to be calculated. Eris has a calculated density of 2520 ± 50 kg/m^3 (Sicardy et al., 2011). It has a very high visible geometric albedo of $0.96^{+0.09}_{-0.04}$ (Sicardy et al., 2011), which is argued to be possibly due to the collapse of a nitrogen atmosphere.

6.4.11 (136472) Makemake

Dwarf planet (136472) Makemake was discovered by Michael Brown, Chad Trujillo, and David Rabinowitz in 2005. The body was nicknamed "Easterbunny" because it was discovered shortly after Easter and also used the code "K50331A" for the object. Makemake is officially named after the Easter Island god of fertility for the same reason. The provisional designation for this body was 2005 FY$_9$. The discoverer of this body was also announced on July 29, 2005 since "K50331A" was also observed using the SMARTS telescope. This name was also searchable on the Internet if you knew what you were looking for and this could permit its position in the sky to be determined

A stellar occultation of Makemake in 2011 allowed many physical parameters to be calculated (Ortiz et al., 2012; Brown, 2013). It was found to have an equatorial diameter of 1434 ± 14 km and a polar diameter 1422 ± 14 km (Brown, 2013). Makemake also had an extremely high visual geometric albedo of 0.81. Brown (2013) argued that Makemake's density could not be accurately determined from just measurements of the size, shape, and albedo of its surface. Makemake has one known moon [S/2015 (136472) 1] (Parker et al., 2016).

Spectra of Makemake have an extremely red spectral slope with absorptions due to methane ice (Licandro et al., 2006; Lorenzi et al., 2015). The spectra of Makemake were found to be very similar to Pluto (Licandro et al., 2006). However, other ices identified in the spectra of Pluto such as CO and N$_2$ could not be identified in the spectra. The ground-based spectral resolution is too low to identify CO absorptions, and the 2.15 μm absorption due to N$_2$ is too weak to be identified with the very strong absorption at 2.2 μm nearby. Rotational visible spectra showed no variation in the CH$_4$/N$_2$ mixing ratio with rotation due to the absence of any change in the depths and wavelength positions of the observed bands (Lorenzi et al., 2015).

6.5 Centaurs

Centaurs have semi-major axes between Jupiter's (5.2 AU) and Neptune's (30.1 AU). A list of some notable centaurs is given in Table 6.5.

Table 6.5 *A list of notable centaurs*

Name	Semi-major axis (AU)	Diameter (km)	Visual geometric albedo
(944) Hidalgo	5.75	30–40	–
(2060) Chiron	13.64	218 ± 20	0.16 ± 0.03
(5145) Pholus	20.37	140 ± 40	$0.08^{+0.07}_{-0.03}$
(10199) Chariklo	15.79	258.6 ± 10.3	$0.0573^{+0.0049}_{-0.0042}$
(60558) Echeclus	10.7	$83.6^{+15.0}_{-15.2}$	$0.0383^{+0.0189}_{-0.0108}$

The semi-major axes are from the Minor Planet Center database. The diameter of Hidalgo is from Davies (1984). The diameter and albedo of Chiron is from Fornasier et al. (2013). The diameter and albedo of Pholus, Chariklo, and Echeclus are from Stansberry et al. (2008).

Centaurs may be transitional bodies between objects in the outer Solar System (TNOs and SDOs) and those in the inner Solar System (JFCs and NEAs). Centaurs with perihelia at 10 AU or smaller tend to show activity, while centaurs that do not tend to show activity have perihelia closer to Neptune's orbit (Jewitt, 2015). Napier et al. (2015) argue that centaurs can be deflected toward Earth. Such objects may fragment due to thermal stresses or the passage through the Roche limit of Jupiter or the Sun. Stellar occultations of (2060) Chiron and (10199) Chariklo imply the presence of rings around these bodies (Braga-Ribas et al., 2014; Duffard et al., 2014; Ortiz et al., 2015; Ruprecht et al., 2015). These stellar occultations were observed by a number of telescopes on the surface of the Earth, which allows the dimming of the star to be measured for the centaur and any material in orbit around it. Hedman (2015) argues that the dense ring systems for bodies between 8 and 20 AU (Saturn, Uranus, Chiron, Chariklo) may be due to the water ice in the rings being at temperatures at ~70 K where the ice may be extremely weak. El Moutamid et al. (2014) argue that these rings could have formed from a moon-forming collision or the disruption of a small moon that has migrated to within the Roche limit.

Centaurs are also divided into "red" and "blue-gray." As with TNOs, "red" centaurs have reddish spectral slopes in the visible, while "blue-gray" centaurs have neutral colors. Extremely red (ultrared) colors are common among non-active centaurs and Kuiper belt objects but not among populations of bodies (e.g., Trojan asteroid, Jupiter-family comets, long-period comets, active centaurs) found closer to the Sun (Jewitt, 2015). Jewitt (2015) argues that cometary activity causes the extremely red surface material to be covered by an optically thick layer of material that was excavated from beneath the surface during the cometary activity.

Centaur (5145) Pholus is one of the reddest observed objects in the Solar System. Pholus' spectrum is best matched by a mixture of a fine-grained tholin, olivine, H_2O and CH_3OH ices, and carbon (Cruikshank et al., 1998). Cruikshank et al. (1998) argue that Pholus is similar in properties of a large comet nucleus that has never been activated.

6.6 Damocloids

Damocloids have Halley-family or long-period comet orbits but do not show any outgassing (Jewitt, 2005). They also tend to have high eccentricities and inclinations. Damocloids have Tisserand parameters with respect to Jupiter of 2 or less. They have spectral slopes in the visible similar to the nuclei of Jupiter-family comets.

6.7 Extrasolar Minor Planets

Evidence for extrasolar minor planets have been found in astronomical observations of white dwarf stars (Jura and Young, 2014). White dwarf stars are primarily composed of carbon and oxygen with thin atmospheres of hydrogen and/or helium. Initially, the atmosphere is very hot (above ~100 000 K). When it cools (below ~20 000 K), the atmosphere is unable to support elements heavier than helium (called metals) by the upward flow of radiation and these metals sink into the interior. So many "cooler" white dwarfs have absorption spectra with only hydrogen (and sometimes helium) lines. However, 25–33% of all "cool" white dwarfs have spectral lines consistent with metals. These metals are consistent with the white dwarfs accreting orbiting minor planets (Jura, 2008, 2014).

Questions

1) Why do comets get much brighter when they approach Earth than would be expected from the inverse square law?
2) Why is comet Halley named after Edmond Halley?
3) If comet Halley has a period of 75.3 years, what is its semi-major axis?
4) What are main-belt comets? What do they tell us about some asteroids in the main asteroid belt?
5) Why are trans-Neptunian objects difficult to discover?
6) Do you think Pluto should still be considered a planet? Explain your answer.
7) Do you think Eris should be considered a planet? Explain your answer.
8) What was the controversy concerning the discovery of Haumea?
9) What are the differences between centaurs and trans-Neptunian objects?
10) How are extrasolar minor planets detected?

7

Near-Earth Asteroids and the Impact Threat

7.1 Near-Earth Asteroids

The first near-Earth asteroid discovered was (433) Eros in 1898 by Gustav Witt (1866–1946) in Berlin. Auguste Charlois (1864–1910) also observed Eros on the same night in Nice. Witt discovered Eros on a single long-exposure photographic plate where Eros appeared as a trailed image (Jedicke et al., 2015). The largest near-Earth asteroid is (1036) Ganymed, which has a diameter of ~30 km. Ganymed was discovered in 1924 by Walter Baade (1893–1960). Over 15 000 NEAs are currently known, and all known taxonomic types are represented among the NEAs (Binzel et al., 2015).

NEAs are broken up into Atens, Apollos, Amors, and Atiras. The Aten and Apollo groups are named after the first discovered member of the group. Atens, named after (2062) Aten, have semi-major axes less than 1 AU and an aphelion greater than 0.983 AU (the perihelion of Earth's orbit). Apollos, named after (1862) Apollo, have semi-major axes greater than Earth's and perihelion distances less than the Earth's aphelion distances. The Amors, named after (1221) Amor, have a perihelion distance between 1.017 and 1.3 AU. Amors can cross Mars' orbit but not the Earth's orbit.

Atiras, named after (163693) Atira, have orbits that are entirely inside the Earth's orbit. Discovered in 2003 by LINEAR, Atira was the first asteroid with this type of orbit and was given a name consistent with Aten, Apollo, and Amor, which all have names beginning with "A" and all are named after gods. Atira is a Pawnee goddess of the Earth and the asteroid was named to honor Native Americans. The Atira group has also been called the Apohele group. Apohele is the Hawaiian word for orbit and was also chosen because it was similar to the word aphelion and helios. Approximately 20 Atiras have been discovered to date.

Vatiras are a proposed population of bodies (Greenstreet et al., 2012) that would have orbits that are completely interior to Venus' orbit. These objects would have aphelia between 0.307 and 0.718 AU. Vulcanoids, which would have orbits that are completely interior to Mercury's orbit, are another proposed population. These objects would have aphelia less than 0.307 AU. No members of the Vatira or Vulcanoid population have been discovered to date. Granvik et al. (2016) has found that there is a deficit of asteroids, especially low-albedo ones, with perihelia close to the Sun due presumably to thermally driven disruption.

7.1.1 Near-Earth Asteroid Surveys

Near-Earth asteroids are important to discover because they could potentially collide with the Earth. However in the 1990s, David Morrison stated that fewer people are searching for near-Earth asteroids than work a shift at a McDonalds (Chandler, 1998). Now Morrison states that the number is up to three McDonalds (Dean, 2015).

The first near-Earth asteroid survey was the Palomar Planet-Crossing Asteroid Survey (PCAS), which ran from 1973 to 1995 (Helin et al., 1997). Eugene Shoemaker and Eleanor Helin started this survey using the 18-inch Palomar Schmidt Telescope and they were joined by Carolyn Shoemaker in the early 1980s (Jedicke et al., 2015). NEOs were discovered manually using blink comparators and stereomicroscopes.

Spacewatch is a discovery survey using two telescopes (0.9 and 1.8 m) located at Kitt Peak, Arizona which was founded in 1980 by Tom Gehrels (1925–2011) and Robert McMillan. In 1984, Spacewatch was the first asteroid survey to use a CCD (Jedicke et al., 2015).

From 1993 to 2008, the Lowell Observatory Near-Earth Object Search (LONEOS) searched for asteroids under the direction of Ted Bowell. LONEOS used a 0.6-meter Schmidt telescope. LONEOS discovered 288 NEOs (Committee to Review Near-Earth Object Surveys and Hazard Mitigation Strategies et al., 2010).

The Near-Earth Asteroid Tracking (NEAT) was a discovery survey jointly run by NASA, the Jet Propulsion Laboratory (JPL), and the U.S. Air Force from 1995 to 2007. NEAT had a cooperative agreement with the Air Force to use a one-meter GEODSS (Ground-based Electro-Optical Deep Space Surveillance) telescope on Haleakalā in Maui, Hawaii. NEAT designed a CCD and computer system for the telescope to discover asteroids. In 2000, NEAT moved from using the GEODSS telescope to a 1.2-meter AMOS (Air Force Maui Optical and Supercomputing) telescope located nearby. In 2001, a 1.2-meter Schmidt telescope at Palomar was also used to discover asteroids. NEAT discovered over approximately 20 000 objects with ~430 being NEOs (Committee to Review Near-Earth Object Surveys and Hazard Mitigation Strategies et al., 2010).

In 1998, Congress mandated that NASA discover in ten years 90% of all NEOs with diameters of 1 km or greater, which is often called the Spaceguard survey. NEOs with diameters of 1 km or greater were initially assumed to have H magnitudes of 18 or less. The term "Spaceguard" is the name of the asteroid detection survey in *Rendezvous with Rama* by Arthur C. Clarke (1917–2008) that was first published in 1973. The Spaceguard survey refers to the telescopes that are funded by NASA to archive this 90% goal. After 10 years, Harris (2008) estimated that 79% of NEOs with diameters greater than 1 km had been discovered. In 2011, NASA achieved its goal. Harris and D'Abramo (2015) estimate that there are 990 ± 20 NEAs larger than 1 km in diameter (H ≤ 17.75). This H magnitude corresponds to a visual geometric albedo of 0.14.

One of the most prolific discovery surveys is the Lincoln Near-Earth Asteroid Research (LINEAR) survey (Evans et al., 2003), which began operating in 1998. LINEAR is a joint project between the United States Air Force, NASA, and the Massachusetts Institute of Technology (MIT) to discover and track near-Earth asteroids. LINEAR is the most prolific discovery survey to date and has discovered over 230 000 asteroids and ~300 comets. As of today, LINEAR had discovered more asteroids than any other asteroid survey.

Also starting in 1998, the Catalina Sky Survey uses two telescopes (one on Mount Lemmon and one near Mount Bigelow) near Tucson, Arizona to discover potentially hazardous asteroids. A third telescope at the Siding Spring Observatory in Australia was in operation between 2004 and 2013. The Catalina Sky Survey has discovered over 5000 NEOs (Johnson et al., 2014). This survey has discovered approximately half of all known NEOS.

In 2005, Congress directed NASA through the George E. Brown, Jr. Near-Earth Object Survey Act to discover, catalog, and characterize 90% of potentially hazardous asteroids (PHAs) larger than 140 meters in 15 years. The sponsor of the bill was California congressman Dana Rohrabacher. The bill honored congressman George E. Brown Jr. (1920–1999), who was a staunch supporter of science.

The Panoramic Survey Telescope and Rapid Response System (Pan-STARRS) is a group of four proposed telescopes to discover NEOs. The first telescope PS1 (Pan-STARRS1) is located on the inactive shield volcano Haleakalā on the island of Maui in Hawaii and began full-time science observations in 2010. The Pan-STARRS CCD cameras are the largest digital cameras ever built. The cameras have about 1.4 billion pixels spread out over an area of 40 cm square. Each camera has a 64 × 64 array of CCDs.

7.1.2 Odds of Dying in an Asteroid Impact

Before all these extensive near-Earth asteroid discovery surveys, the odds of a person dying through the impact of an NEO was estimated to be approximately 1 in

20 000 (Chapman and Morrison, 1994). These odds were similar to the odds of a person in the United States dying in a plane crash.

However, results from discovery surveys have greatly decreased the odds (Chapman, 2008; Harris, 2008). Boslough and Harris (2008) estimate that the odds are much less (1 in 720 000 for individuals with a life expectancy of 80 years) because the number of objects with diameters between tens to hundreds of meters across appears to be less than previously thought. These odds are similar to the odds of dying in a fireworks accident (1 in 600 000) (Harris, 2008).

The problem with comparing these odds is that deaths due to plane crashes and fireworks accidents occur every year, while confirmed deaths due to asteroid impacts have not been conclusively proven in recorded history. However, people have been confirmed to have been struck by meteorites. For example, the Sylacauga meteorite (~6 kg) struck a woman [Ann Hodges (1920–1972)] in Alabama in 1954. Also a small fragment (3.6 g) of the Mbale meteorite did strike a young boy's head in Uganda in 1992 (Jenniskens et al., 1994).

7.2 Cratering

Any object that impacts the Earth or the atmosphere is considered an impactor (Brown et al., 2013). Craters are usually found as circular depressions with elevated rims (Ivanov and Hartmann, 2009). All bodies with solid surfaces in the Solar System contain craters. Even comets appear to have craters.

The energy deposited by an impacting object is

$$E = \frac{1}{2} m_2 v^2 \tag{7.1}$$

where E is the kinetic energy in joules, m_2 is the mass in kilograms of the impactor, and v is the velocity of the impactor in m/s. Energy for an impact is often expressed in kilotons (kT) or megatons (MT) of TNT. One kiloton of TNT is equal to 4.184×10^{12} joules while one megaton of TNT is equal to 4.184×10^{15} joules. TNT is the abbreviation for trinitrotoluene, which is an organic explosive material.

Since the mass of a potential impacting body will probably not be known, it can be estimated by assuming a density from its meteoritic analog from its taxonomic type and a volume for a circular body. The radius for calculating the volume can be determined by knowing (or assuming) an albedo and an absolute magnitude (*Equation 1.30*). The energy will then be

$$E = \frac{2}{3} \pi R^3 \rho v^2 \tag{7.2}$$

where the radius (R) is in meters and the density (ρ) will be in kg/m^3. Therefore for the same size object, an iron asteroid will have more energy than a chondritic or achondritic asteroid due to its higher density.

Example 7.1

What will be the kinetic energy of an S-complex asteroid with an absolute magnitude of +17 and an impact velocity of 17 km/s?

If you assume a visual geometric albedo of 0.22 (Table 5.1), the diameter can be estimated from *Equation 1.30*, resulting in

$$D = \frac{1329}{\sqrt{0.22}} 10^{-0.2(17)} = 1.1 \text{ km} \tag{7.3}$$

The radius of the object will then be 550 meters. Assuming a density of 2700 kg/m^3 (Section 5.5), *Equation 7.2* for the energy becomes

$$E = \frac{2}{3} \pi (550)^3 (2700)(17000)^2 \text{ J} = 2.7 \times 10^{20} \text{ J}. \tag{7.4}$$

The energy of the impactor will be 2.7 × 10^{20} joules (6.5 × 10^4 MT).

The minimum velocity at which an object will impact the Earth is 11.2 km/s (11 200 m/s) (French, 1998), which is the escape velocity of an object from the Earth. Typical impact velocities are 17–20 km/s. The maximum impact velocity at which an object can hit Earth is 72 km/s.

There are many lines of evidence that can be used to confirm the formation of a crater by impact. For example, the presence of coesite and stishovite, polymorphs of SiO$_2$ (quartz), which can only be formed at high pressures such as those that occur during impacts.

The structure of the initial crater that is formed is independent of its diameter, the impact velocity, impact angle (unless at very oblique angles), gravitational acceleration of the target body, and intrinsic properties of the target and impactor (Melosh and Ivanov, 1999). However, the structure of the final crater is a function of the properties of the target body such as the gravitational acceleration, density, and composition.

There are three stages of crater formation (Melosh, 1989; French, 1998). They are the contact/compression stage, the excavation stage, and the modification stage. During the contact and compression stage, the impactor strikes the Earth. There is an extreme increase in pressure (several megabars) and temperature (10 000 K

Figure 7.1 Image of the Zumba crater on Mars taken by the Mars Reconnaissance Orbiter's HiRISE camera. Zumba is a simple crater. The Zumba crater is ~3.3 km in diameter. Credit: NASA.

and higher). The impactor and a volume of the target rock that is comparable to the volume of the impactor is also vaporized. During the excavation stage, shock waves propagate through the target. The shock waves can also be reflected and refracted. The transient crater is formed during the excavation stage. During the modification stage, there is elastic rebound of the crater floor and the crater rim will collapse due to gravity.

There are a few different types of craters. Simple craters are relatively small with a smooth bowl shape (Figure 7.1). Complex craters are larger with a central peak (Figure 7.2). The size transition between simple and complex craters depends on the acceleration of gravity of the impacted parent body and the properties of the impacted material. Basins are extremely large craters with diameters greater than ~300 km. Megabasins have diameters greater than ~1000 km.

A theoretical study done by Rumpf et al. (2016) finds that there is a uniform impact probability across the Earth's surface. However, only ~130 craters have currently been identified on the Earth's surface (Hergarten and Kenkmann, 2015). The identified inventory of craters that are ~6 km in diameter or larger appears relatively complete, and there is no evidence for "missing" craters for this size range. The lack of craters on the Earth's surface compared to other planetary bodies is due

Figure 7.2 Image of the Tycho crater on the Moon taken by the Lunar Reconnaissance Orbiter's Wide Angle Camera. Tycho is a complex crater. The Tycho crater is ~86 km in diameter. This image is a mosaic of a number of images. Credit: NASA.

to erosion due to water and wind and plate tectonics. All of these processes will erase craters on the Earth's surface.

The largest known impact crater on Earth is the Vredefort crater in South Africa, which has a diameter of ~300 km across. The second-largest known crater on Earth is the Sudbury Basin in Canada, with a diameter of ~250 km across. In contrast, the largest crater on the Moon (South Pole–Aitken Basin) is ~2500 km in diameter, while the largest crater on Mercury (Caloris Basin) is ~1550 km in diameter.

One of the most studied craters is Meteor Crater (also known as Barringer crater), a ~1.2 km crater located in Northern Arizona (Figure 7.3). Originally, it was known as the Canyon Diablo crater due to its proximity to the town of Canyon Diablo. The crater was thought to be due to volcanic activity due to presence of a large number of relatively extinct volcanoes in the region (Kring, 2007). The crater is named after Daniel Barringer (1860–1929), who first proposed that the crater was due to an impact. As part of his 1960 PhD thesis, Eugene Shoemaker did a study of Meteor Crater and noted the similarity of geological features of the crater to craters formed by nuclear explosions. The formation of the crater by impact was definitively confirmed by the presence of high-pressure minerals coesite and

Figure 7.3 Image of Meteor Crater in Northern Arizona taken by the NASA Earth Observatory. Meteor Crater is ~1.2 km in diameter. Layers of exposed limestone and sandstone are visible just beneath the crater rim. Credit: NASA. (A black and white version of this figure will appear in some formats. For the color version, please refer to the plate section.)

stishovite (Chao et al., 1960, 1962). The impact occurred ~50 000 years ago. The impactor is known to be an iron asteroid due to the presence of iron–nickel meteorite fragments around the crater. These fragments are known as the Canyon Diablo meteorite, which is a IAB coarse octahedrite.

7.2.1 Major Impact Events

The extinction of the dinosaurs at the K–Pg boundary, which occurred ~66 million years ago (Husson et al., 2011), is generally attributed to the impact of an asteroid. [This boundary was originally called the Cretaceous–Tertiary (K–T) boundary.] Luis (1911–1988) and Walter Alvarez (Alvarez et al., 1980) found that the concentration of iridium at measured K–Pg boundary sites was 20–160 times above the background level. Iridium is a siderophile (iron-loving) element and is relatively rare on the Earth's surface since most of the Earth's iridium traveled to the Earth's core with the metallic iron when the Earth differentiated. However, iridium is relatively concentrated in meteorites compared to the Earth's crust since iridium is a siderophile element. The impactor has been postulated to be similar to carbonaceous chondrites due to finding an extremely altered ~65 Ma old fossil meteorite similar to carbonaceous chondrites (Kyte, 1998) and chromium isotopic compositions of K–Pg sediments are similar to CM2 chondrites (Trinquier et al., 2006).

The ~180-km diameter Chicxulub crater is located underneath the Yucatan Peninsula. Global firestorms would have been caused by this impact that could have caused the mass extinction (e.g., Robertson et al., 2013). Swisher et al. (1992) determined a Ar–Ar age of glassy melt rock from beneath a massive impact breccia yielded a mean plateau age of 64.98 ± 0.05 Ma. This age was similar to Ar–Ar ages for tektite glass found in Haiti and Mexico (Swisher et al., 1992). Tektites are glass that formed during impacts on Earth (Koeberl, 1986). They have a variety of colors (e.g., grey, brown, green). Renne et al. (2013) determined Ar–Ar ages for Haitian tektites and altered volcanic ashes (bentonites) in Montana that were a few centimeters above the largest iridium anomaly. Renne et al. (2013) found that the Ar–Ar age of the impact and the Ar–Ar age of the associated mass extinctions differed by ~32 000 years.

The Nördlinger Ries crater (usually known as the Ries crater) is a 24 km crater in Southern Germany. The age of the Ries impact event is estimated from ~14.6 ± 0.2 Ma from calculated ages from a variety of calculated Ar–Ar ages (Buchner et al., 2010). The Ries crater is believed to be the originating crater for the moldavites.

The Tunguska event was the explosion of an object on June 30, 1908 that occurred near the Podkamennaya Tunguska River in the region of Siberia, Russia. The equivalent energy of the Tunguska event was ~5–15 MT of TNT (Borovička et al., 2013). (Equivalent energy is the energy released by an explosion.)

A bolide exploded over Indonesia on October 8, 2009. The best estimate of the equivalent energy was ~50 kT of TNT. This object was detected using infrasound, which is a low frequency sound (< 20 Hz) below the typical limit of human hearing. The International Monitoring System (IMS) network that is operated by the Comprehensive Nuclear-Test-Ban Treaty Organization (CTBTO) detected this explosion in the atmosphere.

7.2.2 Crater Counting

Crater counting allows the age of a planetary surface to be determined (e.g., Hartmann and Neukum, 2001). The more craters a surface has, the older the surface. Small bodies impact the body more often than larger bodies. Cratering on a surface is assumed to be a stochastic (random) process. Crater counting is usually done by an experienced human crater counter who determines the number of crater for particular size ranges using some type of software. However, since there is a human factor in the crater counting, the number of craters will vary among different researchers. Robbins et al. (2014) have found that the level of agreement

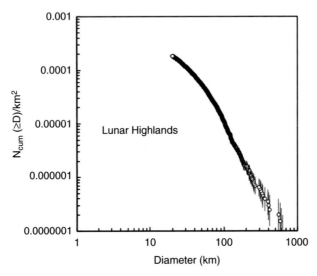

Figure 7.4 Plot of cumulative crater counts versus diameter for the lunar highlands (Head et al., 2010). The slope index is −1.9. Credit: Caleb Fassett, NASA Marshall Space Flight Center.

among "expert" crater counters depends on crater diameter, number of craters per diameter bin, and terrain type with differences that varied from ±10% to ±35% (Robbins et al., 2014). Robbins et al. (2014) also found that the average crater counts of a large number of volunteers were similar to the averages for a smaller number of "experts."

Crater counting results are usually plotted as a crater size-frequency distribution (SFD) (e.g., Giacomini et al., 2015). An SFD plots the frequency of craters of a specified size per unit area. SFD plots have a y-axis that displays the number of craters and an x-axis that displays the crater sizes. Both quantities are usually plotted on a logarithmic scale.

The cumulative SFD (Figure 7.4) is the number N_{cum} of craters per unit area with diameters greater than or equal to a given diameter ($\geq D$) (Melosh, 2011). The diameter is usually given in km. The curves are usually fit with a power law with the form

$$N_{cum}(\geq D) \propto D^b. \tag{7.5}$$

where b is the power law exponent (known as the slope index, the population index, or the cumulative population index) (Neukum et al., 2001). For a power law distribution, the number of objects at a particular diameter varies as the diameter raised to a power (Colwell, 1993). The power law exponent will be negative since there is always more smaller craters than larger ones on a surface. A more negative number indicates a steeper slope. Many different types of objects (e.g., near-Earth asteroids, main-belt asteroids, boulders on asteroids) have power law distributions. Asteroid distributions are usually plotted versus H magnitude, which is related to the diameter and is more precisely known.

The slope index is often given as the differential slope index. If the derivative is taken of both sides, the differential crater size distribution will be proportional to $D^{b-1}dD$. The differential slope index has a value of $b-1$ and is sometimes given as p (Strom et al., 2015). The differential slope index is used in producing R plots. R plots are often considered a better way of displaying crater counts than cumulative plots since the differences between different slope indexes are more obvious when plotted on a R plot (Strom et al., 2015). The R plot is normalized to a differential crater size distribution with a p of -3.

For the cumulative number of craters larger or equal to a particular diameter, the exact formula is given by

$$N_{cum}(\geq D) = c\left(\frac{D}{km}\right)^{b} \tag{7.6}$$

where c is the number density of craters when the crater diameter is 1 km (Melosh, 2011).

For surface with a slope index (b) of -2, the relationship between different crater sizes is relatively simple (Melosh, 2011). For every crater with diameter D, there will be 4 times as many craters with diameter $D/2$ and 16 times as many as diameter $D/4$. The total crater area in each class size is the same for each size distribution.

When a surface is covered with craters, a new crater destroys other craters so the crater density stays approximately the same (Strom and Neukum, 1988). The surface is said to be saturated or reached equilibrium. For a slope index of -2, the crater density is assumed to have reached equilibrium.

Crater counting can tell you the relative age of a surface with the older surface having more craters if both have experienced the same flux of impacting material. The production of craters on a surface is related to scaling laws and the projectile population. Scaling laws relate a crater's diameter and the diameter of the impactor. The crater size produced by an impactor will be a complicated function of the diameter of the impactor, impacting velocity, impact angle, the target and projectile densities, and gravitational acceleration on the body that is being impacted (Werner, 2005).

To determine absolute ages, crater counting must first be done with surfaces of known ages. The only known planetary body where this can be done is the Moon since the Apollo and Luna missions have returned samples from different areas of the Moon (e.g., Neukum et al., 2001; Hiesinger et al., 2012), which can be dated in the laboratory. The ages are usually correlated with cumulative SFDs for craters that are 1 km in diameter and larger. Crater counts can be determined for these areas, and these ages can be extrapolated for crater counts of surface with unknown ages. The formula (lunar chronology function) for extrapolating lunar crater count ages (e.g., Neukum et al., 2001) for the Moon is

$$N_{cum}(D \geq 1 \text{ km}) = \left[5.44 \times 10^{-14} \left(e^{6.93t/Ga} - 1\right) + \left(8.38 \times 10^{-4}\right)t / Ga\right] \text{ km}^{-2} \quad (7.7)$$

where N_{cum} ($D \geq 1$ km) is the number of craters greater than 1 km in diameter per km^2 and t is the age of the surface in Ga. Ages are estimated by calculating the cumulative number of craters greater than a km in diameter per km^2 and reading the age off the curve (Figure 7.5) or solving *Equation 7.7* for t. For example, an area of the Moon with 0.01 craters greater than 1 km in diameter would have an estimated surface age of ~3.8 Ga. The equation is not valid ages greater than ~4.1 Ga (e.g., Schmedemann et al., 2014).

However the impacting flux on the Moon may not be expected to be directly comparable to the impacting flux on other planetary bodies (e.g., Ivanov, 2001). The asteroid belt is often assumed to have a higher impact rate than the lunar environment. The asteroid chronology functions plot more than a magnitude (over a factor of 10) above the lunar chronology function (Schmedemann et al., 2014). One asteroid chronology function for Vesta used with Dawn spacecraft crater counts is

$$N_{cum}(D \geq 1 \text{ km}) = \left[1.3223 \times 10^{-12} \left(e^{6.93t/Ga} - 1\right) + \left(0.02037\right)t / Ga\right] \text{ km}^{-2}. \quad (7.8)$$

Chronology functions derived for the Rosetta target (21) Lutetia and the Galileo targets (243) Ida and (951) Gaspra have similar equations.

Strom et al. (2005) found that main-belt asteroids have a size distribution similar to craters on the lunar highlands (due to the Late Heavy Bombardment that ended ~3.8 Ga), while near-Earth asteroids have a size distribution similar to the Mars young plains (which are younger than ~3.8 Ga). Minton et al. (2015) found that to match the number of ~100 km diameter craters on the lunar highlands, an impactor distribution similar to a main-belt asteroid distribution would produce too many megabasins that are larger than the Imbrium Basin (~1200 km diameter). Minton et al. (2015) finds that the impacting distribution must be depleted in objects with diameters greater than 70 km than the main-belt asteroid distribution.

O'Brien et al. (2006) argue that the cratering records of asteroids observed by spacecraft missions [Galileo observations of (951) Gaspra and (243) Ida and NEAR Shoemaker observations of (253) Mathilde and (433) Eros] are consistent with a

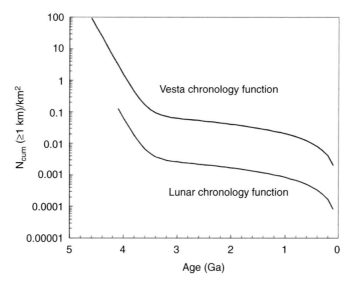

Figure 7.5 Plot of a lunar chronology function (Neukum et al., 2001) and a Vesta chronology function (Schmedemann et al., 2014). Both functions plot cumulative number of craters greater than a km in diameter per km^2 versus surface age in Ga. The lunar chronology greater than 4.1 Ga is not yet well-constrained.

common impacting population. The O'Brien et al. (2006) work is an update of the modeling of Greenberg et al. (1994, 1996) for just Gaspra and Ida. Greenberg et al. (1994, 1996) modeled effects that could erase craters such as global and local seismic jolting, the superposition of new craters on older craters, and seismic shaking over the lifetime of the asteroid. Besides the impacting population, these effects will be a function of the diameter of the asteroid and the time that the asteroid has experienced cratering.

7.3 Discovery of a Potentially Hazardous Asteroid

When a minor planet is detected, observations are reported to the Minor Planet Center. The MPC determines whether or not it is a new discovery. If it is not a new discovery, these observations are used to refine the orbit. If it is a new discovery, a preliminary orbit is then determined. It is then determined by the MPC whether the object is a potentially hazardous asteroid (PHA) (or PHO for potentially hazardous object), which is an object that could potentially strike the Earth in the future. A PHA has an Earth Minimum Orbit Intersection Distance (MOID) of 0.05 AU or less and an absolute magnitude (H) of 22.0 or less. An object with this magnitude and an albedo of 0.13 would have a diameter of ~150 meters. Approximately ~1400 PHAs are known today out of an estimated population of 4700 ± 1450 (Perna et al.,

2016). If not a PHA, routine processing of the observations is done. If it is a PHA, then the NEO Program Office at JPL also determines the probability of a potential impact for the next 100 years. If the impact probability is not 0 (greater than 10^{-10}), follow-up observations are requested from observatories in the world.

When the orbit is initially calculated from just a few observations, the uncertainty in the orbit may allow for a potential impact with the Earth. Further observations will usually allow the impact probability to significantly decrease (often to ~0) as the orbit is refined.

Within the US government, NASA coordinates all NEO detection and threat information (Johnson, 2014). NASA's Planetary Defense Coordination Office (PDCO) has just been established to take a leading role in coordinating interagency and intergovernmental efforts in response to any potential impact threats. For any potential impact, PDCO informs the NASA administrator, who will inform the Executive Office of the President. After acknowledgment by the Executive Office, federal agencies would be notified. For threats that affect the United States territory, the Federal Emergency Management Authority (FEMA) is the lead organization and would notify appropriate federal, state, and local authorities and emergency response agencies. NASA would provide expert input on the threat to FEMA. For threats beyond US territory, the Department of State would notify affected countries through diplomatic channels and to member nations of international organizations such as the UN and NATO. Post-impact, the Department of State would also coordinate relief efforts.

Only two asteroids (2008 TC$_3$ and 2014 AA) have been discovered before they impacted the Earth. The Mount Lemmon Observatory of the Catalina Sky Survey discovered both bodies. Both objects struck the Earth's atmosphere less than a day after their discovery.

The ~4-m diameter asteroid 2008 TC$_3$ exploded in the atmosphere over the Sudan on October 7, 2008 (Jenniskens et al., 2009, 2010). The equivalent energy of the event was ~1 kT of TNT (Chodas et al., 2010). This object was significant for a number of reasons. The object was discovered on October 6, 2008 and predicted to hit the Earth's atmosphere over the Nubian Desert 19 hours later. This is the first object discovered before striking the Earth. Also for the first time, a visible reflectance spectrum was obtained of a body before its impact. The object was identified as an F-class asteroid, an asteroid class first defined by Tholen (1984). Petrus (Peter) Jenniskens and Muawia Shaddad then organized an expedition to recover meteorites from the explosion (Figure 7.6). This was the first time that samples were recovered from a predicted impact. The ~4 kg of recovered meteorites were called Almahata Sitta, which is Arabic for the nearby railway station "Station Six." The meteorite was classified as a polymict ureilite on the basis of oxygen isotopes,

Figure 7.6 Peter Jenniskens, Muawia Shaddad, and several University of Khartoum students, pointing at one of the Almahata Sitta fragments. Credit: NASA.

bulk chemistry and mineralogy (Zolensky et al., 2010; Horstmann and Bischoff, 2014); however, enstatite and ordinary chondrite fragments of Almahata Sitta were also identified.

Asteroid 2014 AA (discovered on January 1, 2014) was the second object to have been predicted to strike the Earth's atmosphere after 2008 TC$_3$. The object struck the Earth's atmosphere approximately 21 hours after its discovery.

A third object (WT1190F) was also discovered at the same observatory on October 3, 2015 before it struck the Earth on November 13, 2015, but is thought to be space "junk" (e.g., rocket booster) (Watson, 2015). The letters ("WTF") in the name was thought to be very fitting. This object had been previously observed twice in 2013 by the Catalina Sky Survey, but then lost. The Catalina Sky Survey automatically assigned this object's designation, but this object was not given a provisional designation by the Minor Planet Center since its orbit was more like an artificial satellite than an asteroid. WT1190F was observed burning up in the atmosphere over the Indian Ocean. The object is thought to have been the translunar injection module of Lunar Prospector that was launched in 1998 (Watson, 2016).

7.3.1 Hazard Scales

The threat of an asteroid impact is primarily a function of the body's orbit and mass. The orbit determines whether the asteroid will hit the Earth, and its mass determines the kinetic energy of the impact. The higher the probability of an impact, the larger the threat; however, a more massive body will create more damage than a less massive one. An impact on land would cause a crater, while an impact in the ocean could possibly cause a tsunami (e.g., Weiss et al., 2006).

Richard Binzel developed a Hazard Scale just based on the impact probability and its kinetic energy. The scale was developed for a United Nations conference in 1995 and revised in 1999 where it was officially called the Torino Impact Hazard Scale (Binzel, 1997, 2000) to commemorate the city where the IAU officially adopted it. The need for such a scale became readily apparent when an IAU Circular released in March 1998 (Marsden, 1998a) mentioned that a near-Earth asteroid (1997 XF_{11}) would pass only 0.00031 AU from the Earth. A press release was then issued on this object by Brian Marsden which stated "The chance of an actual collision is small, but one is not entirely out of the question" (Reichhardt, 1998). This statement caused a huge news story about a potential asteroid impact. Prediscovery observations refined the orbit (Marsden, 1998b) and showed that this body would miss Earth; however, scientists started to talk about better ways of communicating asteroid near-misses with Earth.

The Torino Impact Hazard Scale (or Torino Scale) (Figures 7.7 and 7.8) communicates to the public the impact hazard of asteroids and comets with the Earth using just a number from 0 to 10. The hope was to satisfy the public with a simple-to-understand scale when an object was announced to pass close by but with an extremely low probability of ever striking the Earth. The Torino Scale is only used for potential impacts less than 100 years in the future. Currently, no known asteroid has a Torino Scale value greater than 0.

The Torino Scale plots the collision probability of the object on the x-axis and its kinetic energy (in megatons of TNT) on the y-axis using a logarithmic scale. The Torino Scale is then subdivided into a number of regions with a number between "0" (indicating an object with a negligible chance of hitting the Earth or too small to make it through the Earth's atmosphere intact) to "10" (indicating an object that will certainly strike the Earth and is large enough to cause a global disaster). The scale is color-coded. White (number"0") indicates no hazard; green ("1") indicates a routine discovery that passes near the Earth but there is no unusual danger; yellow ("2," "3," "4") indicates objects meriting more attention from astronomers

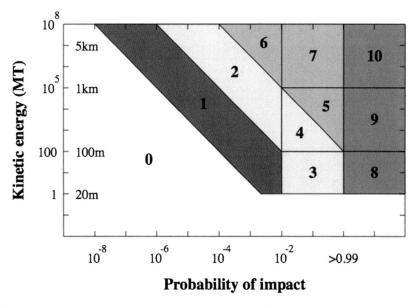

Figure 7.7 Torino Scale values for an impacting object. Kinetic energy in mega-tons is plotted versus impact probability. The estimated diameters in km are also given on the *y*-axis. White (number"0") indicates no hazard, green ("1") indicates a routine discovery, yellow ("2," "3," "4") indicates objects meriting more attention from astronomers, orange ("5," "6," "7") indicates objects that will have a close encounter with the Earth and could possibly cause regional or global devastation, and red ("8," "9," "10") indicates a certain collision. Credit: Richard Binzel, Massachusetts Institute of Technology. (A black and white version of this figure will appear in some formats. For the color version, please refer to the plate section.)

since current observations imply that the object may make a close encounter with the Earth; orange ("5," "6," "7") indicates objects that will have a close encounter with the Earth and could possibly cause regional or global devastation; and red ("8," "9," "10") indicates a certain collision. Green ("1") was initially listed as "events meriting careful monitoring" but was changed to indicate "normal" due to the media coverage that initially occurred for any object listed as a "1" on the Torino Scale even though such an object is relatively common and no "real" impact threat to Earth (Morrison et al., 2004).

A more complex scale is the Palermo Technical Impact Hazard Scale (or Palermo Scale), which is a logarithmic scale that also calculates the potential hazard of near-Earth objects. The scale was developed by Chesley et al. (2002) to aid asteroid researchers in prioritizing objects for future study. Their scale gives impact date, impact energy, and impact probability equal weight. The Palermo Scale formula is

$$P = \log_{10}\left(\frac{p_i}{f_B t}\right) \tag{7.9}$$

THE TORINO SCALE
Assessing Asteroid/Comet Impact Predictions

No Hazard	**0**		The likelihood of collision is zero, or is so low as to be effectively zero. Also applies to small objects such as meteors and bolides that burn up in the atmosphere as well as infrequent meteorite falls that rarely cause damage.
Normal	**1**		A routine discovery in which a pass near the Earth is predicted that poses no unusual level of danger. Current calculations show the chance of collision is extremely unlikely with no cause for public attention or public concern. New telescopic observations very likely will lead to re-assignment to Level 0.
Meriting Attention by Astronomers	**2**		A discovery, which may become routine with expanded searches, of an object making a somewhat close but not highly unusual pass near the Earth. While meriting attention by astronomers, there is no cause for public attention or public concern as an actual collision is very unlikely. New telescopic observations very likely will lead to re-assignment to Level 0.
	3		A close encounter, meriting attention by astronomers. Current calculations give a 1% or greater chance of collision capable of localized destruction. Most likely, new telescopic observations will lead to re-assignment to Level 0. Attention by the public and by public officials is merited if the encounter is less than a decade away.
	4		A close encounter, meriting attention by astronomers. Current calculations give a 1% or greater chance of collision capable of regional devastation. Most likely, new telescopic observations will lead to re-assignment to Level 0. Attention by the public and by public officials is merited if the encounter is less than a decade away.
Threatening	**5**		A close encounter posing a serious, but still uncertain threat of regional devastation. Critical attention by astronomers is needed to determine conclusively whether or not a collision will occur. If the encounter is less than a decade away, governmental contingency planning may be warranted.
	6		A close encounter by a large object posing a serious, but still uncertain threat of a global catastrophe. Critical attention by astronomers is needed to determine conclusively whether or not a collision will occur. If the encounter is less than three decades away, governmental contingency planning may be warranted.
	7		A very close encounter by a large object, which if occurring this century, poses an unprecedented but still uncertain threat of a global catastrophe. For such a threat in this century, international contingency planning is warranted, especially to determine urgently and conclusively whether or not a collision will occur.
Certain Collisions	**8**		A collision is certain, capable of causing localized destruction for an impact over land or possibly a tsunami if close offshore. Such events occur on average between once per 50 years and once per several 1000 years.
	9		A collision is certain, capable of causing unprecedented regional devastation for a land impact or the threat of a major tsunami for an ocean impact. Such events occur on average between once per 10,000 years and once per 100,000 years.
	10		A collision is certain, capable of causing a global climatic catastrophe that may threaten the future of civilization as we know it, whether impacting land or ocean. Such events occur on average once per 100,000 years, or less often.

Figure 7.8 Discussion of each of the Torino Scale values (Morrison et al., 2004). Credit: NASA. (A black and white version of this figure will appear in some formats. For the color version, please refer to the plate section.)

where P is the Palermo Scale value, p_i is the impact probability, f_B is the frequency of the background hazard in yr^{-1}, and t is the number of years before the impact event. The normalized risk part of the formula ($p_i/f_B t$) was first suggested by Binzel (2000). Objects with Palermo Scale values less than -2 rate very little concern, those between -2 and 0 should have some further monitoring, and those with positive values are objects that we need to be concerned with. The background frequency of impacts is assumed to be

$$f_B = \frac{3}{100}\left(\frac{E}{MT}\right)^{-4/5} yr^{-1}. \tag{7.10}$$

This formula estimates the annual probability of an impact with an energy E (in MT of TNT). E for an object can be estimated from $(1/2)m_2v^2$ where m_2 is the mass of the impactor and v is the impact velocity.

Example 7.2

What are the Torino and Palermo Scale values of a NEO that will occur 5 years in the future with an impact probability of 0.011 and an impact energy of 110 MT? Is this an object that we need to be concerned with?

From **Figure 7.7**, the Torino Scale value will be 4. To calculate the Palermo Scale value, the background frequency is first calculated and will be

$$f_B = \frac{3}{100}\left(\frac{110\,MT}{MT}\right)^{-4/5} yr^{-1} = 7\times10^{-4}. \tag{7.11}$$

The Palermo Scale value will then be

$$P = \log_{10}\left[\frac{0.011}{(7\times10^{-4})(5)}\right] = 0.50. \tag{7.12}$$

A Torino Scale value of 4 and a Palermo Scale value of 0.50 indicates an object that the world needs to be concerned with.

The asteroid currently with the highest Palermo Scale value is (29075) 1950 DA, which has a value of −0.83 (Farnocchia and Chesley, 2014). This asteroid has an impact probability of 2.5×10^{-4} in March 2880.

Near-Earth asteroid (99942) Apophis will also pass close to the Earth (e.g., Giorgini et al., 2008). Due to this close approach, this body was named after the Egyptian god of evil and destruction who dwelled in eternal darkness. This god was also known as Apep, the Destroyer.

Apophis has a diameter of ~300 m (Delbó et al., 2007a) and has been classifed as an Sq-type (Binzel et al., 2009), implying an ordinary chondrite-like mineralogy. The visible and near-infrared spectrum of Apophis is best matched by a space-weathered LL chondrite. Apophis' close encounter with the Earth in 2029 may also be able to test the Binzel et al. (2010) model that "seismic shaking" can change the spectral properties on near-Earth asteroids so they appear less space-weathered.

There is no direct conversion between Torino Scale and Palermo Scale numbers. However if the probability of the impact is above the background levels, the Torino Scale value will be 1 or greater and the Palermo Scale value will be above 0.

7.3.2 Asteroid Mitigation

If a NEO were on a collision course, how would the Earth be defended? The type of defense depends on the size of the impactor and the warning time (Committee to Review Near-Earth Object Surveys and Hazard Mitigation Strategies et al., 2010). For a small object (1–10 meters in diameter) predicted to impact with hours to days notice, the body would be expected to explode in the atmosphere before reaching the ground. People near the impact site would be warned to stay indoors and away from windows. For a larger body (10–25 meters) predicted to impact with days to weeks notice, the impacting area would be evacuated if it were on land. If this body were predicted to impact the ocean, tsunami warnings would be given for possible affected areas.

For a larger body (30 meters and larger) predicted to impact in decades, there are a number of postulated ways to prevent the impact. One method is "slow-push-pull" where a small but steady force is continually applied to the NEO (Committee to Review Near-Earth Object Surveys and Hazard Mitigation Strategies et al., 2010). If the acceleration were applied along or against its direction of motion, the NEO would miss the Earth since the body would reach Earth's orbit before or after the Earth does. One possibility would be to coat the surface with a dark material to change the body's albedo. The Yarkovsky effect would then increase due to the hotter surface temperature resulting from the dark coating. Concentrating solar energy at the NEO would create a jet that would change its orbit. A "tugboat" spacecraft attached to the NEO could be used to push the NEO. A "gravity tractor" positions itself very close to the NEO and continually "pulls" the NEO toward the spacecraft. These "slow-push-pull" methods work best on NEOs that are tens of meters to 100 meters in diameter with decades of notice or on a larger (hundreds of meters) NEO that will pass through a keyhole, which is a small region of space where the Earth's gravity will cause the NEO's orbit to change so it hits the Earth.

For larger bodies (up to 0.5 km) with only years of warning time, a kinetic impactor (Bruck et al., 2016) or a series of impactors could be used to strike the NEO and change its orbit. For larger bodies (greater than 0.5 km) with impact times of months to years, the only feasible alternative to deflect the NEO is a nuclear explosion on the surface or slightly below the surface to change the orbit. (Nuclear weapons in space are currently prohibited under the Outer Space Treaty.) Edward Teller (1908–2003), who is known as the "the father of the hydrogen bomb," and Lowell Wood proposed at a conference in 1992 at Los Alamos that the world's arsenal of nuclear weapons could be used to "shoot" down asteroids (Caldicott, 1994). Wood actually yelled out "Nukes forever!"

With decades of warning time, a nuclear standoff explosion could be used. A nuclear standoff explosion occurs in space and radiates the surface with X-rays, gamma rays, and fast-moving neutrons. The exposed area of the surface then evaporates, which propels the NEO in the opposite direction. A report to Congress in 2007 argued that nuclear standoff explosions are 10–100 times more effective than non-nuclear alternatives for diverting an asteroid (NASA, 2007). Aa nuclear standoff explosion also poses less risk of fragmenting the asteroid than a nuclear explosion on the surface.

One problem with almost all of these techniques is that to accurately determine how much the orbit will be changed, some information on the physical properties of the NEO must be known. Rubble piles will dissipate more energy during an impact than solid bodies and will require higher energy impacts for the same velocity change. Volatile-rich surfaces will be easier to vaporize than metallic iron surfaces. The exception is the gravity tractor, which just requires knowing the mass of the NEO.

Questions

1) Why is there such an interest in discovering near-Earth asteroids?
2) Would you expect a 5-km asteroid or a 5-km comet traveling at the same velocity to make a larger crater on the Earth? Why?
3) What will the kinetic energy of a C-complex asteroid with an H magnitude of 16.0 and an impact velocity of 17 km/s?
4) On Earth, what is the best evidence that a circular feature is due to an impact?
5) What was the conclusive evidence that an impact caused the extinction of the dinosaurs?
6) On Vesta, a region has a cumulative crater frequency of 0.5 craters per km^2. What would be its estimated age?
7) What celestial objects have been predicted to strike the Earth? What happened during and after the impact of these objects?
8) What are the Torino and Palermo Scale values for a NEO that will occur 10 years in the future with an impact probability of 0.0015 and an impact energy of 10 000 MT? Is this an object that we need to be concerned with?

9) What are the Torino and Palermo Scale values for a NEO that will occur 1 year in the future with an impact probability of 0.1 and an impact energy of 200 000 MT? Is this an object that we need to be concerned with?

10) An asteroid is on a collision course with the Earth. How would you decide which deflection strategy to use?

8

Spacecraft Missions

8.1 Mission Types

A spacecraft mission to an asteroid allows the body to be studied as a geological body instead of just as a point of light. Due to its proximity to its target, a spacecraft can take high-resolution images of a minor planet's surface. A spacecraft can also measure radiation from all parts of the electromagnetic spectrum for different areas of the surface. These spacecraft missions have shown that minor planets have a wide variety of shapes, surface features, and mineralogies with very different geological histories.

There are many different types of spacecraft missions. A flyby, as the name implies, is when a spacecraft observes a planetary body as it flies by the body. An orbiter goes into orbit around the body. A lander touches or lands on the surface. A sample return mission returns samples of the surface back to Earth. Many missions have had asteroids as their secondary target that they fly by on their way to their primary target. Launched successful and unsuccessful missions that had minor planets as their primary or secondary targets are listed in Table 8.1.

The classifications and bulk properties of each minor planet successfully observed by spacecraft missions are listed in Table 8.2. As can be seen by the table, a wide variety of sizes and taxonomic types have been studied by spacecrafts.

8.2 Naming of Features

The IAU Working Group for Planetary System Nomenclature (WGPSN) approves the names of all planetary features. Different features on different planetary bodies have a naming convention that needs to be followed. For example, craters on the Moon are named after deceased scientists and polar explorers who have made outstanding or fundamental contributions to their field. Wrinkle ridges (dorsa) on

Table 8.1 *A list of spacecraft missions to minor planets*

Spacecraft	Year launched	Primary target(s)	Secondary target(s)
Galileo	1989	Jupiter and its moons (orbiter)	(951) Gaspra (flyby), (243) Ida (flyby)
Clementine	1994	Moon (orbiter)	(1620) Geographos (flyby) (unsuccessful)
NEAR Shoemaker	1996	(433) Eros (orbiter, lander)	(253) Mathilde (flyby)
Cassini–Huygens	1997	Saturn and its moons (orbiter), Titan (lander)	(2685) Masursky (flyby)
Deep Space 1	1998	(9969) Braille (flyby)	107P/Wilson–Harrington (flyby) (unsuccessful), 19P/Borrelly (flyby)
Stardust	1999	81P/Wild (flyby, sample return)	(5535) Annefrank (flyby) 9P/Tempel (flyby)
Hayabusa	2003	(25143) Itokawa (orbiter, lander, sample return)	
Rosetta	2004	67P/Churyumov–Gerasimenko (orbiter, lander)	(2867) Šteins (flyby), (21) Lutetia (flyby)
New Horizons	2006	(134340) Pluto and its moons	(132524) APL (flyby), (15810) 1994 JR_1 (flyby), 2014 MU_{69} (flyby) (planned)
Dawn	2007	(4) Vesta (orbiter), (1) Ceres (orbiter)	
Chang'e 2	2010	Moon (orbiter)	(4179) Toutatis (flyby)
Hayabusa2	2014	(162173) Ryugu (orbiter, lander, sample return) (launched)	
Procyon	2014	(185851) 2000 DP_{107} (flyby) (unsuccessful)	
OSIRIS-REx	2016	(101955) Bennu (orbiter, lander, sample return) (launched)	

Table 8.2 *List of minor planets and minor planet moons that have been successfully observed by a spacecraft mission*

Body	Mission	Spectral type	Dimensions (km³)	Density (kg/m³)	Visual geometric albedo
(1) Ceres	Dawn	C	965 × 961 × 891	2161	0.087
(4) Vesta	Dawn	V	573 × 557 × 446	3456	0.38
(21) Lutetia	Rosetta	Xc	121 × 101 × 75	3400	0.19
(243) Ida	Galileo	Sw	60 × 25 × 19	2600	0.21
Dactyl	Galileo	S	1.6 × 1.4 × 1.2	–	0.20
(253) Mathilde	NEAR Shoemaker	C	66 × 48 × 44	1300	0.047
(433) Eros	NEAR Shoemaker	Sw	34 × 13 × 13	2640	0.23
(951) Gaspra	Galileo	S	18 × 11 × 9	1400	0.22
(2685) Masursky	Cassini– Huygens	–	(15–20 km)	–	–
(2867) Šteins	Rosetta	E	6.7 × 5.8 × 4.5	–	0.41
(4179) Toutatis	Chang'e 2	Sq	4.4 × 1.8 × 2.2	2500	≥0.20
(5535) Annefrank	Stardust	S	6.6 × 5.0 × 3.4	–	0.24
(9969) Braille	Deep Space 1	Q	2 × 1 × 1	–	0.34
(15810) 1994 JR$_1$	New Horizons	–	–	–	–
(25143) Itokawa	Hayabusa	S	0.5 × 0.3 × 0.2	1900	0.53
(132524) APL	New Horizons	S	(2.5 km)	–	–
(134340) Pluto	New Horizons	–	(2374 km)	1860	0.49–0.66
Charon	New Horizons	–	(1212 km)	1702	0.37–0.41
Nix	New Horizons	–	50 × 35 × 33	–	–
Hydra	New Horizons	–	65 × 45 × 25	–	–
Kerberos	New Horizons	–	19 × 10 × 9	–	–
Styx	New Horizons	–	16 × 9 × 8	–	–

The properties of Ceres are from Li et al. (2006), Russell et al. (2015), and Park et al. (2015). The properties of Vesta are from Russell et al. (2012) and Li et al. (2013). The properties of Lutetia are from Sierks et al. (2011) and Pätzold et al. (2011). The properties of Ida and Dactyl are from Belton et al. (1994, 1996). The properties of Mathilde are from Veverka et al. (1999). The properties of Eros are from Thomas et al. (2002) and Li et al. (2004). The properties of Gaspra are from Thomas et al. (1994), Richardson and Bowling (2014), and Helfenstein et al. (1994). The properties of Masursky are from the NASA/JPL/ Cassini Imaging Team (2004). The properties of Šteins are from Keller et al. (2010). The properties of Toutatis are from Howell et al. (1994a), Scheeres et al. (1998), and Huang et al. (2013). The properties of Annefrank are from Newburn et al. (2003) and Duxbury et al. (2004). The properties of Braille are from Oberst et al. (2001) and Buratti et al. (2004). The properties of Itokawa are from Lederer et al. (2005) and Fujiwara et al. (2006). The properties of APL are from Tubiana et al. (2007). The properties of Pluto and its moons are from Lellouch (2011), Stern et al. (2015), and Weaver et al. (2016).

Table 8.3 *List of geological features identified on minor planets*

Feature	Plural of name	Description
crater	craters	circular depression
catena	catenae	crater chain
cavus	cavi	irregular, steep-sided depression
chasma	chasmata	deep, elongated, steep-sided depression
collis	colles	small hill
dorsum	dorsa	wrinkle ridge
fossa	fossae	long, narrow depression
linea	lineae	elongated marking
macula	maculae	dark spot
mons	montes	mountain
planitia	planitiae	smooth, low-elevation terrain
planum	plana	smooth, elevated terrain (plateau)
labes	labēs	landslide
regio	regiones	large area marked by a reflectivity or albedo distinction from adjacent areas or broad geographic region
rima	rimae	fissure
rupes	rupēs	scarp (steep bank or cliff)
terra	terrae	extended land mass
tholus	tholi	hill (small dome-like mountain)
vallis	vallae	valley

the Moon are named after deceased geoscientists. A person being honored with a feature on a planetary body must have been deceased for at least three years. Official names will not be given to features whose longest dimensions are less than 100 meters unless the feature has exceptional scientific interest. Duplication of the same surface feature name on two or more bodies or of the same name for satellites and minor planets is discouraged. Names that have political, military, or religious significance may not be used, except for names of political figures prior to the nineteenth century.

A multitude of geological features have been identified on minor planets (Table 8.3). Cratering is the primary geological feature on all minor planets. Table 8.4 lists which features have been observed on different minor planets. As can be seen in the table, the larger the geological body (e.g., Ceres, Vesta, Lutetia, Pluto), the greater the range of geological features it is likely to have. Planets such as Venus and Mars have over 20 types of geological features identified on their surfaces.

8.3 Missions

Missions to asteroids will be discussed in the following sections. The missions will include successful and unsuccessful ones.

Table 8.4 *Types of features that have been identified on minor planets.*
Plural names are used for features even if only one feature has been
identified on the minor planet.

Body	Features
(1) Ceres	craters, catenae, montes, plana, rupēs, tholi
(4) Vesta	craters, catenae, dorsa, fossae, planitiae, rupēs, terrae, tholi
(21) Lutetia	craters, dorsa, fossae, labēs, regiones, rimae, rupēs
(243) Ida	craters, dorsa, regio
Dactyl	craters
(253) Mathilde	craters
(433) Eros	craters, dorsa, regio
(951) Gaspra	craters, regiones
(2867) Šteins	craters, regiones
(25143) Itokawa	craters, regiones
(134340) Pluto	craters, cavi, colles, dorsa, fossae, lineae, maculae, montes, plana, regiones, rupēs, terrae, valles
Charon	craters, chasmata, maculae, montes
Nix	craters
Hydra	craters

8.4 Galileo

The first mission to fly by an asteroid was Galileo. Galileo was named after the discover of Jupiter's moons since this NASA mission's primary objective was to study the Jupiter system. The project scientist for Galileo was Torrance Johnson. Galileo flew by (951) Gaspra in October 1991 and (243) Ida in August 1993 on its way to Jupiter. Galileo came within ~1600 km of Gaspra and ~2400 km of Ida. Gaspra is a member of the Flora family (Section 5.2.3), while Ida is a member of the Koronis family (Section 5.2.4). The significance of the Galileo flybys is that the obtained images of Gaspra and Ida conclusively showed that asteroids were geological bodies due the observed surface features.

The instruments on Galileo that observed Gaspra and Ida were the camera (the Solid State Imager or SSI), the Near-Infrared Mapping Spectrometer (NIMS), and the magnetometer (MAG). A number of other instruments on Galileo were only used to study Jupiter and its satellite while in orbit around Jupiter and did not observe the asteroids. The Galileo spacecraft (and all other spacecrafts) also have a radio transponder to communicate with Earth.

The SSI had an eight-position filter wheel with filters chosen to optimize spectral coverage of Jupiter and its satellites (Belton et al., 1992a). There was a clear filter (0.611 μm) with a very broad passband (0.440 μm width), three broadband filters at violet (0.404 μm), green (0.559 μm), and red (0.671 μm), two (0.727 and

0.889 μm) for methane absorption bands, one (0.756 μm) to calculate the continuum for the methane absorption bands, and one filter (0.986 μm) that overlaps NIMS measurements. However, the spectral coverage worked well for mineralogically characterizing the asteroids. The detector for the SSI was a CCD. NIMS had a wavelength coverage of ~0.7 to ~5.2 μm and used 17 individual detectors (15 InSb detectors for wavelengths greater than 1 μm and 2 Si detectors for wavelengths less than 1 μm) (Carlson et al., 1992). NIMS observations ranged from radiances at 17 different wavelengths to radiances measured at a much larger (hundreds) number of wavelengths (e.g., Granahan, 2002).

A magnetometer was used to measure the strengths of the magnetic fields of the asteroids. Galileo had a primary boom-mounted triaxial fluxgate magnetometer with ring core sensors plus a second magnetometer located two-thirds as far as the primary magnetometer from the spin axis (Kivelson et al., 1992). The secondary magnetometer was used to measure the large fields of the Jovian magnetosphere and assist in identifying spacecraft fields. A fluxgate magnetometer has a ferromagnetic core that is surrounded by two coils of wire. One of the coils has an alternating current flowing through it, which induces a current in the second coil that is measured by a detector. The alternating current in the coil continually causes the core to be magnetically saturated in one direction and then another over and over. An external field, which will affect the magnetization of the core, will break the symmetry of producing this induced current. How the symmetry changes will be a function of the strength and direction of the external field.

8.4.1 (951) Gaspra

Images of Gaspra (Figure 8.1) show an irregularly shaped body with a heavily cratered surface (Belton et al., 1992b). Gaspra is said to look like a potato. Gaspra's mass could not be determined during the encounter using radio science due to speed and distance of the flyby and Gapra's relatively small size (Richardson and Bowling, 2014). Richardson and Bowling (2014) were able to estimate the density of Gaspra to be ~1400 (range of 930–2800) kg/m^3 from the body shape and spin of a particular region on Gaspra's surface.

Craters on Gaspra are named after spas of the world. Regions (regio) on Gaspra are named after its discoverer Grigori Neujmin (1885–1946) and people [Clayne Yeates (1936–1991) and James Dunne (1934–1992)] who were part of the Galileo mission but had passed away before the encounter. Regio are geologically distinct regions on a body's surface.

Initial crater counts (Chapman et al., 1996b) for Gaspra find that this body is dominated by relatively fresh craters several hundreds of meters in diameter and smaller. The differential slope index for Gaspra's fresh and obvious craters

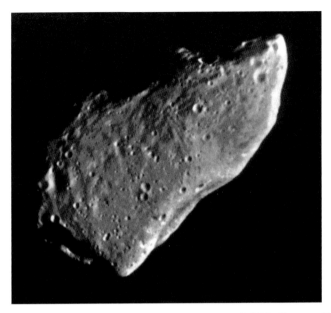

Figure 8.1 Mosaic of two Galileo clear-filter images of (951) Gaspra. Gaspra has dimensions of $18 \times 11 \times 9$ km^3. Credit: NASA.

is -4.3 ± 0.3 (which equals a slope index of -3.3 ± 0.3) (Chapman et al., 1996b). This steep slope was argued to indicate collisions due to a recent breakup event in Gaspra's vicinity. Large craters (~1 km in diameter) appear to be relatively rare and have very subdued features, implying these craters are very old. Gaspra's surface age is estimated to be ~0.2 Ga by Chapman et al. (1996b). Stooke and Ford (2001) reanalyzed images of Gaspra and identified more craters on the surface. Stooke and Ford (2001) found that only the Yeates Regio region of Gaspra is young and the rest of the surface is relatively old (~4 Ga).

Using Band I centers and Band Area Ratios, Granahan et al. (1994) and Granahan (2011) interpreted Galileo's visible and near-infrared data for Gaspra as indicating two different spectral units on the asteroid. Both spectral units have higher interpreted olivine abundances than ordinary chondrites, implying that this body has undergone some igneous differentiation, which is contrary to the LL-like composition interpreted for most analyzed Flora family members (Vernazza et al., 2008).

8.4.2 (243) Ida

Images of Ida (Figure 8.2) also show an irregularly shaped body with a heavily cratered surface (Belton et al., 1994). Craters on Ida are named after caverns and grottos of the world. Dorsa (ridges) are named after people were part of the Galileo mission who had passed away. Regions are named after the discoverer of Ida [Johann Palisa (1848–1925)] and places associated with him.

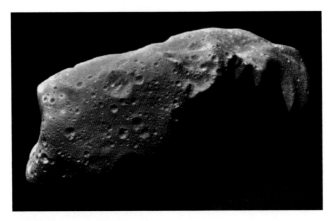

Figure 8.2 Mosaic of five Galileo clear-filter images of (243) Ida. Ida has dimensions of $60 \times 25 \times 19$ km^3. Credit: NASA.

Ida's surface appears saturated with craters with a range of craters from fresh to extremely degraded (Chapman et al., 1996a). If Gaspra and Ida have similar surface strengths and experienced similar projectile populations, Chapman et al. (1996a) estimated that Ida's surface age is ~10 times that of Gaspra's surface (0.2 Ga). Ida's surface age is, therefore, estimated to be ~2 Ga. However, it is possible that Ida's surface is actually younger in age if it underwent an early heavy bombardment after the breakup of the original Koronis parent body.

Both Gaspra and Ida produced magnetic field deflections when Galileo passed by them (Kivelson et al., 1992, 1993; Southwood, 1994; Hood, 1995; Wang et al., 1995). These deflections are interpreted as the interaction of the solar wind with a magnetospheric obstacle (the asteroid). Kivelson et al. (1993) found that the computed magnetic moment per unit mass for Gaspra was consistent with iron meteorites and highly magnetized chondrites, consistent with a significant metallic iron component (Hood, 1995). However, Hood (1995) found that the range of possible values for Gaspra was consistent with a large variety of meteorites (iron meteorites, stony-irons, ordinary chondrites, achondrite, carbonaceous chondrites) measured by Pesonen et al. (1993).

The first moon (Dactyl) (Figure 8.3) around another asteroid was discovered during Galileo's flyby of Ida. Dactyl has dimensions of $1.6 \times 1.4 \times 1.2$ km^3. Craters on Dactyl are named after Idaean Dactyls. Only two craters have been named on Dactyl and they are called Acmon and Celmis. The presence of Dactyl allowed the density of Ida to be determined. The calculated density of Ida was 2600 ± 500 g/cm^3 (Belton et al., 1995).

Using Band I centers and Band Area Ratios, Granahan et al. (1995) and Granahan (2002, 2013) found that Ida and Dactyl had interpreted mineralogies similar to LL chondrites. Evidence for space weathering was also evident from analyses of Ida's spectra. Chapman (1996) found that there was a trend when comparing the visible

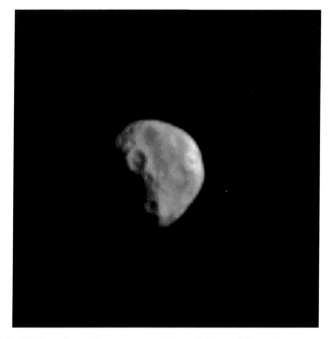

Figure 8.3 Galileo clear-filter image of Dactyl. Dactyl has dimensions of 1.6 × 1.4 × 1.2 km³. The top large crater is Acmon and the lower large crater is Celmis. Credit: NASA.

spectra of "typical" Ida terrains, "fresh" craters, and ordinary chondrites. The spectra of "typical" Ida terrains were redder than "fresh" craters, which were redder than ordinary chondrites.

8.5 Clementine

NASA's Clementine spacecraft was launched in 1994 with a primary mission of studying the Moon. The mission director was Pedro Rustan. Clementine is named after the song "Oh My Darling, Clementine" since the spacecraft will be "lost and gone forever." After mapping the Moon, Clementine was also to do a flyby of (1620) Geographos (Nozette and Shoemaker, 1994). However, a thruster misfired after leaving the Moon's orbit, which exhausted its fuel and caused the spacecraft to tumble at a speed greater than 80 rpm (revolutions per minute). Because of this misfire, Clementine was never able to fly by Geographos.

8.6 NEAR Shoemaker

NASA's Near Earth Asteroid Rendezvous (NEAR) mission was the first spacecraft to orbit an asteroid [(433) Eros] and land on it. Eros is the second-largest known near-Earth asteroid. The spacecraft was launched in February 1996. After launch, the mission was renamed NEAR Shoemaker after Eugene Shoemaker, the pioneering planetary geologist. It was the second Discovery mission, which attempted to do "faster, better, cheaper" planetary missions. NEAR Shoemaker orbited S-type (433) Eros for a year starting in 2000. NEAR Shoemaker also flew by C-type (253) Mathilde in June 1997. The flyby distance of Mathilde was ~1200 km. NEAR Shoemaker also imaged comet Hyakutake in March 1996. The PI of the mission was Andrew Cheng.

Instruments on NEAR Shoemaker included the Multispectral Imager (MSI), an infrared spectrometer (NIS), the NEAR Laser Rangefinder (NLR), the X-Ray/Gamma-Ray Spectrometer (XGRS), and a magnetometer (MAG). The laser rangefinder measures the distance of a spacecraft to an object by measuring the time it takes for an emitted laser pulse to be reflected off the surface and be detected by a spacecraft. By measuring this time, the topography of the surface can be determined since laser pulses reflected off higher surfaces will have shorter transit times. The XGRS contains two distinct instruments: the X-Ray Fluorescence Spectrometer and the Gamma-Ray Spectrometer. Both spectrometers detect high-energy photons from the surface of an asteroid.

8.6.1 (253) Mathilde

Earth-based observations found that Mathilde had a C-type reflectance spectrum and an extremely low albedo. Only ~60% of the surface was observed during the flyby (Veverka et al., 1999). Mathilde is relatively large with dimensions of $66 \times 48 \times 44$ km^3. Not all instruments were used to observe Mathilde during the encounter so as to conserve power for the Eros encounter (Farquhar et al., 2002).

During the flyby, Mathilde was confirmed to have an extremely low albedo (0.047) (Veverka et al., 1999). Mathilde was found to have at least five giant (diameters of ~20–33 km) craters on its surface (Figure 8.4) (Veverka et al., 1997, 1999; Cheng and Barnouin-Jha, 1999; Housen et al., 1999). Four of these craters have diameters that are similar to Mathilde's radius. Craters on Mathilde are named after coal fields of the world because Mathilde is a dark, C-type body and is thought to have a composition similar to carbonaceous chondrites.

Mathilde was only imaged but had its mass determined due to the gravitational perturbation of Mathilde on NEAR Shoemaker which could be measured using the spacecraft tracking data from the radio transponder (Yeomans et al., 1997). Since its mass was determined, the density could be determined using the images to estimate a volume. Mathilde was found to have an extremely low density (~1300 kg/m^3) (Veverka et al., 1997, 1999; Yeomans et al., 1997). This low density indicates a very porous interior since this measured density is much lower than the densities of other

Figure 8.4 Mosaic of four NEAR Shoemaker images of (253) Mathilde. Mathilde has dimensions of 66 × 48 × 44 km³. The extremely large crater apparent on its surface is called Karoo. Credit: NASA/JPL/JHUAPL.

low-albedo carbonaceous chondrites (e.g., CI, CM). Due to its low density, Housen et al. (1999) argued that the craters on Mathilde were due to compaction and not excavation. The porous nature of Mathilde may prevent shock and seismic waves being easily transmitted, allowing Mathilde not to be disrupted by large impacts (Veverka et al., 1999).

8.6.2 (433) Eros

Before the encounter with Eros, Murchie and Pieters (1996) analyzed previously obtained ground-based rotational spectra (~0.3–2.5 μm) and concluded that Eros had distinct rotational spectral variations. They interpreted the spectral data as showing Eros had both a pyroxene-rich hemisphere and an olivine-rich hemisphere. McCoy et al. (2000) argued from the interpretation of the ground-based spectral data that Eros may be a partially differentiated assemblage.

Images showed that Eros was extremely elongated and heavily cratered (Figure 8.5). Yeomans et al. (2000) finds that there is a considerable range in gravitational accelerations and escape velocities across the surface of Eros due to its elongated shape. The surface appeared to be saturated with large craters (> 100 m); however, smaller craters appear depleted in number (Chapman et al., 2002). Richardson et al. (2004) argue that this depletion of small craters is due to impact-induced seismic shaking, which causes the downslope movement of loose regolith that can cover up the craters. Eros had a density of 2640 ± 20 kg/m³

Figure 8.5 Mosaic of six NEAR Shoemaker images of (433) Eros. Eros has dimensions of $34 \times 13 \times 13$ km³. The large crater partially in shadow on the top of Eros in the image is Psyche and the large crater partially in shadow at the bottom of Eros in the image is Himeros. Credit: NASA/JPL/JHUAPL.

(Thomas et al., 2002), implying a porous interior if it has a composition similar to ordinary chondrites.

Images of Eros also showed that its surface had a considerable range in surface roughness (Robinson et al., 2001). Some areas have a high density of boulders up to 100 m in size while others are extremely smooth and flat. These smooth and flat regions are called "pond" deposits. There are ~300 "ponds" on Eros with most being less than 60 meters in width (Dombard et al., 2010).

Craters on Eros are named after mythological and legendary lovers. Regions (Charlois Regio, Witt Regio) on Eros are named after the discoverers of Eros. Dorsa (Finsen Dorsum, Hinks Dorsum) are named after scientists who contributed to the exploration and study of Eros.

Using the XGRS, NEAR Shoemaker was the first spacecraft to determine elemental compositions of an asteroid. X-rays measured from an asteroid are due to fluorescence, which is the emission of light after material has absorbed electromagnetic radiation. This fluorescence is due to the absorption of X-rays or gamma rays by inner orbital electrons of atoms, which causes these electrons to be expelled from these orbitals. As electrons from higher orbitals fill these lower energy electron "holes," X-rays are emitted by the electrons. The strongest lines are those from the transition from the L orbital to the K orbital, which are called Kα lines. Since every element has orbitals with characteristic energies, the energies of the emitted X-rays are characteristic of a particular element. The flux of emitted characteristic X-rays is related to the concentration of the element. For asteroids, the X-ray source is the solar corona. The sampling depth for the measured X-rays is less than 100 μm.

To directly compare NEAR Shoemaker results to meteorite data, bulk chemistries of the meteorites need to be used since that is what is measured on the asteroid. Almost all of the elemental ratios derived from the X-ray measurements for major (Mg/Si, Al/Si, Ca/Si, Fe/Si) (Trombka et al., 2000; Nittler et al., 2001; Lim and Nittler, 2009) and minor (Cr/Fe, Mn/Fe, Ni/Fe) (Foley et al., 2006) elements are consistent with Eros having an ordinary chondrite composition; however, a number of meteorite classes could not be ruled out (e.g., acapulcoites/lodranites).

The exception is the S/Si ratio (0.005 ± 0.008), which is depleted relative to ordinary chondrite values [average of ~0.12 from Nittler et al. (2004) database]. This depletion is attributed to space weathering, which could devolatilize troilite (FeS) on the asteroid surface. This explanation was backed up through experimental work (Loeffler et al., 2008) using the impact of ions to simulate the solar wind and laser irradiation to duplicate micrometeorite impacts.

Gamma rays from an asteroid (Evans et al., 2001) are emitted from the radioactive decay of a number of isotopes (e.g., ^{40}K, ^{232}Th, ^{238}U) or through the excitation of nuclei by galactic cosmic rays or solar particles. To produce gamma rays through excitation, galactic cosmic rays or solar particles strike the nuclei, which then emit neutrons that collide with other nuclei (Prettyman, 2007). These nuclei become excited and emit gamma rays when they return to their normal energy state. The sampling depth for the measured gamma rays is tens of cm. While in orbit, the XRGS had no trouble detecting a statistically significant flux of X-rays from Eros above the background flux the XGRS was not able to detect a statistically significant flux of gamma rays. This high background was due to the gamma-ray spectrometer not being located on a boom (Trombka et al., 2001), which would have reduced the background gamma-ray signal from the spacecraft that is due to cosmic-ray interactions.

A decision was made late in the mission to do something that had never been done before: land on an asteroid. This landing allowed close-up images of the surface to be taken (Veverka et al., 2001) and the gamma-ray spectrometer to obtain a

Figure 8.6 Last image of asteroid (433) Eros received from NEAR Shoemaker taken at a distance of ~120 meters from the surface. The streak lines at the bottom are due to the loss of signal when the spacecraft landed on the surface. Credit: NASA/JPL/JHUAPL.

statistically significant signal from Eros (Evans et al., 2001). Images were taken as close as 120 meters from the surface (Figure 8.6).

The landing on Eros allowed the XRGS to detect a statistically significant number of gamma rays from Eros' surface. Originally Evans et al. (2001) found roughly chondritic Mg/Si and Si/O ratios and the K abundance; however, the Fe/Si and Fe/O ratios were depleted compared to chondritic values. The depleted Fe/Si ratio was not consistent with the X-ray results. Evans et al. (2001) argued that the most probable cause of the mismatch was regolith processes and the "pond" deposit that NEAR Shoemaker landed on was depleted in metallic iron.

Peplowski et al. (2015) did an improved analysis of the gamma-ray data and determined that Eros had a surface composition consistent with L or LL chondrites. This reanalysis of the XGRS data (Peplowski et al., 2015) found that the measured elemental composition derived from the gamma-ray observations was consistent with L and LL chondrites. Peplowski et al. (2015) determined a hydrogen concentration that was consistent with hydrogen concentrations measured in L and LL chondrites falls. They argue that the absence of any measured depletion found for volatiles such as hydrogen and potassium for

Eros indicates that the sulfur depletion is a surface effect, consistent with space weathering.

Using spectral reflectance and X-ray/gamma-ray data, McCoy et al. (2001) found that Eros had a surface mineralogy most similar to a space-weathered ordinary chondrite but a primitive achondritic assemblage could not be ruled out. Later analyses (Foley et al., 2006; Lim and Nittler, 2009; Peplowski et al., 2015) of the X-ray/gamma-ray data were also consistent with ordinary chondrites.

8.7 Deep Space 1

Launched in October 1998, NASA's Deep Space 1 spacecraft was a technology demonstration mission (Rayman et al., 2000) that flew by (9969) Braille at a distance of 26 km from the asteroid during July 1999. Its mission was later extended so it could potentially visit comets Wilson–Harrington [107P/Wilson–Harrington or (4015) Wilson–Harrington] and Borrelly (19P/Borrelly); however, the star tracker failed so it only flew by comet Borrelly during September 2001. Instruments (Rayman et al., 2000) included NSTAR (NASA Solar Technology Application Readiness), Autonav, SDST (Small Deep Space Transponder), MICAS (Miniature Imaging Camera and Spectrometer), SCARLET (Solar Concentrator Array with Refractive Linear Element Technology), Remote Agent (remote intelligent self-repair software), Beacon Monitor, and PEPE (Plasma Experiment for Planetary Exploration). Arguably, the most successful of these technology demonstrations were NSTAR, Autonav, and SDST.

Deep Space 1 was the first spacecraft to use ion propulsion (NSTAR) as its primary propulsion system. Ion propulsion works by ionizing xenon gas atoms. A voltage is then applied, which accelerates the ions out the engine. This action causes a subsequent reaction in the opposite direction, which accelerates the spacecraft. The velocity of the ions is much higher than the exhaust from typical rocket engines that use chemical propellant; however, the number of ions that are accelerated are relatively low. Ion propulsion can accelerate a spacecraft to speeds approximately ten times faster than typical rocket engines; however, the thrust is much smaller due to the small number of ions that are accelerated, so it takes much longer to reach the same speed as the typical rocket engine. NSTAR initially failed after 4.5 minutes of operation, but approximately 30 days later, the propulsion system was restarted and worked successfully for the rest of the mission. After this technology demonstration, ion propulsion was later used on the Dawn spacecraft.

Autonav was designed to navigate the spacecraft through observing bright aster-oids. Usually spacecraft are tracked by communicating with the Deep Space Network (DSN). DSN includes the Goldstone Deep Space Communications Complex near Barstow, California, the Madrid Deep Space Communication Complex near Madrid, Spain, and the Canberra Deep Space Communication Complex near Canberra, Australia. All of the facilities have antennas, transmitters, and receivers. The positions of these facilities approximately 120° from each other allows one of the stations to be in contact with any spacecraft that is located at 30 000 km from Earth. The problem with ground-based systems is that there is always a slight delay in determining the position of the spacecraft due to the two-way light transit time.

Autonav works because asteroids will move in comparison to the relative fixed positions of stars. The movement of two or more asteroids relative to the stars will allow the spacecraft to determine its position. By doing these observations at two or more times, the spacecraft can determine its trajectory. Autonav also is used to track the position of the spacecraft target. Autonav was used on a number of com-etary missions such Stardust and Deep Impact.

SDST was a transponder that combined a number of functions of spacecraft communication into one smaller unit. SDST technology was used on a number of other missions such as Deep Impact and Dawn.

Using telescopic and flyby data in combination, Braille's dimensions were found to be $2.1 \times 1 \times 1$ km^3 (Oberst et al., 2001). Ground-based observations of Braille classified it as a Q-type (Binzel et al., 2001a); however, the ground-based spectral data of Lazzarin et al. (2001) could not rule out a V-type classification. Buratti et al. (2004) found that the visible and near-infrared reflectance spectrum was most similar to Q-type asteroids. The visual geometric albedo of Braille is 0.34 (Buratti et al., 2004).

8.8 Cassini–Huygens

Cassini–Huygens is a mission that was launched to the Saturn system in 1997 and started orbiting Saturn in July 2004. Cassini–Huygens was a joint collaboration between NASA, the European Space Agency (ESA), and the Italian Space Agency (*Agenzia Spaziale Italiana* or ASI). Cassini was the orbiter and Huygens was the probe that successfully landed on Saturn's moon Titan.

A graduate student (Tolis Christou) of one of the Imaging Team members real-ized that Cassini–Huygens would fly reasonably close to asteroid (2685) Masursky. Masursky is named after planetary scientist Harold Masursky (1923–1990). Cassini–Huygens flew by Masursky in January 2000 with a closest approach of 1.6 million km (NASA/JPL/Cassini Imaging Team, 2004). Because Cassini passed Masursky at such a large distance, the asteroid only appeared as a small dot in the images with no resolvable features. Cassini was able to determine that Masursky was ~15–20 km in diameter.

8.9 Stardust

Before the encounter with Wild 2, Stardust imaged (5535) Annefrank at a minimum distance of 3100 km (Duxbury et al., 2004) to test the encounter sequence for Wild 2. Annefrank was named after the Holocaust victim Anne Frank (1929–1945) whose diary chronicled her time in hiding during World War II. Over 70 images were obtained of the object (Newburn et al., 2003). Annefrank was found to have a highly irregular shape with a number of flat surfaces. Annefrank was at least $6.6 \times 5.0 \times 3.4$ km^3 in size. Annefrank has a broadband visual geometric albedo of 0.24 (Newburn et al., 2003). Photometric modeling of Annefrank's phase curve found most parameters consistent with S-complex asteroids (Hillier et al., 2011).

8.10 Hayabusa

The JAXA Hayabusa mission was the first spacecraft mission to return samples of an asteroid back to Earth. Originally the mission was called MUSES-C (Mu Space Engineering Spacecraft C) but was later changed to Hayabusa (Japanese for peregrine falcon). The target was (25143) Itokawa, which was named after the Japanese space program pioneer Hideo Itokawa (1912–1999). Hayabusa launched in May 2003, rendezvoused with Itokawa from mid-September to the beginning of December 2005, and returned to Earth with a sample in June 2010.

Ground-based reflectance spectra (Figure 4.17) (Binzel et al., 2001b; Abell et al., 2007) in the visible and near-infrared were taken of Itokawa and analyzed to estimate Itokawa's mineralogy. The Binzel et al. (2001b) spectra were taken before launch, while the Abell et al. (2007) spectra were taken before samples were returned back to Earth. Binzel et al. (2001b) used MGM to fit Itokawa's absorption bands and interpreted the fits as indicating that Itokawa had a mineralogy similar to LL chondrites.

Scientific instruments on Hayabusa included the Asteroid Multi-band Imaging Camera (AMICA), the Near-Infrared Spectrometer (NIRS), Light Detection and Ranging Altimeter (LIDAR) and the X-Ray Spectrometer (XRS). AMICA took images in four filters (0.430, 0.550, 0.700, and 0.950 μm). NIRS spectral coverage was only from 0.75 to 2.1 μm so it did not cover the full wavelength region of the Band II found in pyroxenes. The X-ray results (Okada et al., 2006) were later retracted. Hayabusa also had a mini-lander (MIcro/Nano Experimental Robot Vehicle for Asteroid or MINERVA). MINERVA was launched toward Itokawa; however, it missed its target. MINERVA would have hopped on the surface of Itokawa, taken images, and measured surface temperatures.

Figure 8.7 Image of asteroid (25143) Itokawa taken by Hayabusa on November 1, 2005. Itokawa has dimensions of 0.5 × 0.3 × 0.2 km³. Note the smooth area on the surface (MUSES-C Regio) where Hayabusa landed to retrieve samples. Also, note all the boulders on the surface. Courtesy of JAXA. Credit: JAXA.

Hayabusa had a number of technical issues during its mission. The mini-lander MINERVA that was deployed on November 12, 2005 was inadvertently released while Hayabusa was ascending and at a higher altitude from Itokawa than planned. MINERVA missed Itokawa's surface and went floating off into space.

Itokawa is a peanut-shaped (or sea otter-shaped) body that is predominately covered with meter-sized boulders (Figure 8.7) (Saito et al., 2006; Michel et al., 2009). One region is often called the head and the other the body. The high abundance of boulders on the surface is much different the surface properties of previous asteroids that were observed. Itokawa is rather small with dimensions of 0.535 × 0.294 × 0.209 km³.

Features (including craters) on Itokawa are primarily named after places associated with astronautics and planetary sciences. The only exception is the large flat region, which is called Muse-C Regio and refers to the original spacecraft name.

Hirata et al. (2009) identified ~40 possible craters on Itokawa's surface. These craters appear very shallow with an average depth/diameter ratio of 0.08±0.03, which is much shallower than the average depth/diameter ratios of other observed asteroids. For large craters (~200 m), the density of the craters is close to the empirical saturation line; however for craters less than 100 meters, the number of craters declines significantly with decreasing diameter. This lack of small craters is proposed to be due to processes such as seismic shaking or due to armoring (Michel et al., 2009; Güttler et al., 2012). Armoring is where an impactor strikes a boulder, which is larger in size than the impactor, and expends most of its energy in crushing the boulder and not forming a crater on the surface. Michel et al. (2009) find that the time needed to accumulate the craters on Itokawa ranges from ~75 Ma to ~1 Ga depending on the scaling law.

Itokawa's bulk density is estimated to be $1900 \pm 130 \text{ kg/m}^3$ and is assumed to be a rubble pile from this low density (Fujiwara et al., 2006). Near-infrared spectral measurements of Itokawa found that it had a reflectance spectrum similar to LL chondrites (Abe et al., 2006).

Two separate "landings" were performed on the smooth MUSES-C Regio to retrieve samples (Yano et al., 2006). These "landings" were designed to be just quick touches on the surface where Hayabusa's sampling horn makes a quick contact with the asteroid and a bullet is fired into the surface to eject particles into the sampling horn. However during these two encounters, technical issues caused the bullets not be fired into the surface. But there was hope that simple contact with the surface would eject material into the sampling container, which contains two compartments.

These hopes were confirmed after Hayabusa's re-entry capsule landed in the Australian desert in June 2010. The sampling container contained thousands of particles but less than a milligram of material. Analyses of the grains showed that they had olivine (Fa_{27-30}) and pyroxene ($Fs_{23.1\pm2.2}$) mineralogies consistent with equilibrated LL (LL4–LL6) chondrites (Nakamura et al., 2011; Mikouchi et al., 2014). Oxygen isotopic ratios of these grains (Yurimoto et al., 2011) are also similar to those of LL4–LL6 chondrites. Noble gas isotopic measurements (Nagao et al., 2011) showed large amounts amounts of solar helium (^4He), neon (^{20}Ne), and argon (^{36}Ar), indicating that these grains resided in the regolith on the very surface of Itokawa where they were exposed to the solar wind. The Ar–Ar age of three Itokawa grains (measured as a single sample) by Park et al. (2015) was interpreted as indicating that found that the body became a rubble pile ~1.3 Ga ago (Figure 3.20). However, Jourdan et al. (2015) determined much different Ar–Ar ages for two separate grains. One of the grains appeared to have been shocked and had an Ar–Ar age of 2.3 ± 0.1 Ga while an unshocked grain had an age of 3.8–4.6 Ga. This wide range of ages is consistent with each measured grain having its own "personal history" (as coined by Tomoki Nakamura) and possibly being exposed to different types of processes (e.g., shock, space weathering) after it formed compared to other collected grains.

Using a scanning transmission electron microscope (STEM), Noguchi et al. (2011) were able to identify nanophase iron particles in a thin layer on grain surfaces. Coupled with the reddening of Itokawa's spectrum relative to LL chondrites, this result proved that space weathering does occur on asteroid surfaces. These results were confirmed by Thompson et al. (2014) who also identified nanophase iron particles on Itokawa grains.

Hayabusa was able to definitively determine that Itokawa had an LL chondrite composition confirming the ground-based prediction of Binzel et al. (2001b). Hayabusa also confirmed that space weathering occurs on S-complex asteroids. These results were only possible with returned samples that could be analyzed in a laboratory on Earth.

8.11 New Horizons

New Horizons also flew by asteroid (132524) APL in 2006 as it flew through the main belt. The asteroid was originally known as 2002 JF_{56} but was later named APL after the Johns Hopkins University Applied Physics Lab (APL). APL was one of the manufacturers of New Horizons and many other spacecraft. New Horizons flew by asteroid APL in June 2006 at a distance of ~100 000 km (Olkin et al., 2006). Ground-based observations found that APL (Tubiana et al., 2007) was an S-type asteroid with a diameter of ~2.3 km. Very little was learned about APL during the flyby.

8.12 Rosetta

On its way to comet Churyumov–Gerasimenko, Rosetta flew by (2867) Šteins during September 2008 and (21) Lutetia during July 2010. Rosetta was the first European mission to fly by an asteroid. Šteins was classified as an E-type before the encounter while Lutetia was classified as either an Xc- or M-type. Šteins is much smaller than Lutetia with dimensions of $6.7 \times 5.8 \times 4.5$ km^3, while Lutetia has dimensions of $121 \times 101 \times 75$ km^3. Closest approach to Šteins was at a distance of 803 km, while the closest approach to Lutetia was at a distance of 3170 km.

Instruments on Rosetta were designed to study a comet nucleus but most could also be used to study an asteroid. These instruments included an ultraviolet imaging spectrograph (ALICE, which is not an acronym), two cameras (Wide-Angle Camera or WAC and Narrow-Angle Camera or NAC) which are part of the Optical, Spectroscopic, and Infrared Remote Imaging System (OSIRIS), imaging spectrometer (Visible and Infrared Thermal Imaging Spectrometer or VIRTIS), microwave detector (Microwave Instrument for the Rosetta Orbiter or MIRO), and radio science (Radio Science Investigation or RSI). The name OSIRIS was chosen to honor the Egyptian god of the underworld, whose name in hieroglyphics includes the all-seeing eye" (Keller et al., 2007).

WAC has a much larger field of view than NAC. WAC was designed for mapping the gas and dust near the comet, while NAC was designed for mapping the nucleus of the comet. WAC uses 14 filters between 0.24 and 0.72 μm, while NAC uses 12 filters between 0.25 and 1.0 μm. These cameras used backside

Figure 8.8 Reprocessed Rosetta image of (2867) Šteins. Šteins has dimensions of 6.7 × 5.8 × 4.5 km³. Image is reprocessed to enhance the differences between bright crater rims and their shadowed floors. Credit: Ted Stryk, Roane State Community College.

illuminated CCD detectors that were 2048 pixels × 2048 pixels in size. In contrast, the camera on Dawn only had a CCD detector that was 1024 pixels × 1024 pixels in size.

8.12.1 (2867) Šteins

Šteins is named after Kārlis Šteins (1911–1983), a Latvian and Soviet astronomer. Craters on Šteins are named after gemstones due to its diamond-like shape. With the montes of Mercury (which are named after the word for "hot" in various languages) and lunar maria (which tend to be named after states of mind), the craters on Šteins are the only planetary features that are not derived from proper nouns. Šteins is an oblate body with a number of linear faults and a large ~2.1 km diameter crater near its south pole (Figure 8.8) (e.g., Keller et al., 2010), which is called the Diamond crater. This crater has a crater diameter to mean asteroid radius ratio of 0.79 (Burchell and Leliwa-Kopystynski, 2010). If this is the largest crater size possible on its surface, Šteins would have a macroporosity of ~20%. There is also a chain (catena) of rimless circular pits going from the north pole to the Diamond crater (Marchi et al., 2010). These circular features are not thought to be craters since the probability of such a crater chain on a low-gravity body is thought to be extremely low.

One region on Šteins is named Chernykh Regio after its discoverer Nikolai Chernykh (1931–2004). Other features are named after places associated with gemstones. Approximately 40 crater-like features were identified on Šteins surface with diameters ranging from 0.2 to 2.1 km; however, there is an absence of small craters (≲0.6 km) on its surface (Marchi et al., 2010). For the distribution of craters larger than 0.5 km in diameter, an age of 154 ± 35 Ma was derived (Keller et al., 2010) using a crater-erasing rate that was scaled from a rate that was derived from Gaspra (O'Brien et al., 2006).

The visual geometric albedo for Šteins is 0.41 ± 0.016 (Keller et al., 2010). OSIRIS multicolor photometry is consistent with ground-based observations. A strong absorption is present at 0.49 μm. OSIRIS did find that Šteins had an extremely strong dropoff shortward of 0.4 μm, which cannot be observed using Earth-based observations. No surface color variations larger than 1% were observed on the surface of Šteins. The high albedo and color photometry for Šteins is consistent with this body having a surface mineralogy similar to the enstatite-rich differentiated aubrites.

8.12.2 (21) Lutetia

Images of Lutetia show it to have a heavily cratered surface (Figure 8.9) (Sierks et al., 2011; Marchi et al., 2012; Massironi et al., 2012). Lutetia is named after the town Lutetia in Roman Gaul (that contained France) that was later renamed Paris. Craters on Lutetia are named after cities in the Roman Empire and adjacent parts of Europe at the time of Lutetia (52 BC – AD 360). The largest crater is the ~55 km Massilia crater (Massironi et al., 2012). Regions on Lutetia are named after the discoverer of Lutetia and provinces of the Roman Empire at the time of Lutetia. Other features are named after rivers in the Roman Empire and adjacent parts of Europe at the time of Lutetia.

Five major regions on the surface of Lutetia were identified using crater densities, cross-cutting and overlapping relationships, and the presence of deformational features such as faults, fractures, and grooves (Sierks et al., 2011).

VIRTIS observations found that Lutetia had a relatively flat spectrum with no absorption bands from 0.4 to 3.5 μm with a thermal inertia in the range of 20–30 J m^{-2} K^{-1} s$^{0.5}$ (Coradini et al., 2011). This thermal inertia value is comparable to a lunar-like powdery regolith. The calculated density for Lutetia is 3400 kg/m^3 (Sierks et al., 2011; Pätzold et al., 2011), which is higher than values for most meteorites. The average visual geometric albedo is 0.19 ± 0.01.

Figure 8.9 Rosetta image of (21) Lutetia. Lutetia has dimensions of 121 × 101 × 75 km³. Note the heavily cratered surface. Credit: ESA 2010 MPS for OSIRIS Team MPS/UPD/LAM/IAA/RSSD/INTA/UPM/DASP/IDA.

The absence of absorption bands and relatively high density for Lutetia was argued by Coradini et al. (2011) to indicate a surface mineralogy that is a mixture of low FeO silicates and metallic iron. Possible compositional analogs for Lutetia include CB, CH, CR, or enstatite chondrites.

8.13 Dawn

Dawn's mission goal was to study the two most massive bodies in the asteroid belt, Vesta and Ceres. Ceres is the largest body in the asteroid belt and is almost twice the diameter of Pallas. Ceres is the only dwarf planet that is interior to the orbit of Neptune. Dawn was the first spacecraft to "visit" a dwarf planet since it reached Ceres a few months before New Horizons flew by Pluto.

Ceres and Vesta experienced two distinct types of differentiation. Vesta appears to have melted and formed a basaltic crust, an olivine mantle, and an iron core. Ceres is believed to have undergone ice–rock differentiation, where an initial undifferentiated mixture of ice and chondritic material reorganizes itself with sufficient heating into a rocky core and an icy mantle or shell. Dawn was launched in September 2007 by NASA and entered orbit around Vesta in July 2011. The PI

Figure 8.10 Dawn image of (4) Vesta. Vesta has dimensions of 573 × 557 × 446 km³. Note the heavily cratered northern hemisphere (top) and smoother southern hemisphere (bottom). The towering mountain at the south pole is visible at the bottom of the image. The set of three craters known as the "snowman" can be seen at the top left. Credit: NASA/JPL-Caltech/UCLA/MPS/DLR/IDA.

of the mission is Christopher Russell. Dawn left orbit around Vesta in September 2012 and reached Ceres in March 2015.

The instruments on Dawn include the Framing Camera (FC), the Visual and Infrared Spectrometer (VIR), and the Gamma-Ray and Neutron Detector (GRaND). The framing camera observed Vesta using eight filters. There were seven filters in the visible and near-infrared (0.438 μm, 0.555 μm, 0.653 μm, 0.749 μm, 0.829 μm, 0.917 μm, 0.965 μm) plus one clear filter (Nathues et al., 2015a). VIR potentially could measure reflectance spectra from ~0.25 and ~5.1 μm.

8.13.1 (4) Vesta

Images of Vesta revealed a heavily cratered northern hemisphere and smoother southern hemisphere (Figure 8.10) (Marchi et al., 2012). Approximately 1900 craters have been identified on Vesta's surface. Two overlapping large craters are present at its south pole with the 500-km wide Rheasilvia crater overlying the 400-km wide Veneneia crater (Figure 8.11). Diogenite regions are more abundant in the southern hemisphere (e.g., Reddy et al., 2012a; Thangjam et al., 2013), which is consistent with deeper excavation in the southern hemisphere due to

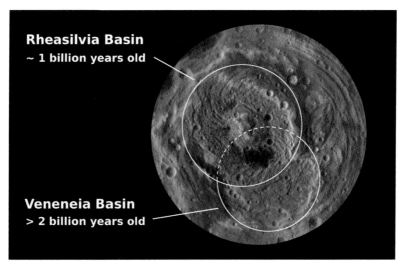

Figure 8.11 Topographic map of the southern hemisphere of (4) Vesta made by the Dawn mission. The map is color-coded by elevation with red showing the higher areas and blue showing the lower areas. Rheasilvia, the largest impact basin on Vesta, is ~500 km in diameter. The other basin, Veneneia, is ~400 kilometers across and lies partially beneath Rheasilvia. The topographic map was constructed from analyzing images from Dawn's framing camera that were taken with varying Sun and viewing angles. Credit: NASA/JPL-Caltech/UCLA/MPS/DLR/IDA/PSI. (A black and white version of this figure will appear in some formats. For the color version, please refer to the plate section.)

these large craters. Numerous troughs are present on Vesta's equator. No moons were discovered around Vesta with a detection limit of 3 meters (McFadden et al., 2015).

Craters on Vesta are given names associated with the Roman goddess Vesta and famous Roman women. The 450-km diameter Rheasilvia crater is named after the vestal virgin Rhea Silvia, who was also the mythical mother of the founders of Rome (Romulus and Remulus). The 400-km diameter Veneneia crater is named after one of the founding vestal virgins. The 18-km diameter Angioletta crater was named after Italian planetary scientist Angioletta Coradini (1946–2011), who helped develop Dawn's VIR instrument. Regions are named after Vesta's discoverer and other scientists who have contributed to the study of Vesta. Other features are named after places and festivals associated with vestal virgins.

Vesta's density is calculated to be 3456 kg/m^3 (Russell et al., 2012). This density is only 5% more than the density (3300 ± 500 kg/m^3) calculated from Earth-based observations (Viateau and Rapport, 2001). The Viateau and Rapport (2001) density was calculated using average dimensions calculated from a variety of Earth-based (including Earth-orbiting spacecraft) observations and a mass determined from gravitational perturbations of Vesta on other asteroids. The density calculated by

Dawn is consistent with Vesta (assuming a metallic iron core) having ~10% porosity in its mantle and crust (Russell et al., 2012; Park et al., 2014).

Spectral reflectance measurements of Vesta taken by Dawn are consistent with HED meteorites with some areas being more eucritic and some more diogenitic (e.g., De Sanctis et al., 2012a, 2013; Reddy et al., 2012a; Thangjam et al., 2013; Ammannito et al., 2013; Zambon et al., 2014). Blewett et al. (2016) found no evidence of lunar-style space weathering on Vesta's surface. Vesta also has a number of terrains that are enriched in low-albedo material (Reddy et al., 2012b; McCord et al., 2012). These dark areas are generally associated with impact craters and are spectrally similar to carbonaceous chondrites. They are believed to be due to an influx of carbonaceous material striking the surface. This was not a surprising result, since HEDs are known to contain CM2-like and CR2-like inclusions (e.g., Buchanan et al., 1993; Zolensky et al., 1996), which have abundant hydrated silicates. Using FC data, dark material on Vesta was also found to have a feature at ~0.7 μm (Nathues et al., 2014), which is consistent with CM chondrite material (Cloutis et al., 2011b).

One surprising result for Vesta is the relatively rare occurrence of olivine-rich (> 40 wt%) areas on the surface on the basis of the analysis of its FC data (e.g., Ammanito et al., 2013; Thangjam et al., 2014; Nathues et al., 2015a). A number of filter reflectance ratios such as the band tilt ($R_{0.92μm}$ / $R_{0.96μm}$), mid ratio [($R_{0.75μm}$/$R_{0.83μm}$) /($R_{0.83μm}$ / $R_{0.92μm}$)], and mid curvature [($R_{0.75μm}$ + $R_{0.92μm}$)/$R_{0.83μm}$], which are based primarily on the work of Isaacson and Pieters (2009), have been used to try to identify olivine-rich regions (Thangjam et al., 2013, 2014). [R_x is the reflectance at a particular wavelength (x)]. Differentiation of an asteroid is thought to produce a basaltic crust, an olivine-dominated mantle, and a metallic iron core. Large craters are present on the surface, which should have broken through the assumed thickness of the basaltic crust and exposed the olivine-rich mantle. It is possible that the crust is relatively thick on Vesta or an olivine-rich mantle did not form on Vesta (Nathues et al., 2015a). Nathues et al. (2015a) argue that most of the olivine on Vesta's surface is endogenic and due to influx of olivine-rich material striking the surface.

Reddy et al. (2013) discussed the accuracy of compositional interpretations for Vesta's surface based on ground-based and Hubble observations. Ground-based rotational spectra (Gaffey, 1997; Reddy et al., 2010) and Hubble Space Telescope observations (Thomas et al., 1997) had previously noted deeper band depths for the southern hemisphere, which is consistent with the Dawn results. An olivine-rich region on the equator identified by Gaffey (1997) was not observed.

Williams et al. (2014) proposed a geological time scale and time-stratigraphy for Vesta based on its geological map. The time periods (Figure 8.12) are named using different impact events on Vesta. The Pre-Veneneian time period covers the time from the formation of Vesta and the Veneneia impact event. The Veneneian

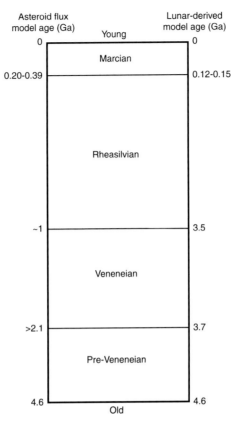

Figure 8.12 The geological time scale of (4) Vesta derived from geological mapping (Williams et al., 2014). The uncertainty in the ages stems from the uncertainty of the flux of impacting small asteroids on Vesta. Credit: NASA/JPL-Caltech/ASU.

time period covers the time between the Veneneia and Rheasilvia impact events. The Rheasilvia time period covers the time between the Rheasilvia and Marcia impact events. The Marcia time period covers the time between the Marcia impact event until the present day.

Without returned samples (such as from the Moon), assumptions must be made to derive absolute ages of Vesta. Two methods were used to determine these ages. One method was to extrapolate crater production rates and ages derived for the Moon, which have been calibrated using ages determined for Apollo samples, to Vesta. The same projectile population and flux that struck the Moon is assumed to have also impacted Vesta. The other method is to use crater production rates and ages derived from models of asteroid belt dynamics.

Global Fe/O (0.30 ± 0.04) and Si/O (0.56 ± 0.06) mass ratios (equivalent to weight ratios) derived from gamma-ray measurements (Prettyman et al., 2012) are consistent with HEDs, which have Fe/O values that range from 0.17 to 0.40

and Si/O values that range from 0.47 to 0.58. These values are also consistent with some angrites, ureilites, and the anomalous Shallowater aubrite; however, none of these meteorites have reflectance spectra similar to Vesta. Measurements of the high-energy gamma-ray flux from smaller areas on Vesta's surface are also consistent with HEDs (Peplowski et al., 2013) and so is the calculated K/Th ratio (900 ± 400) (Prettyman et al., 2015).

Dawn was the first mission to an asteroid to have a neutron detector, which is used to determine H abundances and, therefore, the water content. The neutron detector measures thermal (energies less than 0.1 eV), epithermal (0.1 eV–0.7 MeV), and fast neutrons (> 0.7 MeV) from approximately the top meter of an asteroid's surface (e.g., Prettyman et al., 2011, 2016). As cosmic rays (primarily protons) bombard Vesta's surface, they collide with atoms and dislodge neutrons from their nuclei. These fast-moving neutrons can then collide with other nuclei and lose energy. The lighter the nuclei they strike, the more energy the neutron loses. Since hydrogen has the lowest atomic mass of any element, abundant hydrogen in the subsurface will significantly "slow" down the neutrons. The relative abundances of thermal, epithermal, and fast neutrons will be both a function of the amount of hydrogen and the average atomic mass of the elements in the surface. Fewer neutrons detected from a particular part of the surface indicates less hydrogen in that region (Prettyman et al., 2016). The hydrogen is assumed to be a constituent of H_2O or $-OH$.

Analyses of the fast neutron data (Lawrence et al., 2013) are also consistent with an HED-like mineralogy on Vesta's surface. Also, water contents as high as 400 ppm were calculated for Vesta's surface (Lawrence et al., 2013). This result is consistent with dark carbonaceous chondritic material on the surface. This result also confirmed the Hasegawa et al. (2003) detection of a weak (~1%) 3 µm feature for Vesta, which indicated OH and/or H_2O-bearing minerals on its surface. Dawn also observed a 2.8 µm absorption due to OH on Vesta's surface (De Sanctis et al., 2012b). Curvilinear features on the walls of young craters have been proposed (Scully et al., 2015) to be due to the impact release of water from deeply buried ice deposits that are too deep to be detected by GRaND; however, there is no supporting evidence that such ice deposits exist on Vesta.

Dawn confirmed that Vesta has an HED-like surface composition (e.g., McCoy et al., 2015). Almost certainly, Vesta is the parent body of almost all HEDs. Dawn also confirmed that impacts of carbonaceous chondrite projectiles were common on its surface. However, the presence of an olivine mantle has not been confirmed.

8.13.2 (1) Ceres

After leaving Vesta's orbit, Dawn traveled to Ceres. During its approach, Dawn imaged Ceres. As expected, Ceres appeared heavily cratered (Figure 8.13).

Figure 8.13 Dawn images of (1) Ceres. Ceres has dimensions of 965 × 961 × 891 km³. The images were processed to enhance clarity. The images show the full range of different crater shapes that can be found at Ceres' surface. The crater shapes range from shallow, flattish craters to those with peaks at their centers. Note the white patches that are apparent on the surface. Credit: NASA/JPL-Caltech/UCLA/MPS/DLR/IDA.

A number of high-albedo "white spots" were observed on its surface during the approach. This was not unexpected since a high-albedo spot had been previously observed in Hubble Space Telescope images (Thomas et al., 2005).

Dawn entered orbit around Ceres in March 2015. Framing camera data of the bright areas in the Occator crater had reflectance spectra consistent with a high concentration of sodium carbonate mixed with dark background material (De Sanctis et al., 2016). The sodium carbonate is argued to be the solid residue from the crystallization of brines (salty water). The bright material was first proposed by Nathues et al. (2015b) to be hydrated magnesium sulfates. Craters on Ceres are named for agriculture gods and goddesses, while other features are named for world agricultural festivals. There is a lack of large craters on Ceres' surface compared to Vesta (Marchi et al., 2016). The largest recognizable craters (Kerwan and Yalode) on Ceres are 280 km and 260 km, respectively, in diameter while the smaller Vesta has two craters (Rheasilvia and Veneneia) larger than 400 km in diameter.

VIR spectra of Ceres from 0.4 to 5 µm were consistent with widespread ammoniated phyllosilicates on its surface but no water ice absorption features were apparent (De Sanctis et al., 2015). The presence of ammoniated phyllosilicates on Ceres' surface was first predicted by King et al. (1992) from the analysis of a 3 µm spectrum of Ceres. Since ammonia (NH_3) is only stable in the outer Solar System,

Figure 8.14 Map of neutron counts acquired by the GRaND instrument on Dawn for a portion of the northern hemisphere of (1) Ceres. The color scale of the map is from blue (lowest neutron count) to red (highest neutron count). The color information is based on the number of neutrons detected per second by GRaND. These data reflect the concentration of hydrogen in the upper meter of the regolith. Counts decrease with increasing hydrogen concentration. Lower neutron counts near the pole suggest the presence of water ice within a meter of the surface at high latitudes. Credit: NASA/JPL-Caltech/UCLA/MPS/DLR/IDA/PSI. (A black and white version of this figure will appear in some formats. For the color version, please refer to the plate section.)

Ceres may have formed in the outer Solar System and migrated to its present position or incorporated outer Solar System material when forming. Also apparent in the Dawn spectra of Ceres is an absorption band at 2.72 μm band that is consistent with Mg-rich serpentines (e.g., antigorite, chrysotile, lizardite) (De Sanctis et al., 2015; McSween et al., 2016). Mg-rich serpentines are produced by significant degrees of aqueous alteration and have absorption band positions due to –OH in serpentine near 2.7 μm (e.g., Takir et al., 2013). A feature at ~3.4 μm is consistent with organic material and one at ~4 μm is consistent with carbonates (De Sanctis et al., 2015; McSween et al., 2016). The ~3.4 μm feature for Ceres is weaker than the same feature in CI/CM chondrites while the ~4 μm feature is stronger than the one found in CI/CM chondrites (McSween et al., 2016).

GRaND measurements of Ceres detected fewer neutrons from the northern hemisphere than the equator (Figure 8.14), indicating higher hydrogen abundances in the top one meter of the regolith near the pole (Prettyman et al., 2016). These higher hydrogen abundances could indicate either higher ice or hydrated silicate abundances near the pole. Since the poles should have a lower subsolar surface temperature, these results are more consistent with larger concentrations of subsurface water ice near the poles (Prettyman et al., 2016).

From Dawn observations, Ceres' density was calculated to be 2161 kg/m³ (Park et al., 2015). This density is only 4% greater than the density for Ceres calculated by Thomas et al. (2005) from Earth-based (including an Earth-orbiting spacecraft) observations, giving confidence that asteroid densities can be accurately determined from Earth-based observations. The Thomas et al. (2005) density was calculated using dimensions calculated from Hubble Space Telescope images and an average mass determined from studies of gravitational perturbations of Ceres on other asteroids (e.g., Michalak, 2000). The low density for Ceres is consistent with a body that is ~25–35% water that is found either as ice or in hydrated silicates (Thomas et al., 2005; McSween et al., 2016).

McSween et al., (2016) find that Ceres is somewhat similar mineralogically to CI/CM chondrites. However, there are a number of mineralogical differences. Ceres appears to contain ammoniated clays and there are inferred differences in the relative proportions of different constituents such as carbonates and organic matter compared to CI/CM chondrites. McSween et al. (2016) argue that Ceres is more extensively altered than CI/CM chondrites.

8.14 Chang'e 2

Chang'e 2 (also written as Chang'E-2) was a Chinese spacecraft whose primary goal was to study the Moon but also flew by (4179) Toutatis. Chang'e 2 is part of a series of Chinese Lunar probes with Chang'e 1 being a previously launched lunar orbiter. Both spacecraft are named after an ancient Chinese Moon goddess (Chang'e).

Chang'e 2 launched in October 2010, orbited the Moon for eight months, and then left lunar orbit for the Earth–Sun L2 Lagrangian point. After leaving the L2 point, the spacecraft flew by Toutatis on December 13, 2012. Toutatis was classified as an Sq-type by DeMeo et al. (2009).

At closest approach, Chang'e 2 was 3.2 km from Toutatis. Chang'e 2 only imaged Toutatis and was able to identify a number of craters (Zou et al., 2014). The two-lobed nature of Toutatis, which was first identified using radar observations (Ostro et al., 1995), was confirmed using the images. Over 200 boulders were identified on the surface of Toutatis (Jiang et al., 2015).

8.15 Hayabusa2

Hayabusa2 is the JAXA sample return mission to C-complex NEA (162173) Ryugu. Hayabusa2 was launched at the beginning of December 2014 and is expected to reach its target in June 2018. It will leave Ryugu in December 2019 and return to

Earth in December 2020. Instruments on Hayabusa2 include a the Light Detection and Ranging Altimeter (LIDAR), the multi-band telescopic camera (ONC-T), the wide-angle cameras (ONC-W1 and ONC-W2), the Near-Infrared Spectrometer (NIRS3), the Thermal Infrared Imager (TIR), the Small Carry-on Impactor (SCI), a Deployable Camera (DCAM3), and the sampler (SMP) (Tachibana et al., 2014). NIRS3 will spectroscopically observe from 1.8 to 3.2 μm, which is a much longer wavelength region that the region covered by the original Hayabusa spectrometer. This wavelength region fully covers the 3 μm band due to water.

A lander (MASCOT) and three small rovers (MINERVA-II-1A, -1B, and -2) will also be part of the mission package (Tachibana et al., 2014). MASCOT has a multi-band wide-angle camera (CAM), a six-band thermal radiometer (MARA), a three-axis fluxgate magnetometer (MAG), and a hyperspectral microscope (MicrOmega). MicrOmega obtains images of a ~2 mm area and will obtain spectra from 0.9 to 3.5 μm. MASCOT will hopefully function for ~15 hours.

The Hayabusa2 sampler system is similar to the Hayabusa sampler (Tachibana et al., 2014). The plan is to retrieve at least 100 mg of material. Three projectiles will be used at three different sampling locations. A backup sampling method was designed for the sampler so that it has teeth like a comb to dig into the soil during touchdown and pick up some material. The sample catcher will then be placed into the sample container, which is located in the re-entry capsule that will return to Earth.

Reflectance spectra (Binzel et al., 2001a; Vilas, 2008; Moskovitz et al., 2013) of Ryugu and its albedo (~5%) (Ishiguro et al., 2014) are consistent with C-complex asteroids. Vilas (2008) noted a possible feature between 0.6 and 0.7 μm due to a charge-transfer absorption between Fe^{2+} and Fe^{3+} that may indicate hydrated silicates. This feature is not present in any of the other visible reflectance spectra and may indicate that the surface has a heterogeneous distribution of hydrated silicates.

8.16 PROCYON

PROCYON (Proximate Object Close flyby with Optical Navigation) is a microsatellite that was launched with Hayabusa2 in December 2014. The spacecraft was developed by the University of Tokyo and JAXA. The plan was for the spacecraft to fly by (185851) 2000 DP_{107} in 2016 (Naidu et al., 2015); however, a malfunction of the ion propulsion system did not allow this encounter to happen.

8.17 OSIRIS-REx

OSIRIS-REx (Origins Spectral Interpretation Resource Identification Security Regolith Explorer) is a launched NASA mission to asteroid (101955) Bennu, land on it, and acquire at least 60 g of material that will be returned to Earth for study. Osiris was chosen to honor the Egyptian god of the afterlife who was also the god of resurrection and fertility. Osiris was killed by his brother Seth, chopped into pieces, and then thrown into the Nile; however, Osiris was magically reassembled. The death and rebirth symbolized the annual death and regrowth of crops and the yearly flooding of the Nile. This name was fitting since Bennu should contain organic compounds that are similar to the material that potentially "seeded" the Earth. Rex means "king" in Latin. The PI of the mission is Dante Lauretta. Lauretta became PI after the original PI, Michael Drake (1946–2011), passed away.

OSIRIS-REx was launched on September 8, 2016. The spacecraft will reach Bennu in 2018, orbit this asteroid for a year and a half, briefly touch the surface, and then eject the Sample Return Capsule (SRC) that will return to Earth in 2023. Instruments include the OSIRIS-REx Laser Altimeter (OLA), the OSIRIS-REx Camera Suite (OCAMS), the OSIRIS-REx Thermal Emission Spectrometer (OTES), the OSIRIS-REx Visible and Infrared Spectrometer (OVIRS), and the Regolith X-ray Imaging Spectrometer (REXIS). The Touch-and-Go Sample Acquisition Mechanism (TAGSAM) is a robotic arm with an attached sampler head. When TAGSAM touches Bennu, nitrogen gas will be released, which will cause regolith to be directed into a collector. Also as a backup, contact pads on the sampler head will also trap small particles. Up to three attempts will be made. The sampler head will then be placed in the SRC. The goal is to collect at least 60 grams of regolith.

Bennu is classified as a B-type (e.g., Clark et al., 2011) due to its blue spectral slope in the visible and near-infrared. This object also has a low visual albedo of 5% (Emery et al., 2014). Clark et al. (2011) found that the best spectral match to Bennu was CI and/or CM chondrites. The estimated bulk density of Bennu (1260 ± 70 kg/m^3) (Chesley et al., 2014) is also consistent with a carbonaceous chondritic rubble pile assemblage.

Questions

1) What are the advantages and disadvantages of a flyby spacecraft mission?
2) What asteroids did the Galileo mission study? What are the major accomplishments of the Galileo mission?
3) What was the evidence that Eros had an ordinary chondrite composition?
4) What was the evidence from the Hayabusa mission that Itokawa had an LL chondrite composition?
5) What is the evidence from the Dawn mission that Vesta had a composition similar to HEDs?
6) What are the differences between Vesta and Ceres that were apparent from the results of the Dawn mission?

7) What can spacecraft missions accomplish that Earth-based telescopes cannot accomplish?

8) What is the importance of sample return missions?

9) What type of main-belt asteroid would be the best target for a future spacecraft mission? What five instruments should be part of the mission? Explain why you chose the instruments.

10) What type of near-Earth asteroid would be the best target for a future spacecraft mission? What five instruments should be part of the mission? Explain why you chose the instruments.

References

Abe, M., Takagi, Y., Kitazato, K., *et al.* (2006) Near-infrared spectral results of asteroid Itokawa from the Hayabusa spacecraft. *Science*, 312, 1334–1338.

Abell, P. A. and Gaffey, M. J. (2000) Probable geologic composition, thermal history, and meteorite affinities for mainbelt asteroid 349 Dembowska. Lunar and Planetary Science Conference, XXXI, 1291. www.lpi.usra.edu/meetings/lpsc2000/pdf/1291.pdf.

Abell, P. A., Vilas, F., Jarvis, K. S., Gaffey, M. J. and Kelley, M. S. (2007) Mineralogical composition of (25143) Itokawa 1998 SF_{36} from visible and near-infrared reflectance spectroscopy: Evidence for partial melting. *Meteoritics & Planetary Science*, 42, 2165–2177.

Agee, C. B., Wilson, N. V., McCubbin, F. M., *et al.* (2013) Unique meteorite from early Amazonian Mars: Water-rich basaltic breccia Northwest Africa 7034. *Science*, 339, 780–785.

A'Hearn, M. F., Belton, M. J. S., Delamere, W. A., *et al.* (2005) Deep Impact: Excavating comet Tempel 1. *Science*, 310, 258–264.

Albarède, F. (2011) Oxygen fugacity. In *Encyclopedia of Astrobiology*, eds. Gargaud, M., Amils, R., Quintanilla, J. C., Cleaves, H. J. II, Irvine, W. M., Pinti, D. L. and Viso, M. Berlin: Springer Berlin Heidelberg, p. 1196.

Alexandersen, M., Gladman, B., Greenstreet, S., Kavelaars, J. J., Petit, J.-M. and Gwyn, S. (2013) A Uranian Trojan and the frequency of temporary giant-planet co-orbitals. *Science*, 341, 994–997.

Altshuler, D. R. (2002) The National Astronomy and Ionosphere Center's (NAIC) Arecibo Observatory in Puerto Rico. In *Single–Dish Radio Astronomy: Techniques and Applications*, *ASP Conference Series 278*, eds. Stanimirović, S., Altschuler, D. R., Goldsmith, P. F. and Salter, C. J. San Francisco: Astronomical Society of the Pacific, pp. 1–24.

Alvarez, L. W., Alvarez, W., Asaro, F. and Michel, H. V. (1980) Extraterrestrial cause for the Cretaceous-Tertiary extinction. *Science*, 208, 1095–1108.

Alvarez-Candal, A., Duffard, R., Lazzaro, D. and Michtchenko, T. (2006) The inner region of the asteroid main belt: A spectroscopic and dynamic analysis. *Astronomy & Astrophysics*, 459, 969–976.

Amelin, Y. and Krot, A. (2007) Pb isotopic age of the Allende chondrules. *Meteoritics & Planetary Science*, 42, 1321–1335.

Amelin, Y., Krot, A. N., Hutcheon, I. D. and Ulyanov, A. A. (2002) Lead isotopic ages of chondrules and calcium–aluminum-rich inclusions. *Science*, 297, 1678–1683.

Amelin, Y., Kaltenbach, A., Iizuka, T., Stirling, C. H, Ireland, T. R., Petaev, M. and Jacobsen, S. B. (2010) U–Pb chronology of the Solar System's oldest solids with variable ^{238}U/^{235}U. *Earth and Planetary Science Letters*, 300, 343–350.

American Meteor Society (2016) Meteor shower calendar. www.amsmeteors.org/meteor-showers/meteor-shower-calendar/.

Ammannito, E., De Sanctis, M. C., Capaccioni, F., *et al.* (2013) Vestan lithologies mapped by the visual and infrared spectrometer on Dawn. *Meteoritics & Planetary Science*, 48, 2185–2198.

Anders, E. and Grevesse, N. (1989) Abundances of the elements: Meteoritic and solar. *Geochimica et Cosmochimica Acta*, 53, 197–214.

Appenzeller, I. (2012) *Introduction to Astronomical Spectroscopy (Cambridge Observing Handbooks for Research Astronomers)*. New York: Cambridge University Press, 268 pp.

Asher, D. J., Bailey, M. E., Hahn, G. and Steel, D. I. (1994) Asteroid 5335 Damocles and its implications for cometary dynamics. *Monthly Notices of the Royal Astronomical Society*, 267, 26–42.

Ashley, J. W. (2015) The study of exogenic rocks on Mars – An evolving subdiscipline in meteoritics. *Elements*, 11, 10–11.

Baer, J., Chesley, S. R. and Matson, R. D. (2011) Astrometric masses of 26 asteroids and observations on asteroid porosity. *Astronomical Journal*, 141, 143.

Baker, D. (2015) *NASA Hubble Space Telescope – 1990 Onwards (Including All Upgrades): An Insight into the History, Development, Collaboration, Construction and Role of the Earth-Orbiting Space Telescope (Owners' Workshop Manual)*. Newbury Park: Haynes Publishing UK, 181 pp.

Barkume, K. M., Brown, M. E. and Schaller, E. L. (2006) Water ice on the satellite of Kuiper belt object 2003 EL61. *Astrophysical Journal*, 640, L87–L89.

(2008) Near-infrared spectra of centaurs and Kuiper belt objects. *Astronomical Journal*, 135, 55–67.

Barrios, M. A., George, J. V. and Marschall, L. A. (2004) Photometry and astrometry of asteroids by Gettysburg College students at NURO. *Bulletin of the American Astronomical Society*, 36, 1348.

Barucci, M. A., Cruikshank, D. P., Dotto, E., *et al.* (2005) Is Sedna another Triton? *Astronomy & Astrophysics*, 439, L1–L4.

Barucci, M. A., Dalle Ore, C. M., Alvarez-Candal, A., *et al.* (2010) (90377) Sedna: Investigation of surface compositional variation. *Astronomical Journal*, 140, 2095–2100.

Barucci, M. A., Dalle Ore, C. M., Perna, D., *et al.* (2015) (50000) Quaoar: Surface composition variability. *Astronomy & Astrophysics*, 584, A107.

Basilevsky, A. T. and Keller, H. U. (2006) Comet nuclei: Morphology and implied processes of surface modification. *Planetary and Space Science*, 54, 808–829.

Batygin, K. and Brown, M. E. (2016) Evidence for a distant giant planet in the Solar System. *Astronomical Journal*, 151, 22.

Beck, P., Quirico, E., Montes-Hernandez, G., *et al.* (2010) Hydrous mineralogy of CM and CI chondrites from infrared spectroscopy and their relationship with low albedo asteroids. *Geochimica Cosmochimica Acta*, 74, 4881–4892.

Beck, P., Barrat, J.-A., Grisolle, F., *et al.* (2011a) NIR spectral trends of HED meteorites: Can we discriminate between the magmatic evolution, mechanical mixing and observation geometry effects? *Icarus*, 216, 560–571.

Beck, P., Quirico, E., Sevestre, D., Montes-Hernandez, G., Pommerol, A. and Schmitt, B. (2011b) Goethite as an alternative origin of the 3.1 μm band on dark asteroids. *Astronomy & Astrophysics*, 526, A85.

Beech, M. (2013) Towards an understanding of the fall circumstances of the Hoba meteorite. *Earth, Moon, and Planets*, 111, 15–30.

Beekman, G. (2006) I. O. Yarkovsky and the discovery of 'his' effect. *Journal for the History of Astronomy*, 37, 71–86.

Bell, J. F. (1986) Mineralogical evolution of meteorite parent bodies. *Lunar and Planetary Science Conference*, XVII, 985–986.

(1988) A probable asteroidal parent body for the CO or CV chondrites. *Meteoritics*, 23, 256–257.

Bell, J. F., Davis, D. R., Hartmann, W. K. and Gaffey, M. J. (1989) Asteroids: The big picture. In *Asteroids II*, eds. Binzel, R. P., Gehrels, T. and Matthews, M. S. Tucson: University of Arizona Press, pp. 921–945.

Belton, M. J. S., Klaasen, K. P., Clary, M. C., *et al.* (1992a) The Galileo Solid-State Imaging experiment. *Space Science Reviews*, 60, 413–455.

Belton, M. J. S., Veverka, J., Thomas, P., Helfenstein, P., *et al.* (1992b) Galileo encounter with 951 Gaspra: First pictures of an asteroid. *Science*, 257, 1647–1652.

Belton, M. J. S., Chapman, C. R., Thomas, P. C., *et al.* (1994) Bulk density of asteroid 243 Ida from the orbit of its satellite Dactyl. *Nature*, 374, 785–788.

Belton, M. J. S., Chapman, C. R., Klaasen, K. P., *et al.* (1996) Galileo's encounter with 243 Ida: An overview of the imaging experiment. *Icarus*, 120, 1–19.

Belton, M. J. S., Chapman, C. R., Veverka, J., *et al.* (2004) First images of asteroid 243 Ida. *Science*, 265, 1543–1547.

Bendjoya, Ph. and Zappalà, V. (2002) Asteroid family identification. In *Asteroids III*, eds. Bottke, W. F. Jr., Cellino, A., Paolicchi, P. and Binzel, R. P. Tucson: University of Arizona Press, pp. 613–618.

Benishek, V. and Pilcher, F. (2016) Rotation period and H-G parameters of (57868) 2001 YD. *Minor Planet Bulletin*, 43, 100–101.

Benner, L. A. M., Ostro, S. J., Magri, C., *et al.* (2008) Near-Earth asteroid surface roughness depends on compositional class. *Icarus*, 198, 294–304.

Bennett, C. J., Pirim, C. and Orlando, T. M. (2013) Space-weathering of Solar System bodies: A laboratory perspective. *Chemical Reviews*, 113, 9086–9150.

Bessell, M. S. (1990) UBVRI passbands. *Publications of the Astronomical Society of the Pacific*, 102, 1181–1199.

(2005) Standard photometric systems. *Annual Review of Astronomy and Astrophysics*, 43, 293–336.

Binzel, R. P. (1988) Hemispherical color differences on Pluto and Charon. *Science*, 241, 1070–1072.

(1997) A near-Earth object hazard index. *Annals of the New York Academy of Sciences*, 822, 545–551.

(2000) The Torino Impact Hazard Scale. *Planetary and Space Science*, 48, 297–303.

(2003) Spin control for asteroids. *Nature*, 425, 131–132.

Binzel, R. P. and van Flandern, T. C. (1979) Minor planets: The discovery of minor satellites. *Science*, 203, 903–905.

Binzel, R. P. and Xu, S. (1993) Chips off of asteroid 4 Vesta: Evidence for the parent body of basaltic achondrite meteorites. *Science*, 260, 186–191.

Binzel, R. P., Gehrels, T. and Matthews, M. S., eds. (1989) *Asteroids II*. Tucson: University of Arizona Press, 1258 pp.

Binzel, R. P., Xu, S., Bus, S. J., *et al.* (1993) Discovery of a main-belt asteroid resembling ordinary chondrite meteorites. *Science*, 262, 1541–1543.

Binzel, R. P., Bus, S. J., Xu, S., *et al.* (1995) Rotationally resolved spectra of asteroid 16 Psyche. *Icarus*, 117, 443–445.

Binzel, R. P., Gaffey, M. J., Thomas, P. C., Zellner, B. H., Storrs, A. D. and Wells, E. N. (1997) Geologic mapping of Vesta from 1994 Hubble Space Telescope images. *Icarus*, 128, 95–103.

Binzel, R. P., Harris, A. W., Bus, S. J. and Burbine, T. H. (2001a) Spectral properties of Near-Earth objects: Palomar and IRTF results for 48 objects including spacecraft targets (9969) Braille and (10302) 1989 ML. *Icarus*, 151, 139–149.

Binzel, R. P., Rivkin, A. S., Bus, S. J., Sunshine, J. M. and Burbine, T. H. (2001b) MUSES-C target asteroid (25143) 1998 SF36: A reddened ordinary chondrite. *Meteoritics & Planetary Science*, 36, 1167–1172.

Binzel, R. P., Rivkin, A. S., Stuart, J. S., Harris, A. W., Bus, S. J. and Burbine, T. H. (2004) Observed spectral properties of near-Earth objects: Results for population distribution, source regions, and space weathering processes. *Icarus*, 170, 259–294.

Binzel, R. P., Thomas, C. A., DeMeo, F. E., Tokunaga, A., Rivkin, A. S. and Bus, S. J. (2006) The MIT-Hawaii-IRTF joint campaign for NEO spectral reconnaissance. Lunar and Planetary Science Conference, XXXVII, 1491. www.lpi.usra.edu/meetings/lpsc2006/pdf/1491.pdf.

Binzel, R. P., Rivkin, A. S., Thomas, C. A., *et al.* (2009) Spectral properties and composition of potentially hazardous asteroid (99942) Apophis. *Icarus*, 200, 480–485.

Binzel, R. P., Morbidelli, A., Merouane, S., *et al.* (2010) Earth encounters as the origin of fresh surfaces on near-Earth asteroids. *Nature*, 463, 331–334.

Binzel, R. P., Reddy, V. and Dunn, T. (2015) The near-Earth object population: Connections to comets, main-belt asteroids, and meteorites. In *Asteroids IV*, eds. Michel, P., DeMeo, F. E. and Bottke, W. F. Tucson: University of Arizona Press, pp. 243–256.

Birck, J. L. and Allegre, C. J. (1979) ^{87}Rb-^{87}Sr chronology of the Binda howardite. *Nature*, 282, 288–289.

Birlan, M., Vernazza, P. and Nedelcu, D. A. (2007) Spectral properties of nine M-type asteroids. *Astronomy & Astrophysics*, 475, 747–754.

Bland, P. A., Cressey, G. and Menzies, O. N. (2004) Modal mineralogy of carbonaceous chondrites by X-ray diffraction and Mössbauer spectroscopy. *Meteoritics & Planetary Science*, 39, 3–16.

Bland, P. A., Zolensky, M. E., Benedix, G. K. and Sephton, M. A. (2006) Weathering of chondritic meteorites. In *Meteorites and the Early Solar System,* eds. Lauretta, D. S. and McSween, H. Y. Jr. Tucson: University of Arizona Press, pp. 853–867.

Bland, P. A., Spurný, P., Towner, M. C., *et al.* (2009) An anomalous basaltic meteorites from the innermost main belt. *Science*, 325, 1525–1527.

Blewett, D. T., Denevi, B. W., Le Corre, L., *et al.* (2016) Optical space weathering on Vesta: Radiative-transfer models and Dawn observations. *Icarus*, 265, 161–174.

Bockelée-Morvan, D., Lis, D. C., Wink, J. E., *et al.* (2000) New molecules found in comet C/1995 O1 (Hale-Bopp). Investigating the link between cometary and interstellar material. *Astronomy and Astrophysics*, 353, 1101–1114.

Boesenberg, J. S., Delaney, J. S. and Hewins, R. H. (2012) A petrological and chemical reexamination of Main Group pallasite formation. *Geochimica et Cosmochimica Acta*, 89, 134–158.

Bogard, D. D. (1995) Impact ages of meteorites. A synthesis. *Meteoritics*, 30, 244–268.

Bogard, D. D. and Johnson, P. (1983) Martian gases in an Antarctic meteorite. *Science*, 221, 651–654.

Boice, D. C., Soderblom, L. A., Britt, D. T., *et al.* (2002) The Deep Space 1 encounter with comet 19P/Borrelly. *Earth, Moon, and Planets*, 89, 301–324.

Borovička, J., Spurný, P., Brown, P., *et al.* (2013) The trajectory, structure and origin of the Chelyabinsk asteroidal impactor. *Nature*, 503, 235–237.

Borovička, J., Shrbený, L., Kalenda, P., *et al.* (2016) A catalog of video records of the 2013 Chelyabinsk superbolide. *Astronomy & Astrophysics*, 585, A90.

Boslough, M. B. and Harris, A. W. (2008) Global catastrophes in perspective: Asteroid impacts vs climate change. American Geophysical Union, Fall Meeting 2008, U41D-0034.

Bottke, W. F. Jr., Rubincam, D. P. and Burns, J. A. (2000) Dynamical evolution of main belt meteoroids: Numerical simulations incorporating planetary perturbations and Yarkovsky thermal forces. *Icarus*, 145, 301–331.

Bottke, W. F. Jr., Cellino, A., Paolicchi, P. and Binzel, R. P., eds. (2002) *Asteroids III*. Tucson: University of Arizona Press, 785 pp.

Bottke, W. F. Jr., Vokrouhlický, D., Rubincam, D. P. and Nesvorný, D. (2006) The Yarkovsky and YORP effects: Implications for asteroid dynamics. *Annual Review of Earth and Planetary Sciences*, 34, 157–191.

Bottke, W. F., Vokrouhlický, D. and Nesvorný, D. (2007) An asteroid breakup 160 Myr ago as the probable source of the K/T impactor. *Nature*, 449, 48–53.

Bottke, W., Vokrouhlický, D., Nesvorný, D. and Shrbeny, L. (2010) (6) Hebe really is the H chondrite parent body. *Bulletin of the American Astronomical Society*, 42, 1051.

Bouvier, A. and Wadhwa, M. (2010) The age of the Solar System redefined by the oldest Pb-Pb age of a meteoritic inclusion. *Nature Geoscience*, 3, 637–641.

Bouvier, A., Blichert-Toft, J., Moynier, F., Vervoort, J. D. and Albarède, F. (2007) Pb-Pb dating constraints on the accretion and cooling history of chondrites. *Geochimica et Cosmochimica Acta*, 71, 1583–1604.

Bowell, E. and Lumme, K. (1979) Colorimetry and magnitudes of asteroids. In *Asteroids*, ed. Gehrels, T. Tucson: University of Arizona Press, pp. 132–169.

Bowell, E., Hapke, B., Domingue, D., Lumme, K., Peltoniemi, J. and Harris, A. W. (1989) Application of photometric models to asteroids. In *Asteroids II*, eds. Binzel, R. P., Gehrels, T. and Matthews, M. S. Tucson: University of Arizona Press, pp. 524–556.

Braga-Ribas, F., Sicardy, B., Ortiz, J. L., *et al.* (2013) The size, shape, albedo, density, and atmospheric limit of transneptunian object (50000) Quaoar from multi-chord stellar occultations. *Astrophysical Journal*, 773, 26.

Braga-Ribas, F., Sicardy, B., Ortiz, J. L., *et al.* (2014) A ring system detected around the centaur (10199) Chariklo. *Nature*, 508, 72–75.

Brasil, P. I. O., Roig, F., Nesvorný, D., Carruba, V., Aljbaae, S. and Huaman, M. E. (2016) Dynamical dispersal of primordial asteroid families. *Icarus*, 266, 142–151.

Brearley, A. J. and Jones, R. H. (1998) Chondritic meteorites. In *Reviews in Mineralogy, Vol. 36: Planetary Materials*, ed. Papike, J. J. Washington, DC: Mineralogical Society of America, pp. 3-001–3-398.

Brennecka, G. A. and Wadhwa, M. (2012) Uranium isotope compositions of the basaltic angrite meteorites and the chronological implications for the early Solar System. *Proceedings of the National Academy of Sciences*, 109, 9299–9303.

Britt, D. T., Bell, J. F., Haack, H. and Scott, E. R. D. (1992) The reflectance spectrum of troilite and the T-type asteroids. *Meteoritics*, 27, 207.

Britt, D. T., Yeomans, D., Housen, K. and Consolmagno, G. (2002) Asteroid density, porosity, and structure. In *Asteroids III*, eds. Bottke, W. F. Jr., Cellino, A., Paolicchi, P. and Binzel, R. P. Tucson: University of Arizona Press, pp. 485–500.

Britt, D. T., Boice, D. C., Buratti, B. J., *et al.* (2004) The morphology and surface processes of comet 19/P Borrelly. *Icarus*, 167, 45–53.

Brown, M. (2010) *How I Killed Pluto and Why it had it Coming*. New York: Spiegel & Grau, 288 pp.

Brown, M. E. (2013) On the size, shape, and density of dwarf planet Makemake. *Astrophysical Journal Letters*, 767, L7.

Brown, M. E. and Calvin, W. M. (2000) Evidence for crystalline water and ammonia ices on Pluto's satellite Charon. *Science*, 287, 107–109.

Brown, M. E. and Schaller, E. L. (2007) The mass of dwarf planet Eris. *Science*, 316, 1585.

Brown, M. E., Trujillo, C. and Rabinowitz, D. (2004) Discovery of a candidate inner Oort Cloud planetoid. *Astrophysical Journal*, 617, 645–649.

Brown, M. E., van Dam, M. A., Bouchez, A. H., *et al.* (2006) Satellites of the largest Kuiper belt objects. *Astrophysical Journal*, 639, L43–L46.

Brown, M. E., Barkume, K. M., Ragozzine, D. and Schaller, E. L. (2007) A collisional family of icy objects in the Kuiper belt. *Nature*, 446, 294–296.

Brown, M. E., Ragozzine, D., Stansberry, J. and Fraser, W. C. (2010) The size, density, and formation of the Orcus-Vanth system in the Kuiper belt. *Astronomical Journal*, 139, 2700–2705.

Brown, P. G., Hildebrand, A. R., Zolensky, M. E., *et al.* (2000) The fall, recovery, orbit, and composition of the Tagish Lake meteorite: A new type of carbonaceous chondrite. *Science*, 290, 320–325.

Brown, P. G., Assink, J. D., Astiz, L., *et al.* (2013) A 500-kiloton airburst over Chelyabinsk and an enhanced hazard from small impactors. *Nature*, 503, 238–241.

Brownlee, D. E. (2012) The Stardust comet mission: Studying sediments from the Solar System's frozen attic. *Elements*, 8, 327–328.

Brownlee, D. E., Horz, F., Newburn, R. L., *et al.* (2004) Surface of young Jupiter family comet 81P/Wild 2: View from the Stardust spacecraft. *Science*, 304, 1764–1769.

Brownlee, D., Joswiak, D. and Matrajt, G. (2012) Overview of the rocky component of Wild 2 comet samples: Insight into the early Solar System, relationship with meteoritic materials and the differences between comets and asteroids. *Meteoritics & Planetary Science*, 47, 453–470.

Brozović, M., Showalter, M. R., Jacobson, R. A. and Buie, M. W. (2015) The orbits and masses of satellites of Pluto. *Icarus*, 246, 317–329.

Bruck, S. M., Owen, J. M. and Miller, P. L. (2016) Deflection by kinetic impact: Sensitivity to asteroid properties. *Icarus*, 269, 50–61.

Brunetto, R. (2009) Space weathering of Small Solar system bodies. *Earth, Moon, and Planets*, 105, 249–255.

Brunetto, R., Barucci, M. A., Dotto, E. and Strazzulla, G. (2006) Ion irradiation of frozen methanol, methane, and benzene: Linking to the colors of centaurs and trans-Neptunian objects. *Astrophysical Journal*, 644, 646–650.

Bryson, J. F. J., Nichols, C. I. O., Herrero-Albillos, J., *et al.* (2015) Long-lived magnetism from solidification-driven convection on the pallasite parent body. *Nature*, 517, 472–475.

Buchanan, P. C., Zolensky, M. E. and Reid, A. M. (1993) Carbonaceous chondrite clasts in the howardites Bholghati and EET87513. *Meteoritics*, 28, 659–669.

Buchheim, R. K. (2010) Methods and lessons learned determining the H-G parameters of asteroid phase curves. In *The Society for Astronomical Science 29th Annual Symposium on Telescope Science*. pp. 101–115. www.socastrosci.org/images/SAS_2010_Proceedings.pdf.

Buchner, E., Schwarz, W. H., Schmieder, M. and Trieloff, M. (2010) Establishing a 14.6 ± 0.2 Ma age for the Nördlinger Rie impact (Germany) – A prime example for concordant isotopic ages from various dating materials. *Meteoritics & Planetary Science*, 45, 662–674.

Buie, M. W., Grundy, W. M., Young, E. F., Young, L. A. and Stern, S. A. (2010a) Pluto and Charon with the Hubble Space Telescope. I. Monitoring global change and improved surface properties from light curves. *Astronomical Journal*, 139, 1117–1127.

(2010b) Pluto and Charon with the Hubble Space Telescope. II. Resolving changes on Pluto's surface and a map for Charon. *Astronomical Journal*, 139, 1128–1143.

Bullock, E. S., Gounelle, M., Lauretta, D. S., Grady, M. M. and Russell, S. S. (2005) Mineralogy and texture of Fe-Ni sulfides in CI1 chondrites: Clues to the extent of aqueous alteration on the CI1 parent body. *Geochimica et Cosmochimica Acta*, 69, 2687–2700.

Buratti, B. J., Britt, D. T., Soderblom, L. A., *et al.* (2004) 9969 Braille: Deep Space 1 infrared spectroscopy, geometric albedo, and classification. *Icarus*, 167, 129–135.

Burbine, T. H. (1998) Could G-class asteroids be the parent bodies of the CM chondrites? *Meteoritics & Planetary Science*, 33, 253–258.

Burbine, T. H. and Binzel, R. P. (2002) Small Main-belt Asteroid Spectroscopic Survey in the near-infrared. *Icarus*, 159, 468–499.

Burbine, T. H., Gaffey, M. J. and Bell, J. F. (1992) S-asteroids 387 Aquitania and 980 Anacostia: Possible fragments of the breakup of a spinel-bearing parent body with CO3/CV3 affinities. *Meteoritics*, 27, 424–434.

Burbine, T. H., Meibom, A. and Binzel, R. P. (1996) Mantle material in the main belt: Battered to bits? *Meteoritics & Planetary Science*, 31, 607–620.

Burbine, T. H., Buchanan, P. C., Binzel, R. P., *et al.* (2001) Vesta, Vestoids, and the howardite, eucrite, diogenite group: Relationships and the origin of spectral differences. *Meteoritics & Planetary Science*, 36, 761–781.

Burbine, T. H., McCoy, T. J., Nittler, L. R., Benedix, G. K., Cloutis, E. A. and Dickinson, T. L. (2002) Spectra of extremely reduced assemblages: Implications for Mercury. *Meteoritics & Planetary Science*, 37, 1233–1244.

Burbine, T. H., McCoy, T. J., Jarosewich, E. and Sunshine, J. M. (2003) Deriving asteroid mineralogies from reflectance spectra: Implications for the MUSES-C target asteroid. *Antarctic Meteorite Research*, 16, 185–195.

Burbine, T. H., McCoy, T. J., Hinrichs, J. L. and Lucey, P. G. (2006) Spectral properties of angrites. *Meteoritics & Planetary Science*, 41, 1139–1145.

Burbine, T. H., Buchanan, P. C., Dolkar, T. and Binzel, R. P. (2009) Pyroxene mineralogies of near-Earth Vestoids. *Meteoritics & Planetary Science*, 44, 1331–1341.

Burbine, T. H., Duffard, R., Buchanan, P. C., Cloutis, E. A. and Binzel, R. P. (2011) Spectroscopy of O-type asteroids. Lunar and Planetary Science Conference, 42, 2483. www.lpi.usra.edu/meetings/lpsc2011/pdf/2483.pdf.

Burchell, M. J. and Leliwa-Kopystynski, J. (2010) The large crater on the small asteroid (2867) Steins. *Icarus*, 210, 707–712.

Burns, R. G. (1993) *Mineralogical Applications of Crystal Field Theory (2nd Edition)*. Cambridge: Cambridge University Press, 576 pp.

Burton, A. S., Glavin, D. P., Callahan, M. P., Dworkin, J. P., Jenniskens, P. and Shaddad, M. H. (2011) Heterogeneous distributions of amino acids provide evidence of multiple sources within the Almahata Sitta parent body, asteroid 2008 TC₃. *Meteoritics & Planetary Science*, 46, 1703–1712.

Burton, A. S., Elsila, J. E., Hein, J. E., Glavin, D. P. and Dworkin, J. P. (2013) Extraterrestrial amino acids identified in metal-rich CH and CB carbonaceous chondrites from Antarctica. *Meteoritics & Planetary Science*, 48, 390–402.

Burton, A. S., McLain, H., Glavin, D. P., *et al.* (2015) Amino acid analyses of R and CK chondrites. *Meteoritics & Planetary Science*, 50, 470–482.

Bus, S. J. (1999) Compositional structure in the asteroid belt: Results of a spectroscopic survey. PhD Thesis, Massachusetts Institute of Technology, Cambridge, Massachusetts, 367 pp.

Bus, S. J. and Binzel, R. P. (2002a) Phase II of the Small Main-belt Asteroid Spectroscopic Survey: The observations. *Icarus*, 158, 106–145.

(2002b) Phase II of the Small Main-belt Asteroid Spectroscopic Survey: A feature-based taxonomy. *Icarus*, 158, 146–177.

Buseck, P. R. (1977) Pallasite meteorites – Mineralogy, petrology and geochemistry. *Geochmica et Cosmochimica Acta*, 41, 711–740.

Busfield, A., Turner, G. and Gilmour, J. D. (2008) Testing an integrated chronology: I-Xe analysis of enstatite meteorites and a eucrite. *Meteoritics & Planetary Science*, 43, 883–897.

Caldicott, H. (1994) *Nuclear Madness: What You Can Do* (Revised Edition). New York: W. W. Norton & Company, 240 pp.

Campins, H. and Swindle, T. D. (1998) Expected characteristics of cometary meteorites. *Meteoritics & Planetary Science*, 33, 1201–1211.

Campins, H., Hargrove, K., Pinilla-Alonso, N., *et al.* (2010) Water ice and organics on the surface of the asteroid 24 Themis. *Nature*, 464, 1320–1321.

Canup, R. M. (2005) A giant impact origin of Pluto-Charon. *Science*, 307, 546–550.

Carporzen, L., Weiss, B. P., Elkins-Tanton, L. T., Shuster, D. L., Ebel, D. and Gattacceca, J. (2011) Magnetic evidence for a partially differentiated carbonaceous chondrite parent body. *Proceedings of the National Academy of Sciences*, 108, 6386–6389.

Carlson, R. W., Weissman, P. R., Smythe, W. D., Mahoney, J. C. and the NIMS Science and Engineering Teams (1992) Near-infrared mapping spectrometer experiment on Galileo. *Space Science Reviews*, 60, 457–502.

Carruba, V., Michtchenko, T. A., Roig, F., Ferraz-Mello, S. and Nesvorný, D. (2005) On the V-type asteroids outside the Vesta family. I. Interplay of nonlinear secular resonances and the Yarkovsky effect: The cases of 956 Elisa and 809 Lundia. *Astronomy & Astrophysics*, 441, 819–829.

Carruba, V., Michtchenko, T. A. and Lazzaro, D. (2007) On the V-type asteroids outside the Vesta family. II. Is (21238) 1995 WV7 a fragment of the long-lost basaltic crust of (15) Eunomia? *Astronomy & Astrophysics*, 473, 967–978.

Carry, B. (2012) Density of asteroids. *Planetary and Space Science*, 73, 98–118.

Carry, B., Vernazza, P., Dumas, C. and Fulchignoni, M. (2010) First disk-resolved spectroscopy of (4) Vesta. *Icarus*, 205, 473–482.

Carry, B., Vernazza, P., Dumas, C., *et al.* (2012) The remarkable surface homogeneity of the Dawn mission target (1) Ceres. *Icarus*, 217, 20–26.

Carusi, A., Kresák, L. and Valsecchi, G. B. (1995) Conservation of the Tisserand parameter at close encounters of interplanetary objects with Jupiter. *Earth, Moon, and Planets*, 68, 71–94.

Carvano, J. M., Hasselmann, P. H., Lazzaro, D. and Mothé-Diniz, T. (2010) SDSS-based taxonomic classification and orbital distribution of main belt asteroids. *Astronomy & Astrophysics*, 510, A43.

Casanova, I., Graf, T. and Marti, K. (1995) Discovery of an unmelted H-chondrite inclusion in an iron meteorite. *Science*, 268, 540–542.

Cassidy, W. A. (2003) *Meteorites, Ice, and Antarctica: A Personal Account*. New York: Cambridge University Press, 364 pp.

Castillo-Rogez, J. C. (2011) Ceres – Neither a porous nor salty ball. *Icarus*, 215, 599–602.

Chabot, N. L. and Haack, H. (2006) Evolution of asteroidal cores. In *Meteorites and the Early Solar System*, eds. Lauretta, D. S. and McSween, H. Y. Jr. Tucson: University of Arizona Press, pp. 747–771.

Chambers, J. and Mitton, J. (2013) *From Dust to Life: The Origin and Evolution of our Solar System*. Princeton: Princeton University Press, 320 pp.

Chan, Q. H. S., Chikaraishi, Y., Takano, Y., Ogawa, N. O. and Ohkouchi, N. (2016) Amino acid compositions in heated carbonaceous chondrites and their compound-specific nitrogen isotopic ratios. *Earth, Planets and Space*, 68, 7.

Chandrasekhar, S. (1960) *Radiative Transfer*. New York: Dover Publications, 416 pp.

Chandler, D. (1998) Planetary science: The burger bar that saved the world. *Nature*, 453, 1164–1168.

Chang, K. (2016) You could actually snooze your way through an asteroid belt. www .nytimes.com/2016/04/05/science/you-could-actually-snooze-your-way-through-an-asteroid-belt.html.

Chao, E. C. T., Shoemaker, E. M. and Madsen, B. M. (1960) First natural occurrence of coesite. *Science*, 132, 220–222.

Chao, E. C. T., Fahey, J. J., Littler, J. and Milton, D. J. (1962) Stishovite, SiO_2, a very high pressure new mineral from Meteor Crater, Arizona. *Journal of Geophysical Research*, 67, 419–421.

Chapman, C. R. (1986) Implications of the inferred compositions of asteroids for their collisional evolution. *Memorie della Società Astronomica Italiana*, 57, 103–112.

(1996) S-type asteroids, ordinary chondrites, and space weathering: The evidence from Galileo's fly-bys of Gaspra and Ida. *Meteoritics*, 31, 699–725.

(2004) Space weathering of asteroid surfaces. *Annual Review of Earth and Planetary Sciences*, 32, 539–567.

(2008) Meteoroids, meteors, and the near-Earth object impact hazard. *Earth, Moon, and Planets*, 102, 417–424.

Chapman, C. R. and Morrison, D. (1994) Impacts on the Earth by asteroids and comets: Assessing the hazard. *Nature*, 367, 33–40.

Chapman, C. R. and Salisbury, J. W. (1973) Comparisons of meteorite and asteroid spectral reflectivities. *Icarus*, 19, 507–522.

Chapman, C. R., Morrison, D. and Zellner, B. (1975) Surface properties of asteroids: A synthesis of polarimetry, radiometry, and spectrophotometry. *Icarus*, 25, 104–130.

Chapman, C. R., Ryan, E. V., Merline, W. J., *et al.* (1996a) Cratering on Ida. *Icarus*, 120, 77–86.

Chapman, C. R., Veverka, J., Belton, M. J. S., Neukum, G. and Morrison, D. (1996b) Cratering on Gaspra. *Icarus*, 120, 231–245.

Chapman, C. R., Merline, W. J., Thomas, P. C., Joseph, J., Cheng, A. F. and Izenberg, N. (2002) Impact history of Eros: Craters and boulders. *Icarus*, 155, 104–118.

Chapman, C. R., Enke, B., Merline, W. J., Nesvorný, D., Tamblyn, P. and Young, E. F. (2009) Reflectance spectra of members of very young asteroid families. Lunar and Planetary Science Conference, 40, 2258. www.lpi.usra.edu/meetings/lpsc2009/pdf/2258.pdf.

Chen, M. and El Goresy, A. (2000) The nature of maskelynite in shocked meteorites: Not diaplectic glass but a glass quenched from shock-induced dense melt at high pressures. *Earth and Planetary Science Letters*, 179, 489–502.

Cheng, A. F. and Barnouin-Jha, O. S. (1999) Giant craters on Mathilde. *Icarus*, 140, 34–48.

Chesley, S. R., Chodas, P. W., Milani, A., Valsecchi, G. B. and Yeomans, D. K. (2002) Quantifying the risk posed by potential Earth impacts. *Icarus*, 159, 423–432.

Chesley, S. R., Ostro, S. J., Vokrouhlický, D., *et al.* (2003) Direct detection of the Yarkovsky effect by radar ranging to asteroid 6489 Golevka. *Science,* 302, 1739–1742.

Chesley, S. R., Farnocchia, D., Nolan, M. C., *et al.* (2014) Orbit and bulk density of the OSIRIS-REx target asteroid (101955) Bennu. *Icarus*, 235, 5–22.

Chodas, P. and Chesley, S. R. (2014) The trajectory of the Chelyabinsk impactor. *American Astronomical Society, DDA meeting*, 45, 103.01.

Chodas, P., Chesley, S. and Yeomans, D. (2010) The trajectory and impact circumstances of asteroid 2008 TC3. *Bulletin of the American Astronomical Society*, 42, 931.

Choe, W. H., Huber, H., Rubin, A. E., Kallemeyn, G. W. and Wasson, J. T. (2010) Compositions and taxonomy of 15 unusual carbonaceous chondrites. *Meteoritics and Planetary Science*, 45, 531–554.

Christou, A. A. (2013) Orbital clustering of Martian Trojans: An asteroid family in the inner Solar System? *Icarus*, 224, 144–153.

Ciarniello, M., Capaccioni, F. and Filacchione, G. (2014) A test of Hapke's model by means of Monte Carlo ray-tracing. *Icarus*, 237, 293–305.

Ciesla, F. J. (2005) Chondrule-forming processes–An overview. In *Chondrites and the Protoplanetary Disk, ASP Conference Series, Volume 341*, eds. Krot, A. N., Scott, E. R. D. and Reipurth, B. San Francisco: Astronomical Society of the Pacific, pp. 811–820.

(2010) The distributions and ages of refractory objects in the solar nebula. *Icarus*, 208, 455–467.

Clark, B. E. (1995) Spectral mixing models of S-type asteroids. *Journal of Geophysical Research*, 100, 14443–14456.

Clark, B. E., Bus, S. J., Rivkin, A. S., *et al.* (2004) E-type asteroid spectroscopy and compositional modeling. *Journal of Geophysical Research*, 109, E02001.

Clark, B. E., Ockert-Bell, M. E., Cloutis, E. A., Nesvorný, D., Mothé-Diniz, T. and Bus, S. J. (2009) Spectroscopy of K-complex asteroids: Parent bodies of carbonaceous meteorites? *Icarus*, 202, 119–133.

Clark, B. E., Ziffer, J., Nesvorný, D., *et al.* (2010) Spectroscopy of B-type asteroids: Subgroups and meteorite analogs. *Journal of Geophysical Research*, 115, E06005.

Clark, B. E., Binzel, R. P., Howell, E., *et al.* (2011) Asteroid (101955) 1999 RQ36: Spectroscopy from 0.4 to 2.4 µm and meteorite analogs. *Icarus*, 216, 462–475.

Clark, R. N. (1999) Spectroscopy of rocks and minerals, and principles of spectroscopy. In *Remote Sensing for the Earth Sciences: Manual of Remote Sensing (3rd Edition), Volume 3*, ed. Rencz, A. N. New York: John Wiley & Sons, Inc., pp. 3–52.

Clark, R. N. and Roush, T. L. (1984) Reflectance spectroscopy: Quantitative analysis techniques for remote sensing applications. *Journal of Geophysical Research*, 89, 6329–6340.

Clark, R. N., Swayze, G. A., Wise, R., *et al.* (2007) *USGS Digital Spectral Library splib06a*, U.S. Geological Survey Data Series 231. Tucson: U.S. Geological Survey.

Clayton, R. N. and Mayeda, T. K. (1978) Genetic relations between iron and stony meteorites. *Earth and Planetary Science Letters*, 40, 168–174.

(1988) Formation of ureilites by nebular processes. *Geochimica et Cosmochimica Acta*, 52, 1313–1318.

Clayton, R. N., Onuma, N., Grossman, L. and Mayeda, T. K. (1977) Distribution of the presolar component in Allende and other carbonaceous chondrites. *Earth and Planetary Science Letters*, 34, 209–224.

Cloutis, E. A. (2001) H_2O/OH-associated absorption band depth relationships in mineral reflectance spectra. Lunar and Planetary Science Conference, XXXII, 1146. www.lpi.usra.edu/meetings/lpsc2001/pdf/1146.pdf.

Cloutis, E. A. and Gaffey, M. J. (1993) Accessory phases in aubrites: Spectral properties and implications for asteroid 44 Nysa. *Earth, Moon, and Planets*, 63, 227–243.

Cloutis, E. A., Gaffey, M. J., Jackowski, T. L. and Reed, K. L. (1986) Calibrations of phase abundance, composition, and particle size distribution for olivine-orthopyroxene mixtures from reflectance spectra. *Journal of Geophysical Research*, 91, 11641–11653.

Cloutis, E. A., Gaffey, M. J., Smith, D. G. W. and Lambert, R. St. J. (1990) Reflectance spectra of "featureless" materials and the surface mineralogies of M- and E-class asteroids. *Journal of Geophysical Reseearch*, 95, 281–293.

Cloutis, E. A., Binzel, R. P., Burbine, T. H., Gaffey, M. J. and McCoy, T. J. (2006) Asteroid 3628 Božněmcová: Covered with angrite-like basalts? *Meteoritics & Planetary Science*, 41, 1147–1161.

Cloutis, E. A., Hardersen, P. S., Bish, D. L., Bailey, D. T., Gaffey, M. J. and Craig, M. A. (2010a) Reflectance spectra of iron meteorites: Implications for spectral identification of their parent bodies. *Meteoritics and Planetary Science*, 45, 304–332.

Cloutis, E. A., Hudon, P., Romanek, C. S., *et al.* (2010b) Spectral reflectance properties of ureilites. *Meteoritics & Planetary Science*, 45, 1668–1694.

Cloutis, E. A., Klima, R. L., Kaletzke, L., *et al.* (2010c) The 506 nm absorption feature in pyroxene spectra: Nature and implications for spectroscopy-based studies of pyroxene-bearing targets. *Icarus*, 207, 295–313.

Cloutis, E. A., Hiroi, T., Gaffey, M. J., Alexander, C. M. O'D. and Mann, P. (2011a) Spectral reflectance properties of carbonaceous chondrites: 1. CI chondrites. *Icarus*, 212, 180–209.

Cloutis, E. A., Hudon, P., Hiroi, T., Gaffey, M. J. and Mann, P. (2011b) Spectral reflectance properties of carbonaceous chondrites: 2. CM chondrites. *Icarus*, 216, 309–346.

Cloutis, E. A., Hudon, P., Hiroi, T. and Gaffey, M. J. (2012a) Spectral reflectance properties of carbonaceous chondrites: 3. CR chondrites. *Icarus*, 217, 389–407.

(2012b) Spectral reflectance properties of carbonaceous chondrites. 4: Aqueously altered and thermally metamorphosed meteorites. *Icarus*, 220, 586–617.

Cloutis, E. A., Hudon, P., Hiroi, T., Gaffey, M. J. and Mann, P. (2012c) Spectral reflectance properties of carbonaceous chondrites – 5: CO chondrites. *Icarus*, 220, 466–486.

Cloutis, E. A., Hudon, P., Hiroi, T., Gaffey, M. J., Mann, P. and Bell, J. F. (2012d) Spectral reflectance properties of carbonaceous chondrites: 6. CV chondrites. *Icarus*, 221, 328–358.

Cloutis, E. A., Hudon, P., Hiroi, T. and Gaffey, M. J. (2012e) Spectral reflectance properties of carbonaceous chondrites: 7. CK chondrites. *Icarus*, 221, 911–924.

Cloutis, E. A., Hudon, P., Hiroi, T., Gaffey, M. J. and Mann, P. (2012f) Spectral reflectance properties of carbonaceous chondrites: 8. "Other" carbonaceous chondrites: CH, ungrouped, polymict, xenolithic inclusions, and R chondrites. *Icarus*, 221, 984–1001.

Colwell, J. E. (1993) Power-law confusion: You say incremental, I say differential. *Lunar and Planetary Science Conference*, XXIV, 325–326.

Comelli, D., D'orazio, M., Folco, L., *et al.* (2016) The meteoritic origin of Tutankhamun's iron dagger blade. *Meteoritics & Planetary Science*, 51, 1301–1309.

Committee to Review Near-Earth Object Surveys and Hazard Mitigation Strategies, Space Studies Board, Aeronautics and Space Engineering Board, Division on Engineering and Physical Sciences and National Research Council (2010) *Defending Planet Earth: Near-Earth Object Surveys and Hazard Mitigation Strategies*. Washington, DC: National Academies Press, 134 pp.

Condon, D. J., McLean, N., Noble, S. R. and Bowring, S. A. (2010) Isotopic composition (^{238}U/^{235}U) of some commonly used uranium reference materials. *Geochimica et Cosmochimica Acta*, 74, 7127–7143.

Connelly, J. N., Bizzarro, M., Krot, A. N., Nordlund, Å., Wielandt, D. and Ivanova, M. A. (2012) The absolute chronology and thermal processing of solids in the solar protoplanetary disk. *Science*, 338, 651–655.

Connolly, H. C. Jr. (2005) Refractory inclusions and chondrules: Insights into a protoplanetary disk and planet formation. In *Chondrites and the Protoplanetary Disk, ASP Conference Series, Volume 341*, eds. Krot, A. N., Scott, E. R. D. and Reipurth, B. San Francisco: Astronomical Society of the Pacific, pp. 215–224.

Consolmagno, G. J. and Drake, M. J. (1977) Composition and evolution of the eucrite parent body: Evidence from rare earth elements. *Geochimica et Cosmochimica Acta*, 41, 1271–1282.

Consolmagno, G. J., Britt, D. T. and Macke, R. J. (2008) The significance of meteorite density and porosity. *Chemie der Erde*, 68, 1–29.

Consolmagno, G. J., Schaefer, M. W., Schaefer, B. E., *et al.* (2013) The measurement of meteorite heat capacity at low temperatures using liquid nitrogen vaporization. *Planetary and Space Science*, 87, 146–156.

Connors, M., Wiegert, P. and Veillet, C. (2011) Earth's Trojan asteroid. *Nature*, 475, 481–483.

Coradini, A., Capaccioni, F., Erard, S., *et al.* (2011) The surface composition and temperature of asteroid 21 Lutetia as observed by Rosetta/VIRTIS. *Science*, 334, 492–494.

Cournede, C., Gattacceca, J., Gounelle, M., Rochette, P., Weiss, B. P. and Zanda, B. (2015) An early Solar System magnetic field recorded in CM chondrites. *Earth and Planetary Science Letters*, 410, 62–74.

Cruikshank, D. P. and Brown, R. H. (1987) Organic matter on asteroid 130 Elektra. *Science*, 238, 183–184.

Cruikshank, D. P. and Hartmann, W. K. (1984) The meteorite-asteroid connection: Two olivine-rich asteroids. *Science*, 223, 281–283.

Cruikshank, D. P., Tholen, D. J., Bell, J. F., Hartmann, W. K. and Brown, R. H. (1991) Three basaltic earth-approaching asteroids and the source of the basaltic meteorites. *Icarus*, 89, 1–13.

Cruikshank, D. P., Roush, T. L., Bartholomew, M. J., *et al.* (1998) The composition of centaur 5145 Pholus. *Icarus*, 135, 389–407.

Cruikshank, D. P., Geballe, T. R., Owen, T. C., *et al.* (2002) Search for the 3.4-μm C-H spectral bands on low-albedo asteroids. *Icarus*, 156, 434–441.

Ćuk, M. and Burns, J. A. (2005) Effects of thermal radiation on the dynamics of binary NEAs. *Icarus,* 176, 418–431.

Cunningham, C. J. (1988) *Introduction to Asteroids: The Next Frontier.* Richmond: Willmann-Bell. 208 pp.

 (2016) *Discovery of the First Asteroid, Ceres: Historical Studies in Asteroid Research.* New York: Spring International Publishing, 333 pp.

Cunningham, C. J. and Orchiston, W. (2011) Who invented the word asteroid: William Herschel or Stephen Weston? *Journal of Astronomical History and Heritage*, 14, 230–234.

Curtis, H. D. (2014) *Orbital Mechanics for Engineering Students (Third Edition).* Waltham: Butterworth-Heinemann, 768 pp.

Dalrymple, G. B. (1991) *The Age of the Earth.* Stanford: Stanford University Press, 474 pp.

Davies, J. (1984) Asteroids-the comet connection. *New Scientist*, 1435/1436, 46–48.

Davis, A. M. (2011) Stardust in meteorites. *Proceeding of the National Academy of Sciences*, 108, 19142–19146.

Davis, A. M., Richter, F. M., Mendybaev, R. A., Janney, P. E., Wadhwa, M. and McKeegan, K. D. (2015) Isotopic mass fractionation laws for magnesium and their effects on ^{26}Al–^{26}Mg systematics in Solar System materials. *Geochimica et Cosmochimica Acta*, 158, 245–261.

Dean, J. (2015) The asteroid hunters. www.popularmechanics.com/space/a17822/the-asteroid-hunters/.

De Fourestier, J. (2002) The naming of mineral species approved by the commission on new minerals and mineral names of the International Mineralogical Association: A brief history. *Canadian Mineralogist*, 40, 1721–1735.

de la Fuente Marcos, C. and de la Fuente Marcos, R. (2014) Asteroid 2013 ND$_{15}$: Trojan companion to Venus, PHA to the Earth. *Monthly Notices of the Royal Astronomical Society*, 439, 2970–2977.

Delaney, J. S., Takeda, H., Prinz, M., Nehru, C. E. and Harlow, G. E. (1983) The nomenclature of polymict basaltic achondrites. *Meteoritics*, 18, 103–111.

Delbó, M. (2004) The nature of near-Earth asteroids from the study of their thermal emission. PhD thesis. Free University of Berlin, Berlin, Germany, 210 pp.

Delbó, M. and Harris, A. W. (2002) Physical properties of near-Earth asteroids from thermal infrared observations and thermal modeling. *Meteoritics & Planetary Science*, 37, 1929–1936.

Delbó, M. and Tanga, P. (2009) Thermal inertia of main belt asteroids smaller than 100 km from IRAS data. *Planetary and Space Science*, 57, 259–265.

Delbó, M., Gai, M., Lattanzi, M. G., *et al.* (2006) MIDI observations of 1459 Magnya: First attempt of interferometric observations of asteroids with the VLTI. *Icarus*, 181, 618–622.

Delbó, M., Cellino, A. and Tedesco, E. F. (2007a) Albedo and size determination of potentially hazardous asteroids: (99942) Apophis. *Icarus*, 188, 266–269.

Delbó, M., dell'Oro, A., Harris, A. W., Mottola, S. and Mueller, M. (2007b) Thermal inertia of near-Earth asteroids and implications for the magnitude of the Yarkovsky effect. *Icarus*, 190, 236–249.

DeMeo, F. E. and Carry, B. (2013) The taxonomic distribution of asteroids from multi-filter all-sky photometric surveys. *Icarus*, 226, 723–741.

(2014) Solar System evolution from compositional mapping of the asteroid belt. *Nature*, 505, 629–634.

DeMeo, F. E., Binzel, R. P., Slivan, S. M. and Bus, S. J. (2009) An extension of the Bus asteroid taxonomy into the near-infrared. *Icarus*, 202, 160–180.

DeMeo, F. E., Binzel, R. P., Carry, B., Polishook, D. and Moskovitz, N. A. (2014) Unexpected D-type interlopers in the inner main belt. *Icarus*, 229, 392–399.

De Sanctis, M. C., Ammannito, E., Migliorini, A., Lazzaro, D., Capria, M. T. and McFadden, L. (2011) Mineralogical characterization of some V-type asteroids, in support of the NASA Dawn mission. *Monthly Notices of the Royal Astronomical Society*, 412, 2318–2332.

De Sanctis, M. C., Ammannito, E., Capria, M. T., *et al.* (2012a) Spectroscopic characterization of mineralogy and its diversity across Vesta. *Science*, 336, 697–700.

De Sanctis, M. C., Combe, J.-Ph., Ammannito, E., *et al.* (2012b) Detection of widespread hydrated materials on Vesta by the VIR imaging spectrometer on board the Dawn mission. *Astrophysical Journal Letters*, 758, L36.

De Sanctis, M. C., Ammannito, E., Capria, M. T., *et al.* (2013) Vesta's mineralogical composition as revealed by the visible and infrared spectrometer on Dawn. *Meteoritics & Planetary Science*, 48, 2166–2184.

De Sanctis, M. C., Ammannito, E., Raponi, A., *et al.* (2015) Ammoniated phyllosilicates with a likely outer Solar System origin on (1) Ceres. *Nature*, 528, 241–244.

De Sanctis, M. C., Raponi, A., Ammannito, E., *et al.* (2016) Bright carbonate deposits as evidence of aqueous alteration on (1) Ceres. Nature, 536, 54–57.

Despois, D., Biver, N., Bockelée-Morvan, D. and Crovisier, J. (2005) Observations of molecules in comets. *In Astrochemistry – Recent Successes and Current Challenges*, eds. Lis, D. C., Blake, G. A. and Herbst, E. Cambridge: Cambridge University Press, pp. 119–128.

Di Cicco, D. (1996) Hunting asteroids from your backyard. *CCD Astronomy*, Spring, 8–13.

Dickin, A. P. (2005) *Radiogenic Isotope Geology*. Cambridge: Cambridge University Press, 512 pp.

Dombard, A. J., Barnouin, O. S., Prockter, L. M. and Thomas, P. C. (2010) Boulders and ponds on the asteroid 433 Eros. *Icarus*, 210, 713–721.

Dones, L., Brasser, R., Kaib, N. and Rickman, H. (2015) Origin and evolution of the cometary reservoirs. *Space Science Reviews*, 197, 191–269.

Doressoundiram, A., Boehnhardt, H., Tegler, S. C. and Trujillo, C. (2008) Color properties and trends of the transneptunian objects. In *The Solar System Beyond Neptune,* eds. Barucci, M. A., Boehnhardt, H., Cruikshank, D. P. and Morbidelli, A. Tucson: University of Arizona Press, pp. 91–104.

Drummond, J. D., Carry, B., Merline, W. J., *et al.* (2014) Dwarf planet Ceres: Ellipsoid dimensions and rotational pole from Keck and VLT adaptive optics images. *Icarus*, 236, 28–37.

Duerbeck, H. W. (2009) Principles of photometry. In *Handbook of Practical Astronomy*, ed. Roth, G. D. New York: Springer-Verlag, pp. 205–238.

Duffard, R. and Roig, F. (2009) Two new V-type asteroids in the outer main belt? *Planetary and Space Science*, 57, 229–234.

Duffard, R., Pinilla-Alonso, N., Ortiz, J. L., *et al.* (2014) Photometric and spectroscopic evidence for a dense ring system around centaur Chariklo. *Astronomy & Astrophysics*, 568, A79.

Dunham, D. W. (1978) Satellite of minor planet 532 Herculina discovered during occultation. *Minor Planet Bulletin*, 6, 13–14.

Dunn, T. L., McCoy, T. J., Sunshine, J. M. and McSween, H. Y. (2010) A coordinated spectral, mineralogical, and compositional study of ordinary chondrites. *Icarus*, 208, 789–797.

Dykhuis, M. J. and Greenberg, R. (2015) Collisional family structure within the Nysa-Polana complex. *Icarus*, 252, 199–211.

Dymock, R. (2007) The H and G magnitude system for asteroids. *Journal of the British Astronomical System*, 117, 342–343.

Duxbury, T. C., Newburn, R. L., Acton, C. H., *et al.* (2004) Asteroid 5535 Annefrank size, shape, and orientation: Stardust first results. *Journal of Geophysical Research*, 109, E02002.

Ebel, D. S. (2006) Condensation of rocky material in astrophysical environments. In *Meteorites and the Early Solar System,* eds. Lauretta, D. S. and McSween, H. Y. Jr. Tucson: University of Arizona Press, pp. 253–277.

Edgeworth, K. E. (1943) The evolution of our planetary system. *Journal of the British Astronomical Association*, 53, 181–188.

Elisa, J. E., de Leon, N. P., Buseck, P. R. and Zare, R. N. (2005) Alkylation of polycyclic aromatic hydrocarbons in carbonaceous chondrites. *Geochimica et Cosmochimica Acta*, 69, 1349–1357.

Elkins-Tanton, L. T. (2010) *Asteroids, Meteorites, and Comets (Revised Edition)*. New York: Facts on File, Inc., 270 pp.

El Moutamid, M., Kral, Q., Sicardy, B., *et al.* (2014) How can we explain the presence of rings around the centaur Chariklo? American Astronomical Society, DDA meeting, 45, 402.05.

Emery, J. P. and Brown, R. H. (2003) Constraints on the surface composition of Trojan asteroids from near-infrared (0.8–4.0 μm) spectroscopy. *Icarus*, 164, 104–121.

(2004) The surface composition of Trojan asteroids: Constraints set by scattering theory. *Icarus*, 170, 131–152.

Emery, J. P., Cruikshank, D. P. and van Cleve, J. (2006) Thermal emission spectroscopy (5.2–38 μm) of three Trojan asteroids with the Spitzer Space Telescope: Detection of fine-grained silicates. *Icarus*, 182, 496–512.

Emery, J. P., Burr, D. M. and Cruikshank, D. P. (2011) Near-infrared spectroscopy of Trojan asteroids: Evidence for two compositional groups. *Astronomical Journal*, 141, 25.

Emery, J. P., Fernández, Y. R., Kelley, M. S. P., *et al.* (2014) Thermal infrared observations and thermophysical characterization of OSIRIS-REx target asteroid (101955) Bennu. *Icarus*, 234, 17–35.

Emery, J. P., Marzari, F., Morbidelli, A., French, L. M. and Grav, T. (2015) The complex history of Trojan asteroids. In *Asteroids IV*, eds. Michel, P., DeMeo, F. E. and Bottke, W. F. Tucson: University of Arizona Press, pp. 203–220.

Eugster, O., Herzog, G. F., Marti, K. and Caffee, M. W. (2006) Irradiation records, cosmic-ray exposure ages, and transfer times of meteorites. In *Meteorites and the Early Solar System*, eds. Lauretta, D. S. and McSween, H. Y. Jr. Tucson: University of Arizona Press, pp. 829–851.

Evans, J. B., Shelly, F. C. and Stokes, G. H. (2003) Detection and discovery of near-Earth asteroids by the LINEAR program. *Lincoln Laboratory Journal*, 14, 199–220.

Evans, L. G., Starr, R. D., Brückner, J., *et al.* (2001) Elemental composition from gamma-ray spectroscopy of the NEAR-Shoemaker landing site on 433 Eros. *Meteoritics & Planetary Science*, 36, 1639–1660.

Evatt, G. W., Coughlan, M. J., Joy, K. H., Smedley, A. R. D., Connolly, P. J. and Abrahams, I. D. (2016) A potential hidden layer of meteorites below the ice surface of Antarctica. *Nature Communications*, 7, 10679.

Farinella, P., Gonczi, R., Froeschlé, Ch. and Froeschlé, C. (1993a) The injection of asteroid fragments into resonances. *Icarus*, 101, 174–187.

Farinella, P., Froeschlé, C. and Gonczi, R. (1993b) Meteorites from the asteroid 6 Hebe. *Celestial Mechanics and Dynamical Astronomy*, 56, 287–305.

Farnocchia, D. and Chesley, S. R. (2014) Assessment of the 2880 impact threat from asteroid (29075) 1950 DA. *Icarus*, 229, 321–327.

Farquhar, R., Kawaguchi, J., Russell, C., Schwehm, G., Veverka, J. and Yeomans, D. (2002) Spacecraft exploration of asteroids: The 2001 perspective. In *Asteroids II*, eds. Binzel, R. P., Gehrels, T. and Matthews, M. S. Tucson: University of Arizona Press, pp. 367–376.

Feierberg, M. A., Lebofsky, L. A. and Larson, H. P. (1981) Spectroscopic evidence for aqueous alteration products on the surfaces of low-albedo asteroids. *Geochimica et Cosmochimica Acta*, 45, 971–981.

Feierberg, M. A., Larson, H. P. and Chapman, C. R. (1982) Spectroscopic evidence for un-differentiated S-type asteroids. *Astrophysical Journal*, 257, 361–372.

Feierberg, M. A., Lebofsky, L. A. and Tholen, D. J. (1985) The nature of C-class asteroids from 3-micron spectrophotometry. *Icarus*, 63, 183–191.

Fernández, J. A. (1980) On the existence of a comet belt beyond Neptune. *Monthly Notices of the Royal Astronomical Society*, 192, 481–491.

Fernández, J. A. and Ip, W.-H. (1984) Some dynamical aspects of the accretion of Uranus and Neptune: The exchange of orbital angular momentum with planetesimals. *Icarus*, 58, 109–120.

Fernández, Y. R., Jewitt, D. C. and Sheppard, S. S. (2005) Albedos of asteroids in comet-like orbits. *Astronomical Journal*, 130, 308–318.

Florczak, M., Lazzaro, D. and Duffard, R. (2002) Discovering new V-type asteroids in the vicinity of 4 Vesta. *Icarus*, 159, 178–182.

Foderà Serio, G., Manara, A. and Sicoli, P. (2002) Giuseppe Piazzi and the discovery of Ceres. In *Asteroids III*, eds. Bottke, W. F. Jr., Cellino, A., Paolicchi, P. and Binzel, R. P. Tucson: University of Arizona Press, pp. 17–24.

Foley, C. N., Nittler, L. R., McCoy, T. J., *et al.* (2006) Minor element evidence that asteroid 433 Eros is a space-weathered ordinary chondrite parent body. *Icarus*, 184, 338–343.

Foote, A. E. (1891) A new locality for meteoric iron with a preliminary notice of the discovery of diamonds in the iron. *American Journal of Science*, 42, 413–417.

Fornasier, S., Dotto, E., Hainaut, O., *et al.* (2007) Visible spectroscopic and photometric survey of Jupiter Trojans: Final results on dynamical families. *Icarus,* 190, 622–642.

Fornasier, S., Clark, B. E., Dotto, E., Migliorini, A., Ockert-Bell, M. and Barucci, M. A. (2010) Spectroscopic survey of M-type asteroids. *Icarus,* 210, 655–673.

Fornasier, S., Clark, B. E. and Dotto, E. (2011) Spectroscopic survey of X-type asteroids. *Icarus,* 214, 131–146.

Fornasier, S., Lellouch, E., Müller, T., *et al.* (2013) TNOs are cool: A survey of the trans-Neptunian region. VIII. Combined Herschel PACS and SPIRE observations of nine bright targets at 70–500 μm. *Astronomy & Astrophysics,* 555, A15.

Fowler, J. W. and Chillemi, J. R. (1992) IRAS asteroid data processing. In *The IRAS Minor Planet Survey,* eds. Tedesco, E. F., Veeder, G. J., Fowler, J. W. and Chillemi, J. R. Bedford: Hanscom Air Force Base, pp. 17–43.

Franchi, I. A., Wright, I. P., Sexton, A. S. and Pillinger, C. T. (1999) The oxygen-isotopic composition of Earth and Mars. *Meteoritics & Planetary Science,* 34, 657–661.

French, B. (1998) *Traces of Catastrophe: A Handbook of Shock-Metamorphic Effects in Terrestrial Meteorite Impact Structures.* Houston: Lunar and Planetary Institute, 120 pp.

Friedrich, J. M., Weisberg, M. K., Ebel, D. S., *et al.* (2015) Chondrule size and related physical properties: A compilation and evaluation of current data across all meteorite groups. *Chemie der Erde,* 75, 419–443.

Fries, M., Le Corre, L., Hankey, M., *et al.* (2014) Detection and rapid recovery of the Sutter's Mill meteorite fall as a model for future recoveries worldwide. *Meteoritics & Planetary Science,* 49, 1989–1996.

Froeschlé, Ch. and Scholl, H. (1987) Orbital evolution of asteroids near the secular resonance v_6. *Astronomy and Astrophysics,* 179, 294–303.

Fu, R. R., Weiss, B. P., Shuster, D. L., *et al.* (2012) An ancient core dynamo in asteroid Vesta. *Science,* 338, 238–241.

Fu, R. R., Weiss, B. P., Lima, E. A., *et al.* (2014) Solar nebula magnetic fields recorded in the Semarkona meteorite. *Science,* 346, 1089–1092.

Fujiwara, A., Kawaguchi, J., Yeomans, D. K., *et al.* (2006) The rubble-pile asteroid Itokawa as observed by Hayabusa. *Science,* 312, 1330–1334.

Gaffey, M. J. (1974) A systematic study of the spectral reflectivity characteristics of the meteorite classes with applications to the interpretation of asteroid spectra for mineralogical and petrological information. PhD thesis, Massachusetts Institute of Technology, Cambridge, Massachusetts, 355 pp.

(1976) Spectral reflectance characteristics of the meteorite classes. *Journal of Geophysical Research,* 81, 905–920.

(1980) Mineralogically diagnostic features in the visible and near-infrared reflectance spectra of carbonaceous chondrite assemblages. *Lunar and Planetary Science Conference,* XI, 312–313.

(1984) Rotational spectral variations of asteroid (8) Flora: Implications for the nature of the S-type asteroids and for the parent bodies of the ordinary chondrites. *Icarus,* 60, 83–114.

(1995) The S(IV)-type asteroids as ordinary chondrite parent body candidates: Implications for the completeness of the meteorite sample of asteroids. *Meteoritics,* 30, 507–508.

(1997) Surface lithologic heterogeneity of asteroid 4 Vesta. *Icarus,* 127, 130–157.

Gaffey, M. J. and Gilbert, S. L. (1998) Asteroid 6 Hebe: The probable parent body of the H-type ordinary chondrites and the IIE iron meteorites. *Meteoritics and Planetary Science,* 33, 1281–1295.

Gaffey, M. J. and McCord, T. B. (1978) Asteroid surface materials: Mineralogical characterizations from reflectance spectra. *Space Science Reviews,* 21, 555–628.

Gaffey, M. J., Bell, J. F. and Cruikshank, D. P. (1989) Reflectance spectroscopy and asteroid surface mineralogy. In *Asteroids II*, eds. Binzel, R. P., Gehrels, T. and Matthews, M. S. Tucson: University of Arizona Press, pp. 98–127.

Gaffey, M. J., Reed, K. L. and Kelley, M. S. (1992) Relationship of E-type Apollo asteroid 3103 (1982 BB) to the enstatite achondrite meteorites and the Hungaria asteroids. *Icarus*, 100, 95–109.

Gaffey, M. J., Bell, J. F., Brown, *et al.* (1993) Mineralogic variations within the S-type asteroid class. *Icarus*, 106, 573–602.

Gaffey, M. J., Cloutis, E. A., Kelley, M. S. and Reed, K. L. (2002) Mineralogy of asteroids. In *Asteroids III*, eds. Bottke, W. F. Jr., Cellino, A., Paolicchi, P. and Binzel, R. P. Tucson: University of Arizona Press, pp. 183–204.

Gastineau-Lyons, H. K., McSween, H. Y. Jr. and Gaffey, M. J. (2002) A critical evaluation of oxidation versus reduction during metamorphism of L and LL group chondrites, and implications for asteroid spectroscopy. *Meteoritics & Planetary Science*, 37, 75–89

Gehrels T., ed. (1979) *Asteroids*. Tucson: University of Arizona Press, 1182 pp.

Genge, M. J. (2008) Koronis asteroid dust within Antarctic ice. *Geology*, 36, 687–690.

Ghosh, A., Weidenschilling, S. J., McSween, H. Y. Jr. and Rubin, A. (2002) Asteroidal heating and thermal stratification of the asteroid belt. In *Asteroids III*, eds. Bottke, W. F. Jr., Cellino, A., Paolicchi, P. and Binzel, R. P. Tucson: University of Arizona Press, pp. 555–566.

Giacomini, L., Massironi, M., Marchi, S., Fassett, C. I., Di Achille, G. and Cremonese, G. (2015) Age dating of an extensive thrust system on Mercury: Implications for the planet's thermal evolution. *Geological Society, London, Special Publications*, 401, 291–311.

Gielas, H. L. (2000) *The 13-inch Pluto Discovery Telescope*. Flagstaff: Lowell Observatory, 17 pp.

Gil-Hutton, R. and Brunini, A. (2008) Surface composition of Hilda asteroids from the analysis of the Sloan Digital Sky Survey colors. *Icarus*, 93, 567–571.

Gil-Hutton, R. and Licandro, J. (2010) Taxonomy of asteroids in the Cybele region from the analysis of the Sloan Digital Sky Survey colors. *Icarus*, 206, 729–734.

Gingerich, O. (2006) The inside story of Pluto's demotion. *Sky and Telescope*, 112, 34–39.

Giorgini, J. D., Benner, L. A. M., Ostro, S. J., Nolan, M. C. and Busch, M. W. (2008) Predicting the Earth encounters of (99942) Apophis. *Icarus*, 193, 1–19.

Gladman, B. and Coffey, J. (2009) Mercurian impact ejecta: Meteorites and mantle. *Meteoritics & Planetary Science*, 44, 285–291.

Gladman, B. J., Burns, J. A., Duncan, M., Lee, P. and Levison, H. F. (1996) The exchange of impact ejecta between terrestrial planets. *Science*, 271, 1387–1392.

Gladstone, G. R., Stern, S. A., Ennico, K., *et al.* (2016) The atmosphere of Pluto as observed by New Horizons. *Science*, 351, aad8866-1–aad8866-6.

Goesmann, F., Rosenbauer, H., Bredehöft, J. H., *et al.* (2015) Organic compounds on comet 67P/Churyumov-Gerasimenko revealed by COSAC mass spectrometry. *Science*, 349, aab0689-1–aab0689-3.

Goldberg, E., Uchiyama, A. and Brown, H. (1951) The distribution of nickel, cobalt, gallium, palladium, and gold in iron meteorites. *Geochimica et Cosmochimica Acta*, 2, 1–25.

Goldreich, P. and Tremaine, S. (1980) Disk-satellite interactions. *Astrophysical Journal*, 241, 425–441.

Goldstein, J. I., Scott, E. R. D. and Chabot, N. L. (2009) Iron meteorites: Crystallization, thermal histories, parent bodies, and origin. *Chemie der Erde*, 69, 293–325.

Gomes, R., Levison, H. F., Tsiganis, K. and Morbidelli, M. (2005). Origin of the cataclysmic Late Heavy Bombardment period of the terrestrial planets. *Nature*, 435, 466–469.

Gomes, R. S., Fernández, J. A., Gallardo, T. and Brunini, A. (2008) The scattered disk: Origins, dynamics, and end states. In *The Solar System Beyond Neptune*, eds. Barucci, M. A., Boehnhardt, H., Cruikshank, D. P. and Morbidelli, A. Tucson: University of Arizona Press, pp. 259–273.

Gould, B. A. (1852) On the symbolic notation of the asteroids. *Astronomical Journal*, 2, 80.

Gounelle, M., Morbidelli, A., Bland, P. A., Spurný, P., Young, E. D. and Sephton, M. (2008) Meteorites from the outer Solar System? In *The Solar System Beyond Neptune*, eds. Barucci, M. A., Boehnhardt, H., Cruikshank, D. P. and Morbidelli, A. Tucson: University of Arizona Press, pp. 525–541.

Gounelle, M., Chaussidon, M., Morbidelli, A., *et al.* (2009) A unique basaltic micrometeorite expands the inventory of Solar System planetary crusts. *Proceedings of the National Academy of Sciences*, 106, 6904–6909.

Gradie, J. and Tedesco, E. (1982) Compositional structure of the asteroid belt. *Science*, 216, 1405–1407.

Gradie, J. and Veverka, J. (1980) The composition of the Trojan asteroids. *Nature*, 283, 840–842.

Grady, M. M., Pratesi, G. and Cecchi, V. M. (2014) *Atlas of Meteorites*. Cambridge: Cambridge University Press, 384 pp.

Granahan, J. C. (1993) Investigations of asteroid family geology. PhD thesis, University of Hawaii, Honolulu, Hawaii, 187 pp.

Granahan, J. (2002) A compositional study of asteroid 243 Ida and Dactyl from Galileo NIMS and SSI observations. *Journal of Geophysical Research*, 107, 201-1–201-10.

Granahan, J. C. (2011) Spatially resolved spectral observations of asteroid 951 Gaspra. *Icarus*, 213, 265–272.

   (2013) A comparison of ordinary chondrites with 243 Ida and Dactyl. Lunar and Planetary Science Conference, 44, 1045. www.lpi.usra.edu/meetings/lpsc2013/pdf/1045.pdf.

Granahan, J. C. and Bell, J. F. (1991) On the geologic reality of asteroid families. *Lunar and Planetary Science Conference*, XXII, 477–478.

Granahan, J. C., Fanale, F. P., Robinson, M. S., *et al.* (1994) A Galileo multi-instrument spectral analysis of 951 Gaspra. *Lunar and Planetary Science Conference*, XXV, 453–454.

Granahan, J. C., Fanale, F. P., Carlson, R., *et al.* (1995) Galileo multi-instrument spectral observations of 243 Ida and Dactyl. *Lunar and Planetary Science Conference*, XXVI, 489–490.

Granvik, M., Virtanen, J., Oszkiewicz, D. and Muinonen, K. (2009) OpenOrb: Open-source asteroid orbit computation software including statistical ranging. *Meteoritics & Planetary Science*, 44, 1853–1861.

Granvik, M., Morbidelli, A., Jedicke, R., *et al.* (2016) Super-catastrophic disruption of asteroids at small perihelion distances. *Nature*, 530, 303–306.

Grav, T., Mainzer, A. K., Bauer, J., *et al.* (2011) WISE/NEOWISE observations of the Jovian Trojans: Preliminary results. *Astrophysical Journal*, 742, 40

Grav, T., Mainzer, A. K., Bauer, J., *et al.* (2012a) WISE/NEOWISE observations of the Hilda population: Preliminary results. *Astrophysics Journal*, 744, 197.

Grav, T., Mainzer, A. K., Bauer, J., Masiero, J. and Nugent, C. R. (2012b) WISE/NEOWISE observations of the Jovian Trojan population: Taxonomy. *Astrophysics Journal*, 759, 49.

Green, D. W. E. (2010) (596) Scheila. *International Astronomical Union Circular*, 9188.

Greenberg, R., Nolan, M. C., Bottke, W. F., Kolvoord, R. A. and Veverka, J. (1994) Collisional history of Gaspra. *Icarus*, 107, 84–97.

Greenberg, R., Bottke, W. F., Nolan, M., *et al.* (1996) Collisional and dynamical history of Ida. *Icarus*, 120, 106–118.

Greenstreet, S., Ngo, H. and Gladman, B. (2012) The orbital distribution of near-Earth objects inside Earth's orbit. *Icarus*, 217, 355–366.

Greenwood, R. C., Franchi, I. A., Kearsley, A. T. and Alard, O. (2010) The relationship between CK and CV chondrites. *Geochimica et Cosmochimica Acta*, 74, 1684–1705.

Grossman, J. N. and Brearley, A. J. (2005) The onset of metamorphism in ordinary and carbonaceous chondrites. *Meteoritics & Planetary Science*, 40, 87–122,

Grossman, J. N. and Wasson, J. T. (1981) Compositional study of chondrules from the highly unequilibrated (LL3.0) Semarkona chondrite. *Lunar and Planetary Science Conference*, XII, 371–373.

Grossman, L. (1972) Condensation in the primitive solar nebula. *Geochimica et Cosmochimica Acta*, 36, 597–619.

Grundy, W. M., Binzel, R. P., Buratti, B. J., *et al.* (2016) Surface compositions across Pluto and Charon. *Science*, 351, aad9189-1–aad9189-8.

Guinot, B. (2011) Solar time, legal time, time in use. *Metrologia*, 48, S181–S185.

Gulbis, A. A. S., Elliot, J. L., Person, M. J., *et al.* (2006) Charon's radius and atmospheric constraints from observations of a stellar occultation. *Nature*, 439, 48–51.

Gupta, R. P. (2003) *Remote Sensing Geology (Second Edition)*. Berlin: Springer-Verlag, 656 pp.

Güttler, C., Hirata, N. and Nakamura, A. M. (2012) Cratering experiments on the self armoring of coarse-grained granular targets. *Icarus*, 220, 1040–1049.

Halliday, A. N. (2000) Hf-W chronometry and inner Solar System accretion rates. *Space Science Reviews*, 92, 355–370.

Hapke, B. (1981) Bidirectional reflectance spectroscopy. 1. Theory. *Journal of Geophysical Research*, 86, 3039–3054.

(1984) Bidirectional reflectance spectroscopy. 3. Correction for macroscopic roughness. *Icarus*, 59, 41–59.

(1986) Bidirectional reflectance spectroscopy. 4. The extinction coefficient and the opposition effect. *Icarus*, 67, 264–280.

(1993) *Theory of Reflectance and Emittance Spectroscopy*. Cambridge: Cambridge University Press, 469 pp.

(1999) Scattering and diffraction of light by particles in planetary regoliths. *Journal of Quantitative Spectroscopy and Radiative Transfer*, 61, 565–581.

(2001) Space weathering from Mercury to the asteroid belt. *Journal of Geophysical Research*, 106, 10039–10073.

(2002) Bidirectional reflectance spectroscopy. 5. The coherent backscatter opposition effect and anisotropic scattering. *Icarus*, 157, 523–534.

(2008) Bidirectional reflectance spectroscopy. 6. Effects of porosity. *Icarus*, 195, 918–926.

(2012a) Bidirectional reflectance spectroscopy. 7. The single particle phase function hockey stick relation. *Icarus*, 221, 1079–1083.

(2012b) *Theory of Reflectance and Emittance Spectroscopy (2nd Edition)*. Cambridge: Cambridge University Press, 528 pp.

(2013) Comment on "A critical assessment of the Hapke photometric model" by Y. Shkuratov et al. *Journal of Quantitative Spectroscopy & Radiative Transfer*, 116, 184–190.

Hapke, B. and Wells, E. (1981) Bidirectional reflectance spectroscopy 2. – Experiments and observations. *Journal of Geophysical Research*, 86, 3055–3060.

Hapke, B., Cassidy, W. and Wells, E. (1975) Effects of vapor-phase deposition processes on the optical, chemical, and magnetic properties of the lunar regolith. *Moon*, 13, 339–353.

Hardersen, P. S., Gaffey, M. J. and Abell, P. A. (2004) Mineralogy of asteroid 1459 Magnya and implications for its origin. *Icarus*, 167, 170–177.

(2005) Near-IR spectral evidence for the presence of iron-poor orthopyroxenes on the surfaces of six M-type asteroids. *Icarus*, 175, 141–158.

Hardersen, P. S., Cloutis, E. A., Reddy, V., Mothé-Diniz, T. and Emery, J. P. (2011) The M-/X-asteroid menagerie: Results of an NIR spectral survey of 45 main-belt asteroids. *Meteoritics & Planetary Science*, 46, 1910–1938.

Hardersen, P. S., Reddy, V., Roberts, R. and Mainzer, A. (2014) More chips off of asteroid (4) Vesta: Characterization of eight Vestoids and their HED meteorite analogs. *Icarus*, 242, 269–282.

Hardersen, P. S., Reddy, V. and Roberts, R. (2015) Vestoids, part II: The basaltic nature and HED meteorite analogs for eight V_p-type asteroids and their associations with (4) Vesta. *Astrophysical Journal Supplement Series*, 221, 19.

Harrington, J., de Pater, I., Brecht, S. H., *et al.* (2004) Lessons from Shoemaker-Levy 9 about Jupiter and planetary impacts. In *Jupiter. The Planet, Satellites and Magnetosphere*, eds. Bagenal, F., Dowling, T. E. and McKinnon, W. B. Cambridge: Cambridge University Press, 159–184.

Harris, A. W. (1998) A thermal model for near-Earth asteroids. *Icarus*, 131, 291–301.

Harris, A. (2008) What Spaceguard did. *Nature*, 453, 1178–1179.

Harris, A. W. and D'Abramo, G. (2015) The population of near-Earth asteroids. *Icarus*, 257, 302–312.

Harris, A. W. and Lagerros, J. S. V. (2002) Asteroids in the thermal infrared. In *Asteroids III*, eds. Bottke, W. F. Jr., Cellino, A., Paolicchi, P. and Binzel, R. P. Tucson: University of Arizona Press, pp. 205–218.

Hartmann, W. K. and Neukum, G. (2001) Cratering chronology and the evolution of Mars. *Space Science Reviews*, 96, 165–194.

Hartmann, W. K., Farinella, P., Vokrouhlický, D., *et al.* (1999) Reviewing the Yarkovsky effect: New light on the delivery of stone and iron meteorites from the asteroid belt. *Meteoritics & Planetary Science*, 34, A161–A167.

Hasegawa, S., Murakawa, K., Ishiguro, M., *et al.* (2003) Evidence of hydrated and/or hydroxylated minerals on the surface of asteroid 4 Vesta. *Geophysical Research Letters*, 30, 2123.

Hayatsu, R., Anders, E., Studier, M. H. and Moore, L. P. (1975) Purines and triazines in the Murchison meteorite. *Geochimica et Cosmochimica Acta*, 39, 471–488.

Hayatsu, R., Winans, R. E., Scott, R. G., McBeth, R. L., Moore, L. P. and Studier, M. H. (1980) Phenolic ethers in the organic polymer of the Murchison meteorite. *Science*, 207, 1202–1204.

Head, J. W., Fassett, C. I., Kadish, S. J., *et al.* (2010) Global distribution of large lunar craters: Implications for resurfacing and impactor populations. *Science*, 329, 1504–1507.

Hedman, M. M. (2015) Why are dense planetary rings only found between 8 AU and 20 AU? *Astrophysical Journal Letters*, 801, L33.

Helfenstein, P., Veverka, J., Thomas, P. C., *et al.* (1994) Galileo photometry of asteroid 951 Gaspra. *Icarus*, 107, 37–60.

Helin, E. F., Pravdo, S. H., Rabinowitz, D. L. and Lawrence, K. J. (1997) Near-Earth Asteroid Tracking (NEAT) Program. *Annals of the New York Academy of Sciences*, 822, 6–25.

Henning, T. (2010) Cosmic silicates. *Annual Review of Astronomy and Astrophysics*, 48, 21–46.

Hergarten, S. and Kenkmann, T. (2015) The number of impact craters on Earth: Any room for further discoveries? *Earth and Planetary Science Letters*, 425, 187–192.

Herschel, W. (1802) Observations on the two lately discovered celestial bodies. *Philosophical Transactions of the Royal Society of London*, 92, 213–232.

Herzog, G. F., Caffee, M. W. and Jull, A. J. T. (2015) Cosmogenic nuclides in Antarctic meteorites. In *35 Seasons of U.S. Antarctic Meteorites: A Pictorial Guide To The Collection,* eds. Righter, K., Corrigan, C., McCoy, T. and Harvey, R. Washington, DC: American Geophysical Union, pp. 153–172.

Hevey, P. J. and Sanders, I. S. (2006) A model for planetesimal meltdown by ^{26}Al and its implications for meteorite parent bodies. *Meteoritics & Planetary Science*, 41, 95–106.

Hewins, R. H. (1997) Chondrules. *Annual Review of Earth and Planetary Sciences*, 25, 61–83.

Hezel, D. C. and Russell, S. S. (2008) Comment on "Ancient asteroids enriched in refractory inclusions." *Science*, 322, 1050.

Hiesinger, H., van der Bogert, C. H., Pasckert, J. H., *et al.* (2012) How old are young lunar craters? *Journal of Geophysical Research*, 117, E00H10.

Hillier, J. K., Bauer, J. M. and Buratti, B. J. (2011) Photometric modeling of asteroid 5535 Annefrank from Stardust observations. *Icarus*, 211, 546–552.

Hilton, J. L. (2006) When did the asteroids become minor planets? http://aa.usno.navy.mil/faq/docs/minorplanets.php.

Hinrichs, J. L., Lucey, P. G., Robinson, M. S., Meibom, A. and Krot, A. N. (1999) Implications of temperature-dependent near-IR spectral properties of common minerals and meteorites for the remote sensing of asteroids. *Geophysical Research Letters*, 26, 1661–1664.

Hirata, N., Barnouin-Jha, O. S., Honda, C., *et al.* (2009) A survey of possible impact structures on 25143 Itokawa. *Icarus,* 200, 486–502.

Hirayama, K. (1918) Groups of asteroids probably of common origin. *Astronomical Journal*, 31, 185–188.

(1928) Families of asteroids. *Japanese Journal of Astronomy and Geophysics*, 5, 137–162.

(1933) Present state of the families of asteroids. *Proceedings of the Imperial Academy of Japan*, 9, 482–485.

Hiroi, T. and Hasegawa, S. (2003) Revisiting the search for the parent body of the Tagish Lake meteorite – Case of a T/D asteroid 308 Polyxo. *Antarctic Meteorite Research*, 16, 176–184.

Hiroi, T. and Pieters, C. M. (1994) Estimation of grain sizes and mixing ratios of fine powder mixtures of common geologic minerals. *Journal of Geophysical Research*, 99, 10867–10879.

Hiroi, T. and Sasaki, S. (2001) Importance of space weathering simulation products in compositional modeling of asteroids: 349 Dembowska and 446 Aeternitas as examples. *Meteoritics & Planetary Science,* 36, 1587–1596.

Hiroi, T., Pieters, C. M., Zolensky, M. E. and Lipschutz, M. E. (1993) Evidence of thermal metamorphism on the C, G, B, and F asteroids. *Science*, 261, 1016–1018.

Hiroi, T., Pieters, C. M. and Takeda, H. (1994) Grain size of the surface regolith of asteroid 4 Vesta estimated from its reflectance spectrum in comparison with HED meteorites. *Meteoritics*, 29, 394–396.

Hiroi, T., Binzel, R. P., Sunshine, J. M., Pieters, C. M. and Takeda, H. (1995) Grain sizes and mineral compositions of surface regoliths of Vesta-like asteroids. *Icarus*, 115, 374–386.

Hiroi, T., Zolensky, M. E., Pieters, C. M. and Lipschutz, M. E. (1996) Thermal metamorphism of the C, G, B, and F asteroids seen from the 0.7 μm, 3 μm and UV absorption strengths in comparison with carbonaceous chondrites. *Meteoritics & Planetary Science*, 31, 321–327.

Hiroi, T., Pieters, C. M., Vilas, F., Sasaki, S., Hamabe, Y. and Kurahashi, E. (2001a) The mystery of 506.5 nm feature of reflectance spectra of Vesta and Vestoids: Evidence for space weathering? *Earth, Planets and Space*, 53, 1071–1075.

Hiroi, T., Zolensky, M. E. and Pieters, C. M. (2001b) The Tagish Lake meteorite: A possible sample from a D-type asteroid. *Science*, 293, 2234–2236.

Hiroi, T., Pieters, C. M., Rutherford, *et al.* (2004) What are the P-type asteroids made of? Lunar and Planetary Science Conference, XXXV, 1616. www.lpi.usra.edu/meetings/lpsc2004/pdf/1616.pdf.

Hollis, A. J. (1994) Classifying asteroids. *Journal of the British Astronomical Association*, 104, 112–122.

Hood, L. L. (1995) Frozen fields. *Earth, Moon, and Planets*, 67, 131–142.

Horner, J. and Lykawka, P. S. (2011) The Neptune Trojans: A window on the birth of the Solar System. *Astronomy & Geophysics*, 52, 4.24–4.30.

Horstmann, M. and Bischoff, A. (2014) The Almahata Sitta polymict breccia and the late accretion of asteroid 2008 TC_3. *Chemie der Erdie*, 74, 149–183.

Housen, K. R., Holsapple, K. A. and Voss, M. E. (1999) Compaction as the origin of the unusual craters on the asteroid Mathilde. *Nature*, 402, 155–157.

Howard, K. T., Alexander, C. M. O'D. and Dyl, K. A. (2014) PSD-XRD modal mineralogy of type 3.0 CO chondrites: Initial asteroidal water mass fractions and implications for CM chondrites. Lunar and Planetary Science Conference, 45, 1830. www.hou.usra.edu/meetings/lpsc2014/pdf/1830.pdf.

Howell, E. S., Britt, D. T., Bell, J. F., Binzel, R. P. and Lebofsky, L. A. (1994a) Visible and near-infrared spectral observations of 4179 Toutatis. *Icarus*, 111, 468–474.

Howell, E. S., Merényi, E. and Lebofsky, L. A. (1994b) Classification of asteroid spectra using a neural network. *Journal of Geophysical Research*, 99, 10847–10865.

Howell, E. S., Rivkin, A. S., Vilas, F. and Soderberg, A. M. (2001) Aqueous alteration in low albedo asteroids. Lunar and Planetary Science Conference, XXXII, 2058. www.lpi.usra.edu/meetings/lpsc2001/pdf/2058.pdf.

Howell, S. B. (2006) *Handbook of CCD Astronomy (Second Edition)*. Cambridge: Cambridge University Press, 220 pp.

Hsieh, H. H., Jewitt, D. C. and Fernández, Y. R. (2004) The strange case of 133P/Elst-Pizarro: A comet among the asteroids. *Astronomical Journal*, 127, 2997–3017.

Huang, J., Ji. J., Ye, P., *et al.* (2013) The ginger-shaped asteroid 4179 Toutatis: New observations from a successful flyby of Chang'e-2. *Scientific Reports*, 3, 3411.

Hubbard, A. and Ebel, D. S. (2015) Semarkona: Lessons for chondrule and chondrite formation. *Icarus*, 245, 32–37.

Hubbard, W. B., Hunten, D. M., Dieters, S. W., Hill, K. M. and Watson, R. D. (1988) Occultation evidence for an atmosphere on Pluto. *Nature*, 336, 452–454.

Huber, H., Rubin, A. E., Kallemeyn, G. W. and Wasson, J. T. (2006) Siderophile-element anomalies in CK carbonaceous chondrites: Implications for parent-body aqueous alteration and terrestrial weathering of sulfides. *Geochimica et Cosmochimica Acta*, 70, 4019–4037.

Hughes, D. W. (1994) The historical unravelling of the diameters of the first four asteroids. *Quarterly Journal of the Royal Astronomical Society*, 35, 331–344.

Hughes, D. W. and Green, D. W. E. (2007) Halley's first name: Edmond or Edmund. *International Comet Quarterly*, 29, 7–14.

Huss, G. R., Rubin, A. E. and Grossman, J. N. (2006) Thermal metamorphism in chondrites. In *Meteorites and the Early Solar System,* eds. Lauretta, D. S. and McSween, H. Y. Jr. Tucson: University of Arizona Press, pp. 567–586.

Husson, D., Galbrun, B., Laskar, J., *et al.* (2011) Astronomical calibration of the Maastrichtian (Late Cretaceous). *Earth and Planetary Science Letters*, 305, 328–340.

Hutchison, R. (2004) *Meteorites: A Petrologic, Chemical and Isotopic Synthesis.* Cambridge: Cambridge University Press, 524 pp.

Irwin, M. J. (1997) Detectors and data analysis techniques for wide field optical imaging. In *Instrumentation for Large Telescopes*, eds. Rodriquez Espinosa, J. M., Herrero, A. and Sánchez, F. Cambridge: Cambridge University Press, pp. 35–74.

Isa, J., Ma, C. and Rubin, A. E. (2015) Joegoldsteinite: A new sulfide mineral ($MnCr_2S_4$) from the IVA iron meteorite, Social Circle. Lunar and Planetary Science Conference, 47, 1813. www.hou.usra.edu/meetings/lpsc2016/pdf/1813.pdf.

Isaacson, P. J. and Pieters, C. M. (2009) Northern Imbrium noritic anomaly. *Journal of Geophysical Research*, 114, E09007.

Ishiguro, M., Kuroda, D., Hasegawa, S., *et al.* (2014) Optical properties of (162173) 1999 JU3: In preparation for the JAXA Hayabusa 2 sample return mission. *Astrophysical Journal*, 792, 74.

Ivanov, B. A. (2001) Mars/Moon cratering rate ratio estimates. *Space Science Reviews*, 96, 87–104.

Ivanov, B. A. and Hartmann, W. K. (2009) Exogenic dynamics, cratering and surface ages. In *Treatise on Geophysics, Volume 10: Planets and Moons*, ed. Schubert, G. Amsterdam: Elsevier, pp. 207–242.

Ivanov, B. A., Neukum, G., Bottke, W. F. Jr. and Hartmann, W. K. (2002) The comparison of size-frequency distributions of impact craters and asteroids and the planetary cratering rate. In *Asteroids III*, eds. Bottke, W. F. Jr., Cellino, A., Paolicchi, P. and Binzel, R. P. Tucson: University of Arizona Press, pp. 89–101.

Ivezić, Ž., Tabachnik, S., Rafikov, R., *et al.* (2001) Solar System objects observed in the Sloan Digital Sky Survey commissioning data. *Astronomical Journal,* 122, 2749–2784.

Ivezić, Ž., Lupton, R. H., Jurić, M., *et al.* (2002) Color confirmation of asteroid families. *Astronomical Journal*, 124, 2943–2948.

Jacobsen, B., Yin, Q.-z., Moynier, F., *et al.* (2008) ^{26}Al-^{26}Mg and ^{207}Pb-^{206}Pb systematics of Allende CAIs: Canonical solar initial $^{26}Al/^{27}Al$ ratio reinstated. *Earth and Planetary Science Letters*, 272, 353–364.

Jacobsen, S. B. (2005) The Hf-W isotopic system and the origin of the Earth and Moon. *Annual Review of Earth and Planetary Sciences*, 33, 531–570.

Jansen, W. and Slaughter, M. (1982). Elemental mapping of minerals by electron microprobe. *American Mineralogist*, 67, 521–533.

Jarosewich, E. (1990) Chemical analyses of meteorites: A compilation of stony and iron meteorite analyses. *Meteoritics*, 25, 323–337.

Jedicke, R., Gravnik, M., Micheli, M., Ryan, E., Spahr, T. and Yeomans, D. K. (2015) Surveys, astrometric follow-up and population statistics. In *Asteroids IV*, eds. Michel, P., DeMeo, F. E. and Bottke, W. F. Tucson: University of Arizona Press, pp. 795–813.

Jenniskens, P. (2004) 2003 EH$_1$ is the Quadrantid shower parent comet. *Astronomical Journal*, 127, 3018–3022.

(2006) *Meteor Showers and their Parent Comets.* Cambridge: Cambridge University Press, 804 pp.

Jenniskens, P., Betlem, H., Betlem, J., *et al.* (1994) The Mbale meteorite shower. *Meteoritics*, 29, 246–254.

Jenniskens, P., Shaddad, M. H., Numan, D., *et al.* (2009) The impact and recovery of asteroid 2008 TC$_3$. *Nature*, 458, 485–488.

Jenniskens, P., Vaubaillon, J., Binzel, R. P., *et al.* (2010) Almahata Sitta (=asteroid 2008 TC$_3$) and the search for the ureilite parent body. *Meteoritics & Planetary Science*, 45, 1590–1617.

Jewitt, D. (1999) The Kuiper belt. *Physics World*, 12, 37–41.

   (2005) A first look at the damocloids. *Astronomical Journal*, 129, 530–538.

   (2010) The discovery of the Kuiper belt. *Astronomy Beat*, 48, 1–5.

   (2012) The active asteroids. *Astronomical Journal*, 143, 66.

   (2015) Color systematics of comets and related bodies. *Astronomical Journal*, 150, 201.

Jewitt, D. C. and Luu, J. X. (1990) CCD spectra of asteroids. II. The Trojans as spectral analogs of cometary nuclei. *Astronomical Journal*, 100, 933–944.

Jewitt, D. and Luu, J. (1993) Discovery of the candidate Kuiper belt object 1992 QB$_1$. *Nature*, 362, 730–732.

Jewitt, D. C. and Luu, J. (2004) Crystalline water ice on the Kuiper belt object (50000) Quaoar. *Nature*, 432, 731–733.

Jewitt, D. and Guilbert-Lepoutre, A. (2012) Limits to ice on asteroids (24) Themis and (65) Cybele. *Astronomical Journal*, 143, 21.

Jewitt, D., Aussel, H. and Evans, A. (2001) The size and albedo of the Kuiper-belt object (20000) Varuna. *Nature*, 411, 446–447.

Jewitt, D., Hsieh, H. and Agarwal, J. (2015) The active asteroids. In *Asteroids IV*, eds. Michel, P., DeMeo, F. E. and Bottke, W. F. Tucson: University of Arizona Press, pp. 221–241.

Jiang, Y., Ji, J., Huang, J., Marchi, S., Li, Y. and Ip, W.-H. (2015) Boulders on asteroid Toutatis as observed by Chang'e-2. *Scientific Reports*, 5, 16029.

Jilly, C. E., Huss, G. R., Krot, A. N., Nagashima, K., Yin, Q.-z. and Sugiura, N. (2014) ^{53}Mn-^{53}Cr dating of aqueously formed carbonates in the CM2 lithology of the Sutter's Mill carbonaceous chondrite. *Meteoritics & Planetary Science*, 49, 2104–2117.

Johnson, B. C., Minton, D. A., Melosh, H. J. and Zuber, M. T. (2015) Impact jetting as the origin of chondrules. *Nature*, 517, 339–341.

Johnson, H. L. and Morgan, W. W. (1953) Fundamental stellar photometry for standards of spectral type on the revised system of the Yerkes spectral atlas. *Astrophysical Journal*, 117, 313–352.

Johnson, J. A., Christensen, E. J., Gibbs, A. R., *et al.* (2014) The Catalina Sky Survey: Status, discoveries and the future. *American Astronomical Society, DPS meeting*, 46, 414.09.

Johnson, L. (2014) Finding near Earth objects before they find us! http://sservi.nasa.gov/wp-content/uploads/2014/03/Johnson_NASA-NEO-Program-SA-Workshop.pdf.

Johnson, T. V. and Fanale, F. P. (1973) Optical properties of carbonaceous chondrites and their relationship to asteroids. *Journal of Geophysical Research*, 78, 8507–8518.

Jones, T. D., Lebofsky, L.A., Lewis, J. S. and Marley, M. S. (1990) The composition and origin of the C, P, and D asteroids: Water as a tracer of thermal evolution in the outer belt. *Icarus*, 88, 172–192.

Jourdan, F., Timms, N., Eroglu, E., *et al.* (2015) Collisional history of asteroid Itokawa. *Goldschmidt Abstracts*, 2015, 1479.

Jull, A. J. T. (2006) Terrestrial ages of meteorites. In *Meteorites and the Early Solar System*, eds. Lauretta, D. S. and McSween, H. Y. Jr. Tucson: University of Arizona Press, pp. 889–905.

Jura, M. (2008) Pollution of single white dwarfs by accretion of many small asteroids. *Astronomical Journal*, 135, 1785–1792.

   (2014) The elemental compositions of extrasolar planetesimals. In *Formation, Detection and Characterization of Extrasolar Habitable Planets*, ed. Haghighipour, N. Cambridge: Cambridge University Press, pp. 219–228.

Jura, M. and Young, E. D. (2014) Extrasolar cosmochemistry. *Annual Review of Earth and Planetary Sciences*, 42, 45–67.

Jurewicz, A. J. G., Mittlefehldt, D. W. and Jones, J. H. (1991) Partial melting of the Allende (CV3) meteorite: Implications for origins of basaltic meteorites. *Science*, 252, 695–698.

(1993) Experimental partial melting of the Allende (CV) and Murchison (CM) chondrites and the origin of asteroidal basalt. *Geochimica et Cosmochimica Acta*, 57, 2123–2139.

Jurić, M., Ivezić, Ž., Lupton, R. H., *et al.* (2002) Comparison of positions and magnitudes of asteroids observed in the Sloan Digital Sky Survey with those predicted for known asteroids. *Astronomical Journal*, 124, 1776–1787.

Kallemeyn, G. W., Rubin, A. E. and Wasson, J. T. (1991) The compositional classification of chondrites: V. The Karoonda (CK) group of carbonaceous chondrites. *Geochimica et Cosmochimica Acta*, 55, 881–892.

Karczemska, A. T. (2010) Diamonds in meteorites – Raman mapping and cathodoluminescence studies. *Journal of Achievements in Materials and Manufacturing Engineering*, 43, 94–107.

Kargel, J. S. (1994) Metalliferous asteroids as potential sources of precious metals. *Journal of Geophysical Research*, 99, 21129–21141.

Kaasalainen, M., Mottola, S. and Fulchignoni, M. (2002) Asteroid models from disk-integrated data. In *Asteroids III*, eds. Bottke, W. F. Jr., Cellino, A., Paolicchi, P. and Binzel, R. P. Tucson: University of Arizona Press, pp. 139–150.

Kebukawa, Y., Zolensky, M. E., Kilcoyne, A. L. D., Rahman, Z., Jenniskens, P. and Cody, G. D. (2014) Diamond xenolith and matrix organic matter in the Sutter's Mill meteorite measured by C-XANES. *Meteoritics & Planetary Science*, 49, 2095–2103.

Keil, K. (1989) Enstatite meteorites and their parent bodies. *Meteoritics*, 24, 195–208.

(2014) Brachinite meteorites: Partial melt residues from an FeO-rich asteroid. *Chemie der Erde*, 74, 311–329.

Keller, H. U., Barbieri, C., Lamy, P., *et al.* (2007) OSIRIS the scientific camera system onboard Rosetta. *Space Science Reviews*, 128, 433–506.

Keller, H. U., Barbieri, C., Koschny, D., *et al.* (2010) E-type asteroid (2867) Steins as imaged by OSIRIS on board Rosetta. *Science*, 327, 190–193.

Kelley, M. S. and Gaffey, M. J. (2002) High-albedo asteroid 434 Hungaria: Spectrum, composition and genetic connections. *Meteoritics & Planetary Science*, 37, 1815–1827.

Kelley, M. S., Vilas, F., Gaffey, M. J. and Abell, P. A. (2003) Quantified mineralogical evidence for a common origin of 1929 Kollaa with 4 Vesta and the HED meteorites. *Icarus*, 165, 215–218.

Kerr, R. A. (1996) New source proposed for most common meteorites. *Science*, 273, 1337.

Kim, S, Lee, H. M., Nakagawa, T. and Hasegawa, S. (2003) Thermal models and far infrared emission of asteroids. *Journal of the Korean Astronomical Society*, 36, 21–31.

King, T. V. V., Clark, R. N., Calvin, W. M., Sherman, D. M. and Brown, R. H. (1992) Evidence for ammonium-bearing minerals on Ceres. *Science*, 255, 1551–1553.

Kirkwood, D. (1867) *Meteoric Astronomy: A Treatise on Shooting-Stars, Fire-Balls and Aerolites*. Philadelphia: J.B. Lippincott & Co., 129 pp.

Kita, N. T. and Ushikubo, T. (2012) Evolution of protoplanetary disk inferred from ^{26}Al chronology of individual chondrules. *Meteoritics & Planetary Science*, 47, 1108–1119.

Kita, N. T., Nagahara, H., Togashi, S. and Morishita, Y. (2000) A short duration of chondrule formation in the solar nebula: Evidence from ^{26}Al in Semarkona ferromagnesian chondrules. *Geochimica et Cosmochimica Acta*, 64, 3913–3922.

Kita, N. T., Huss, G. R., Tachibana, S., Amelin, Y., Nyquist, L. E. and Hutcheon, I. D. (2005) Constraints on the origin of chondrules and CAIs from short-lived and long-Lived radionuclides chondrites and the protoplanetary disk. In *Chondrites and the*

Protoplanetary Disk, ASP Conference Series, Volume 341, eds. Krot, A. N., Scott, E. R. D. and Reipurth, B. San Francisco: Astronomical Society of the Pacific, pp. 558–587.

Kita, N. T., Yin, Q.-z., MacPherson, G. J., *et al.* (2013) ^{26}Al-^{26}Mg isotope systematics of the first solids in the early Solar System. *Meteoritics & Planetary Science*, 48, 1383–1400.

Kivelson, M. G., Khurana, K. K., Means, J. D., Russell, C. T. and Snare, R. C. (1992) The Galileo magnetic field investigation. *Space Science Reviews*, 60, 357–383.

Kivelson, M. G., Bargatze, L. F., Khurana, K. K., Southwood, D. J., Walker, R. J. and Coleman, P. J. (1993) Magnetic field signatures near Galileo's closest approach to Gaspra. *Science*, 261, 331–334.

Kleine, T. and Rudge, J. F. (2011) Chronometry of meteorites and the formation of the Earth and Moon. *Elements*, 7, 41–46.

Kleine, T., Touboul, M., Bourdon, B., *et al.* (2009) Hf-W chronology of the accretion and early evolution of asteroids and terrestrial planets. *Geochimica et Cosmochimica Acta*, 73, 5150–5188.

Klima, R. L., Pieters, C. M. and Dyar, M. D. (2007) Spectroscopy of synthetic Mg-Fe pyroxenes I: Spin-allowed and spin-forbidden crystal field bands in the visible and near-infrared. *Meteoritics & Planetary Science*, 42, 235–253.

Klima, R. L., Dyar, M. D. and Pieters, C. M. (2011) Near-infrared spectra of clinopyroxenes: Effects of calcium content and crystal structure. *Meteoritics & Planetary Science*, 46, 379–395.

Kminek, G., Botta, O., Glavin, D. P. and Bada, J. L. (2002) Amino acids in the Tagish Lake meteorite. *Meteoritics & Planetary Science*, 37, 697–701.

Knežević, Z., Lemaître, A. and Milani, A. (2002) The determination of asteroid proper elements. In *Asteroids III,* eds. Bottke, W. F. Jr., Cellino, A., Paolicchi, P. and Binzel, R. P. Tucson: University of Arizona Press, pp. 603–612.

Koeberl, C. (1986) Geochemistry of tektites and impact glasses. *Annual Review of Earth and Planetary Sciences*, 14, 323–350.

Kojima, H. (2006) The history of Japanese Antarctic meteorites. In *The History of Meteoritics and Key Meteorite Collections: Fireballs, Falls & Finds, Geological Society Special Publication, no. 256*, eds. McCall, G. J. H., Bowden, A. J. and Bowden A. J. Bath: Geological Society Publishing House, pp. 291–303.

Komatsu, M., Krot, A. N., Petaev, M. I., Ulyanov, A. A., Keil, K. and Miyamoto, M. (2001) Mineralogy and petrography of amoeboid olivine aggregates from the reduced CV3 chondrites Efremovka, Leoville and Vigarano: Products of nebular condensation, accretion and annealing. *Meteoritics & Planetary Science*, 36, 629–641.

Kornacki, A. S. and Cohen, R. E. (1983) The nature of coarse-grained CAI's. *Lunar and Planetary Science*, XIV, 395–396.

Korochantseva, E. V., Trieloff, M., Lorenz, C. A., *et al.* (2007) L-chondrite asteroid breakup tied to Ordovician meteorite shower by multiple isochron ^{40}Ar–^{39}Ar dating. *Meteoritics & Planetary Science*, 42, 113–130.

Kring, D. A. (2007) *Guidebook to the Geology of Barringer Meteorite Crater, Arizona (a.k.a. Meteor Crater)*. Houston: Lunar and Planetary Institute, 150 pp.

Krot, A. N., Meibom, A. and Keil, K. (2000) A clast of Bali-like oxidized CV material in the reduced CV chondrite breccia Vigarano. *Meteoritics & Planetary Science*, 35, 817–825.

Krot, A. N., Meibom, A., Weisberg, M. K. and Keil, K. (2002) The CR chondrite clan: Implications for early Solar System processes. *Meteoritics & Planetary Science*, 37, 1451–1490.

Krot, A. N., Amelin, Y., Cassen, P. and Meibom, A. (2005) Young chondrules in CB chondrites from a giant impact in the early Solar System. *Nature*, 436, 989–992.

Krot, A. N., Hutcheon, I. D., Brearley, A. J., Pravdivtseva, O. V., Petaev, M. I. and Hohenberg, C. M. (2006) Timescales and settings for alteration of chondritic meteorites. In *Meteorites and the Early Solar System,* eds. Lauretta, D. S. and McSween, H. Y. Jr. Tucson: University of Arizona Press, pp. 525–553.

Krot, A. N., Amelin, Y., Bland, P., *et al.* (2009) Origin and chronology of chondritic components: A review. *Geochimica et Cosmochimica Acta*, 73, 4963–4997.

Krot, A. N., Nagashima, K., Ciesla, F. J., *et al.* (2010) Oxygen isotopic composition of the Sun and mean oxygen isotopic composition of the protosolar silicate dust: Evidence from refractory inclusions. *Astrophysical Journal*, 713, 1159–1166.

Kryszczyńska, K. (2013) Do Slivan states exist in the Flora family? II. Fingerprints of the Yarkovsky and YORP effects. *Astronomy & Astrophysics*, 551, A102.

Kuchynka, P. and Folkner, W. M. (2013) A new approach to determining asteroid masses from planetary range measurements. *Icarus*, 222, 243–253.

Kuiper, G. P. (1951) On the origin of the Solar System. In *Astrophysics: A Topical Symposium*, ed. Hynek, J. A. New York: McGraw-Hill, pp. 357–424.

Kuiper, Y. D. (2002) The interpretation of inverse isochron diagrams in ^{40}Ar/^{39}Ar geochronology. *Earth and Planetary Science Letters*, 203, 499–506.

Kumar, A., Gopalan, K. and Bhandari, N. (1999) ^{147}Sm–^{143}Nd and ^{87}Rb–^{87}Sr ages of the eucrite Piplia Kalan. *Geochimica et Cosmochimica Acta*, 63, 3997–4001.

Kunz, G. F. (1888) Diamonds in meteorites. *Science*, 266, 118–119.

Kurahashi, E., Yamanaka, C., Nakamura, K. and Sasaki, S. (2002) Laboratory simulation of space weathering: ESR measurements of nanophase metallic iron in laser-irradiated materials. *Earth, Planets and Space*, 54, e5–e7.

Kvenvolden, K., Lawless, J., Pering, K., *et al.* (1970) Evidence for extraterrestrial amino-acids and hydrocarbons in the Murchison meteorite. *Nature*, 228, 923–926.

Kyte, F. T. (1998) A meteorite from the Cretaceous/Tertiary boundary. *Nature*, 396, 237–239.

Lagerkvist, C.-I. and Magnusson, P. (1990) Analysis of asteroid lightcurves. II. Phase curves in a generalized HG-system. *Astronomy & Astrophysics Supplement Series*, 86, 119–165.

Lamy, P. L., Toth, I. and Weaver, H. A. (1998) Hubble Space Telescope observations of the nucleus and inner coma of comet 19P/1904 Y2 (Borrelly). *Astronomy & Astrophysics*, 337, 945–954.

Larson, H. P. and Fink, U. (1975) Infrared spectral observations of asteroid 4 Vesta. *Icarus*, 26, 420–427.

Larson, H. P., Feierberg, M. A., Fink, U. and Smith, H. A. (1979) Remote spectroscopic identification of carbonaceous chondrite mineralogies: Applications to Ceres and Pallas. *Icarus*, 39, 257–271.

Larson, H. P., Feierberg, M. A. and Lebofsky, L. A. (1983) The composition of asteroid 2 Pallas and its relation to primitive meteorites. *Icarus*, 56, 398–408.

Lawrence, D. J., Peplowski, P. N., Prettyman, T. H., *et al.* (2013) Constraints on Vesta's elemental composition: Fast neutron measurements by Dawn's gamma ray and neutron detector. *Meteoritics & Planetary Science*, 48, 2271–2288.

Lawrence, S. J. and Lucey, P. G. (2007) Radiative transfer mixing models of meteoritic assemblages. *Journal of Geophysical Research*, 112, E07005.

Lazzarin, M., Fornasier, S., Barucci, M. A. and Birlan, M. (2001) Groundbased investigation of asteroid 9969 Braille, target of the spacecraft mission Deep Space 1. *Astronomy & Astrophysics*, 375, 281–284.

Lazzaro, D., Michtchenko, T., Carvano, J. M., *et al.* (2000) Discovery of a basaltic asteroid in the outer main belt. *Science*, 288, 2033–2035.

Lebofsky, L. and Spencer, J. (1989) Radiometry and a thermal modeling of asteroids. In *Asteroids II*, eds. Binzel, R. P., Gehrels, T. and Matthews, M. S. Tucson: University of Arizona Press, pp. 128–147.

Lebofsky, L. A., Sykes, M. V., Tedesco, E. F., *et al.* (1986) A refined "standard" thermal model for asteroids based on observations of 1 Ceres and 2 Pallas. *Icarus*, 68, 239–251.

Lederer, S. M., Domingue, D. L., Vilas, F., *et al.* (2005) Physical characteristics of Hayabusa target asteroid 25143 Itokawa. *Icarus*, 173, 153–165.

Lee, D.-C. and Halliday, A. (1995) Hafnium-tungsten chronometry and the timing of terrestrial core formation. *Nature*, 378, 771–774.

Lee, T., Papanastassiou, D. A. and Wasserburg, G. J. (1976) Demonstration of ^{26}Mg excess in Allende and evidence for ^{26}Al. *Geophysical Research Letters*, 3, 41–44.

Lellouch, E. (2011) Pluto. In *Encyclopedia of Astrobiology*, eds. Gargaud, M., Amils, R., Quintanilla, J. C., Cleaves, H. J. II, Irvine, W. M., Pinti, D. L. and Viso, M. Berlin: Springer Berlin Heidelberg, pp. 1301–1303.

Lellouch, E., Sicardy, B., de Bergh, C., Käufl, H.-U., Kassi, S. and Campargue, A. (2009) Pluto's lower atmosphere structure and methane abundance from high-resolution spectroscopy and stellar occultations. *Astronomy & Astrophysics*, 495, L17-L21.

Lellouch, E., Kiss, C., Santos-Sanz, P., *et al.* (2010) "TNOs are cool": A survey of the trans-Neptunian region. II. The thermal lightcurve of (136108) Haumea. *Astronomy & Astrophysics*, 518, L147.

Lemaitre, A. (1993) Proper elements: What are they? *Celestial Mechanics and Dynamical Astronomy*, 56, 103–119.

Leonard, F. C. (1930) The new planet Pluto. *Astronomical Society of the Pacific Leaflets*, 30, 121–124.

Leverington, D. (2007) *Babylon to Voyager and Beyond: A History of Planetary Astronomy*. Cambridge: Cambridge University Press, 572 pp.

Lewis, J. S. (2004) *Physics and Chemistry of the Solar System (2nd Edition)*. Burlington: Elsevier Academic Press, 655 pp.

Li, J., A'Hearn, M. F. and McFadden, L. (2004) A photometric analysis of Eros from NEAR data. *Icarus*, 172, 415–431.

Li, J. Y., McFadden, L. A., Parker, J. Wm., *et al.* (2006) Photometric analysis of 1 Ceres and surface mapping from HST observations. *Icarus*, 182, 143–160.

Li, J.-Y., Bodewits, D., Feaga, L. M., *et al.* (2011) Ultraviolet spectroscopy of asteroid (4) Vesta. *Icarus*, 216, 640–649.

Li, J.-Y., Le Corre, L., Schröder, S. E., Reddy, V., Denevi, B. W., Buratti, B. J., Mottola, S., Hoffmann, M., Gutierrez-Marques, P., Nathues, A., Russell, C. T. and Raymond, C. A. (2013) Global photometric properties of asteroid (4) Vesta observed with Dawn framing camera. *Icarus*, 226, 1252–1274.

Li, J.-Y., Helfenstein, P., Buratti, B. J., Takir, D. and Clark, B. E. (2015) Asteroid photometry. In *Asteroids IV*, eds. Michel, P., DeMeo, F. E. and Bottke, W. F. Tucson: University of Arizona Press, pp. 129–150.

Licandro, J., Pinilla-Alonso, N., Pedani, M., Oliva, E., Tozzi, G. P. and Grundy, W. M. (2006) The methane ice rich surface of large TNO 2005 FY$_9$: A Pluto-twin in the trans-Neptunian belt? *Astronomy & Astrophysics*, 445, L35–L38.

Licandro, J., Campins, H., Kelley, M., *et al.* (2011) (65) Cybele: Detection of small silicate grains, water-ice, and organics. *Astronomy & Astrophysics*, 525, A34.

Lim, L. F. and Nittler, L. R. (2009) Elemental composition of 433 Eros: New calibration of the NEAR-Shoemaker XRS data. *Icarus*, 200, 129–146.

Lim, L. F., Emery, J. P. and Moskovitz, N. A. (2011) Mineralogy and thermal properties of V-type asteroid 956 Elisa: Evidence for diogenitic material from the Spitzer IRS (5–35 μm) spectrum. *Icarus*, 213, 510–523.

Lindsay, S. S., Dunn, T. L., Emery, J. P. and Bowles, N. E. (2016) The Red Edge Problem in asteroid band parameter analysis. *Meteoritics & Planetary Science, 51,* 806–817.

Lockwood, A. C., Brown, M. E. and Stansberry, J. (2014) The size and shape of the oblong dwarf planet Haumea. *Earth, Moon, and Planets,* 111, 127–137.

Lodders, K. (2003) Solar System abundances and condensation temperatures of the elements. *Astrophysical Journal,* 591, 1220–1247.

Lodders, K. and Fegley, B. Jr. (2010) *Chemistry of the Solar System.* Cambridge: Royal Society of Chemistry, 496 pp.

Loeffler, M. J., Dukes, C. A., Chang, W. Y., McFadden, L. A. and Baragiola, R. A. (2008) Laboratory simulations of sulfur depletion at Eros. *Icarus,* 195, 622–629.

Loeffler, M., Dukes, C. and Baragiola, R. (2009) Irradiation of olivine by 4 keV He$^+$: Simulation of space weathering by the solar wind. *Journal of Geophysical Research,* 114, E03003.

Loeffler, M. J., Dukes, C. A., Christoffersen, R. and Baragiola, R. A. (2016) Space weathering of silicates simulated by successive laser irradiation: In situ reflectance measurements of Fo$_{90}$, Fo$_{99+}$, and SiO$_2$. *Meteoritics & Planetary Science, 51,* 261–275.

Lorenzi, V., Pinilla-Alonso, N. and Licandro, J. (2015) Rotationally resolved spectroscopy of dwarf planet (136472) Makemake. *Astronomy & Astrophysics,* 577, A86.

Lovering, J. F., Nichiporuk, W., Chodos, A. and Brown, H. (1957) The distribution of gallium, germanium, cobalt, chromium, and copper in iron and stony-iron meteorites in relation to nickel content and structure. *Geochimica et Cosmochimica Acta,* 11, 263–278.

Lowry, S. C., Fitzsimmons, A., Pravec, P., *et al.* (2007) Direct detection of the asteroidal YORP effect. *Science,* 316, 272–274.

Lowry, S., Fitzsimmons, A., Lamy, P. and Weissman, P. (2008) Kuiper belt objects in the planetary region: The Jupiter-family comets. In *The Solar System Beyond Neptune,* eds. Barucci, M. A., Boehnhardt, H., Cruikshank, D. P. and Morbidelli, A. Tucson: University of Arizona Press, pp. 397–410.

Lucey, P. G. (1998) Model near-infrared optical constants of olivine and pyroxene as a function of iron content. *Journal of Geophysical Research,* 103, 1703–1713.

Lucey, P. G., Keil, K. and Whitely, R. (1998) The influence of temperature on the spectra of the A-asteroids and implications for their silicate chemistry. *Journal of Geophysical Research,* 103, 5865–5871.

Lugmair, G. W. and Shukolyukov, A. (1998) Early Solar System timescales according to ^{53}Mn–^{53}Cr systematics. *Geochimica et Cosmochimica Acta,* 62, 2863–2886.

Lugmair, G. W. and Shukolyukov, A. (2001) Early Solar System events and timescale. *Meteoritics & Planetary Science,* 36, 1017–1026.

Luu, J. X. and Jewitt, D. (1988) A two-part search for slow-moving objects. *Astronomical Journal,* 95, 1256–1262.

Macke, R. J. (2010) Survey of meteorite physical properties: Density, porosity, and magnetic susceptibility. PhD thesis, University of Central Florida, Orlando, Florida, 311 pp.

Mackinnon, I. D. R. and Zolensky, M. E. (1984) Proposed structures for poorly characterized phases in C2M carbonaceous chondrite meteorites. *Nature,* 309, 240–242.

MacPherson, G. J., Davis, A. M. and Zinner, E. K. (1995) The distribution of aluminum-26 in the early Solar System – A reappraisal. *Meteoritics,* 30, 365–386.

MacPherson, G. J., Simon, S. B., Davis, A. M., Grossman, L. and Krot, A. N. (2005) Calcium-aluminum-rich Inclusions: Major unanswered questions. In *Chondrites and the Protoplanetary Disk, ASP Conference Series, Volume 341,* eds. Krot, A. N., Scott, E. R. D. and Reipurth, B. San Francisco: Astronomical Society of the Pacific, pp. 225–250.

MacRobert, A. (2006) What are celestial coordinates? www.skyandtelescope.com/astronomy-resources/what-are-celestial-coordinates/.

Magri, C., Ostro, S. J., Rosema, K. D., *et al.* (1999) Mainbelt asteroids: Results of Arecibo and Goldstone radar observations of 37 objects during 1980–1995. *Icarus*, 140, 379–407.

Magri, C., Nolan, M. C., Ostro, S. J. and Giorgini, J. D. (2007) A radar survey of mainbelt asteroids: Arecibo observations of 55 objects during 1999–2003. *Icarus*, 186, 126–151.

Mainzer, A., Grav, T., Masiero, J., *et al.* (2011) NEOWISE studies of spectrophotometrically classified asteroids: Preliminary results. *Astrophysical Journal*, 741, 90.

Mainzer, A., Usui, F. and Trilling, D. E. (2015) Space-based thermal infrared studies of asteroids. In *Asteroids IV*, eds. Michel, P., DeMeo, F. E. and Bottke, W. F. Tucson: University of Arizona Press, pp. 89–106.

Mallama, A. (2011) Planetary magnitudes. *Sky and Telescope,* 121, 51–56.

Mamajek, E. E., Barenfeld, S. A., Ivanov, V. D., *et al.* (2015) The closest known flyby of a star to the Solar System. *Astrophysical Journal Letters*, 800, L17.

Marchi, S., Barbieri, C., Küppers, M., *et al.* (2010) The cratering history of asteroid (2867) Steins. *Planetary and Space Science*, 58, 1116–1123.

Marchi, S., Massironi, M., Vincent, J.-B., *et al.* (2012) The cratering history of asteroid 21 Lutetia. *Planetary and Space Science*, 66, 87–95.

Marchi, S., O'Brien, D. P., Schenk, P., *et al.* (2016) Cratering on Ceres: The puzzle of the missing large craters. Lunar and Planetary Science Conference, 47, 1281. www.hou.usra.edu/meetings/lpsc2016/pdf/1281.pdf.

Margot, J.-L. (2015) A quantitative criterion for defining planets. *Astronomical Journal*, 150, 185.

Marsden, B. G. (1973) The next return of the comet of the Perseid meteors. *Astronomical Journal*, 78, 654–662.

  (1977) Carl Friedrich Gauss, Astronomer. *Journal of the Royal Astronomical Society of Canada*, 71, 309–323.

  (1982) How to reduce plate measurements. *Sky and Telescope*, 64, 284.

  (1992) (4015) 1979 VA = Comet Wilson-Harrington (1949 III). *International Astronomical Union Circular*, 5585.

  (1993a) Comet Shoemaker-Levy (1993e). *International Astronomical Union Circular*, 5725.

  (1993b) Comet Shoemaker-Levy (1993e). *International Astronomical Union Circular*, 5730.

  (1993c) Comet Shoemaker-Levy (1993e). *International Astronomical Union Circular*, 5744.

  (1993d) Periodic comet Shoemaker-Levy 9 (1993e). *International Astronomical Union Circular*, 5800.

  (1993e) Periodic comet Shoemaker-Levy 9 (1993e). *International Astronomical Union Circular*, 5801.

  (1996) Comet P/1996 N2 (Elst-Pizarro). *International Astronomical Union Circular*, 6456.

  (1998a) 1997 XF11. *International Astronomical Union Circular*, 6837.

  (1998b) 1997 XF11. *International Astronomical Union Circular*, 6839.

  (2003) Editorial Notice. *Minor Planet Circular*, 49221.

Marti, K. and Graf, T. (1992) Cosmic-ray exposure history of ordinary chondrite meteorites. *Annual Review of Earth and Planetary Sciences*, 20, 221–243.

Martins, Z. (2011) Organic chemistry of carbonaceous meteorites. *Elements*, 7, 35–40.

Martins, Z., Botta, O., Fogel, M. L., *et al.* (2008) Extraterrestrial nucleobases in the Murchison meteorite. *Earth & Planetary Science Letters*, 270, 130–136.

Marvin, U. B. (1983) The discovery and initial characterization of Allan Hills 81005: The first lunar meteorite. *Geophysical Research Letters*, 10, 775–778.

(1992) The meteorite of Ensisheim: 1492 to 1992. *Meteoritics*, 27, 28–72.

(1996) Ernst Florens Friedrich Chladni (1756–1827) and the origins of modern meteorite research. *Meteoritics & Planetary Science*, 31, 545–588.

(2015) The origin and early history of the U.S. Antarctic search for meteorites program (ANSMET). In *35 Seasons of U.S. Antarctic Meteorites: A Pictorial Guide To The Collection,* eds. Righter, K., Corrigan, C., McCoy, T. and Harvey, R. Washington, DC: American Geophysical Union, pp. 1–22.

Marzari, F., Davis, D. and Vanzani, V. (1995) Collisional evolution of asteroid families. *Icarus*, 113, 168–187.

Masiero, J. R., Mainzer, A. K., Bauer, J. M., Grav, T., Nugent, C. R. and Stevenson, R. (2013) Asteroid family identification using the hierarchical clustering method and WISE/NEOWISE physical properties. *Astrophysical Journal*, 770, 7.

Mason, B. and Taylor, S. R. (1982) Inclusions in the Allende meteorite. *Smithsonian Contributions to the Earth Science*, 25, 30 pp.

Massironi, M., Marchi, S., Pajola, M., *et al.* (2012) Geological map and stratigraphy of asteroid 21 Lutetia. *Planetary and Space Science*, 66, 125–136.

Mayne, R. G., Sunshine, J. M., McSween, H. Y., Bus, S. J. and McCoy, T. J. (2011) The origin of Vesta's crust: Insights from spectroscopy of the Vestoids. *Icarus*, 214, 147–160.

McCanta, M. C., Treiman, A. H., Dyar, M. D., Alexander, C. M. O'D., Rumble, D. III and Essene, E. J. (2008) The LaPaz Icefield 04840 meteorite: Mineralogy, metamorphism, and origin of an amphibole- and biotite-bearing R chondrite. *Geochimica et Cosmochimica Acta*, 72, 5757–5780.

McCord, T. B., Adams, J. B. and Johnson, T. V. (1970) Asteroid Vesta: Spectral reflectivity and compositional implications. *Science*, 168, 1445–1447.

McCord, T. B., Castillo-Rogez, J. and Rivkin, A. (2011) Ceres: Its origin, evolution and structure and Dawn's potential contribution. *Space Science Reviews*, 163, 63–76.

McCord, T. B., Li, J.-Y., Combe, J.-P., *et al.* (2012) Dark material on Vesta from the infall of carbonaceous volatile-rich material. *Nature*, 491, 83–86.

McCoy, T. J. (1994) Partial melting on the acapulocite-lodranite meteorite parent body. PhD thesis, University of Hawaii, Honolulu, Hawaii, 146 pp.

McCoy, T. J., Keil, K., Clayton, R. N., *et al.* (1996) A petrologic, chemical, and isotopic study of Monument Draw and comparison with other acapulcoites: Evidence for formation by incipient partial melting. *Geochimica et Cosmochimica Acta*, 60, 2681–2708.

McCoy, T. J., Keil, K., Clayton, R. N., *et al.* (1997) A petrologic and isotopic study of lodranites: Evidence for early formation as partial melt residues from heterogeneous precursors. *Geochimica et Cosmochimica Acta*, 61, 623–637.

McCoy, T. J., Nittler, L. R., Burbine, T. H., Trombka, J. I., Clark, P. E. and Murphy, M. E. (2000) Anatomy of a partially differentiated asteroid: A "NEAR"-sighted view of acapulcoites and lodranites. *Icarus*, 148, 29–36.

McCoy, T. J., Burbine, T. H., McFadden, L. A., *et al.* (2001) The composition of 433 Eros: A mineralogical-chemical synthesis. *Meteoritics & Planetary Science*, 36, 1661–1672.

McCoy, T. J., Beck, A. W., Prettyman, T. H. and Mittlefehldt, D. W. (2015) Asteroid (4) Vesta II: Exploring a geologically and geochemically complex world with the Dawn mission. *Chemie der Erde*, 75, 273–285.

McDermott, K., Greenwood, R. C., Franchi, I. A., Anand, M. and Scott, E. R. D. (2011) Oxygen isotopic and petrological constraints on the origin and relationship of IIE iron meteorites and H chondrites. Lunar and Planetary Science, 42, 2763. http://www.lpi.usra.edu/meetings/lpsc2011/pdf/2763.pdf.

McFadden, L. A. (1999) The importance of comet Hale-Bopp: An astronomical perspective. *Bulletin of the American Astronomical Society*, 31, 1523.

McFadden, L. A., Gaffey, M. J. and McCord, T. B. (1985) Near-Earth asteroids: Possible sources from reflectance spectroscopy. *Science*, 229, 160–163.

McFadden, L. A., Skillman, D. R, Memarsadeghi, N., *et al.* (2015) Vesta's missing moons: Comprehensive search for natural satellites of Vesta by the Dawn spacecraft. *Icarus*, 257, 207–216.

McKay, D. S., Gibson, E. K. Jr., Thomas-Keptra, K. L., *et al.* (1996) Search for past life on Mars: Possible relic biogenic activity in Martian meteorite ALH84001. *Science*, 273, 924–930.

McKeegan, K. D., Aléon, J., Bradley, J., *et al.* (2006) Isotopic compositions of cometary matter returned by Stardust. *Science*, 314, 1724–1728.

McMahon, J. H. (1978) The discovery of a satellite of an asteroid. *Minor Planet Bulletin*, 6, 14–17.

McSween, H. Y. Jr. and Huss, G. R. (2010) *Cosmochemistry*. Cambridge: Cambridge University Press. 549 pp.

McSween, H. Y, Castillo-Rogez, J, Emery, J. P., De Sanctis, M. C. and the Dawn Science Team (2016) Rationalizing the composition and alteration of Ceres. Lunar and Planetary Science Conference, 47, 1258. www.hou.usra.edu/meetings/lpsc2016/pdf/1258.pdf.

Meech, K. J. and Svoreň, J. (2004) Using cometary activity to trace the physical and chemical evolution of cometary nuclei. In *Comets II*, eds. Festou, M., Keller, H. U. and Weaver, H. A. Tucson: University of Arizona Press, pp. 317–335.

Melosh, H. J. (1989) *Impact Catering: A Geologic Process*. Oxford: Oxford University Press, 245 pp.

  (2011) *Planetary Surface Processes*. Cambridge: Cambridge University Press, 534 pp.

Melosh, H. J. and Ivanov, B. A. (1999) Impact crater collapse. *Annual Review of Earth and Planetary Sciences,* 27, 385–415.

Merényi, E., Földy, L., Szegő, K., Tóth, I. and Kondor, A. (1990) The landscape of comet Halley. *Icarus*, 86, 9–20.

Merlin, F. (2015) New constraints on the surface of Pluto. *Astronomy & Astrophysics*, 582, A39.

Merlin, F., Barucci, M. A., de Bergh, C., *et al.* (2010) Chemical and physical properties of the variegated Pluto and Charon surfaces. *Icarus*, 210, 930–943.

Michalak, G. (2000) Determination of asteroid masses I. (1) Ceres, (2) Pallas and (4) Vesta. *Astronomy and Astrophysics*, 360, 363–374.

Michel, P., Tanga, P., Benz, W. and Richardson, D. C. (2002) Formation of asteroid families by catastrophic disruption: Simulations with fragmentation and gravitational reaccumulation. *Icarus*, 160, 10–23.

Michel, P., Benz, W. and Richardson, D. (2003) Disruption of fragmented parent bodies as the origin of asteroid families. *Nature*, 421, 608–611.

Michel, P., O'Brien, D. P., Abe, S. and Hirata, N. (2009) Itokawa's cratering record as observed by Hayabusa: Implications and collisional history. *Icarus*, 200, 503–513.

Mickel, P., DeMeo F. E. and Bottke, W. F. Jr., eds. (2015) *Asteroids IV*. Tucson: University of Arizona Press, 952 pp.

Michtchenko, T. A., Lazzaro, D., Ferraz-Mello, S. and Roig, F. (2002) Origin of the basaltic asteroid 1459 Magnya: A dynamical and mineralogical study of the outer main belt. *Icarus*, 158, 343–359.

Michtchenko, T. A., Lazzaro, D., Carvano, J. M. and Ferraz-Mello, S. (2010) Dynamic picture of the inner asteroid belt: Implications for the density, size and taxonomic distributions of real objects. *Monthly Notices of the Royal Astronomical Society*, 401, 2499–2516.

Mikouchi, T., Komatsu, M., Hagiya, K., *et al.* (2014) Mineralogy and crystallography of some Itokawa particles returned by the Hayabusa asteroidal sample return mission. *Earth, Planets and Space*, 66, 82.

Milliken, R. E. and Rivkin, A. S. (2009) Brucite and carbonate assemblages from altered olivine-rich materials on Ceres. *Nature Geosciences*, 2, 258–261.

Millis, R. L. and Elliot, J. L. (1979) Direct determination of asteroid diameters from occultation observations. In *Asteroids*, ed. Gehrels, T. Tucson: University of Arizona Press, pp. 98–118.

Minton, D. A., Richardson, J. E. and Fassett, C. I. (2015) Re-examining the main asteroid belt as the primary source of ancient lunar craters. *Icarus*, 247, 172–190.

Misawa, K., Yamaguchi, A. and Kaiden, H. (2005) U-Pb and ^{207}Pb-^{206}Pb ages of zircons from basaltic eucrites: Implications for early basaltic volcanism on the eucrite parent body. *Geochimica Cosmochimica Acta*, 69, 5847–5861.

Mittlefehldt, D. W. (2015) Asteroid (4) Vesta: I. The howardite-eucrite-diogenite (HED) clan of meteorites. *Chemie der Erde*, 75, 155–183.

Mittlefehldt, D. W., McCoy, T. J., Goodrich, C. A. and Kracher, A. (1998) Non-chondritic meteorites from asteroidal bodies. In *Reviews in Mineralogy, Vol. 36: Planetary Materials*, ed. Papike, J. J. Washington D.C.: Mineralogical Society of America, pp. 4-1–4-195.

Mittlefehldt, D. W., Killgore, M. and Lee, M. T. (2002) Petrology and geochemistry of D'Orbigny, geochemistry of Sahara 99555, and the origin of angrites. *Meteoritics & Planetary Science*, 37, 345–369.

Moore, J. M., Howard, A. D., Schenk, P. M., *et al.* (2015) Geology before Pluto: Pre-encounter considerations. *Icarus*, 246, 65–81.

Moore, J. M., McKinnon, W. B., Spencer, J. R., *et al.* (2016) The geology of Pluto and Charon through the eyes of New Horizons. *Science*, 351, 1284–1293.

Morbidelli, A. and Gladman, B. (1998) Orbital and temporal distributions of meteorites originating in the asteroid belt. *Meteoritics & Planetary Science*, 33, 999–1016.

Morbidelli, A., Levison, H. F., Tsiganis, K. and Gomes, R. (2005) Chaotic capture of Jupiter's Trojan asteroids in the early Solar System. *Nature*, 435, 462–465.

Moroz, L. V., Arnold, G., Korochantsev, A. V. and Wäsch, R. (1998) Natural solid bitumens as possible analogs for cometary and asteroid organics: 1. Reflectance spectroscopy of pure bitumens. *Icarus*, 134, 253–268.

Moroz, L., Schade, U. and Wäsch, R. (2000) Reflectance spectra of olivine–orthopyroxene-bearing assemblages at decreased temperatures: Implications for remote sensing of asteroids. *Icarus*, 147, 79–93.

Moroz, L. V., Baratta, G., Distefano, E., *et al.* (2003) Ion irradiation of asphaltite: Optical effects and implications for trans-Neptunian objects and centaurs. *Earth, Moon, and Planets*, 92, 279–289.

Moroz, L. V., Baratta, G., *et al.* (2004) Optical alteration of complex organics induced by ion irradiation: 1. Laboratory experiments. *Icarus,* 170, 214–228.

Morrison, D., Chapman, C. R., Steel, D. and Binzel, R. P. (2004) Impacts and the public: Communicating the nature of the impact hazard. In *Mitigation of Hazardous Comets and Asteroids*, eds. Belton, M. J. S., Morgan, T. H., Samarasinha, N.H. and Yeomans, D.K. Cambridge: Cambridge University Press, pp. 353–390.

Moskovitz, N. A., Jedicke, R., Gaidos, E., *et al.* (2008a) The distribution of basaltic asteroids in the main belt. *Icarus*, 198, 77–90.

Moskovitz, N. A., Lawrence, S., Jedicke, R., *et al.* (2008b) A spectroscopically unique main-belt asteroid: 10537 (1991 RY16). *Astrophysical Journal*, 682, L57–L60.

Moskovitz, N. A., Willman, M., Burbine, T. H., Binzel, R. P. and Bus, S. J. (2010) A spectroscopic comparison of HED meteorites and V-type asteroids in the inner main belt. *Icarus*, 208, 773–788.

Moskovitz, N. A., Abe, S., Pan, K.-S., *et al.* (2013) Rotational characterization of Hayabusa II target asteroid (162173) 1999 JU3. *Icarus*, 224, 24–31.

Mothé-Diniz, T. and Carvano, J. M. (2005) 221 Eos: A remnant of a partially differentiated parent body? *Astronomy & Astrophysics*, 442, 727–729.

Mothé-Diniz, T. and Nesvorný, D. (2008) Visible spectroscopy of extremely young asteroid families. *Astrononomy & Astrophysics*, 486, L9–L12.

Mothé-Diniz, T., Carvano, J. M. and Lazzaro, D. (2003) Distribution of taxonomic classes in the main belt of asteroids. *Icarus*, 162, 10–21.

Mothé-Diniz, T., Roig, F. and Carvano, J. M. (2005) Reanalysis of asteroid families structure through visible spectroscopy. *Icarus*, 174, 54–80.

Mothé-Diniz, T., Carvano, J. M., Bus, S. J., Duffard, R. and Burbine, T. H. (2008) Mineralogical analysis of the Eos family from near-infrared spectra. *Icarus*, 195, 277–294.

Mothé-Diniz, T., Jasmin, F. L., Carvano, J. M., Lazzaro, D., Nesvorný, D. and Ramirez, A. C. (2010) Re-assessing the ordinary chondrites paradox. *Astronomy & Astrophysics*, 514, A86.

Muinonen, K., Belskaya, I. N., Cellino, A., *et al.* (2010) A three-parameter magnitude phase function for asteroids. *Icarus*, 209, 542–555.

Müller, T. G. and Blommaert, J. A. D. L. (2004) 65 Cybele in the thermal infrared: Multiple observations and thermophysical analysis. *Astronomy & Astrophysics*, 418, 347–356.

Murchie, S. L. and Pieters, C. M. (1996) Spectral properties and rotational spectral heterogeneity of 433 Eros. *Journal of Geophysical Research*, 101, 2201–2214.

Mustard, J. F. and Pieters, C. M. (1989) Photometric phase functions of common geologic minerals and applications to quantitative analysis of mineral mixture reflectance spectra. *Journal of Geophysical Research*, 94, 13619–13634.

Nagao, K., Okazaki, R., Nakamura, T., *et al.* (2011) Irradiation history of Itokawa regolith material deduced from noble gases in the Hayabusa samples. *Science*, 333, 1128–1131.

Nagashima, K., Nara, M. and Matsuda, J.-i. (2012) Raman spectroscopic study of diamond and graphite in ureilites and the origin of diamonds. *Meteoritics & Planetary Science*, 47, 1728–1737.

Naidu, S. P., Margot, J. L., Taylor, P. A., *et al.* (2015) Radar imaging and characterization of the binary near-Earth asteroid (185851) 2000 DP107. *Astronomical Journal*, 150, 54.

Nakamura, T., Noguchi, T., Tanaka, M., *et al.* (2011) Itokawa dust particles: A direct link between S-type asteroids and ordinary chondrites. *Science*, 333, 1113–1116.

Napier, B., Asher, D., Bailey, M. and Steel, D. (2015) Centaurs as a hazard to civilization. *Astronomy & Geophysics*, 56, 6.24–6.30.

NASA (2007) *Near-Earth Object Survey and Deflection Analysis of Alternatives*. Report to Congress. 27 pp. www.nasa.gov/pdf/171331main_NEO_report_march07.pdf.

NASA/JPL/Cassini Imaging Team (2004) Asteroid Masursky imaged by Cassini. http://sci.esa.int/cassini-huygens/12080-asteroid-masursky/.

Nathues, A., Hoffmann, M., Cloutis, E. A., *et al.* (2014) Detection of serpentine in exogenic carbonaceous chondrite material on Vesta from Dawn FC data. *Icarus*, 239, 222–237.

Nathues, A., Hoffmann, M., Schäfer, M., *et al.* (2015a) Exogenic olivine on Vesta from Dawn framing camera color data. *Icarus*, 258, 467–482.

Nathues, A., Hoffmann, M., Schaefer, M., *et al.* (2015b) Sublimation in bright spots on (1) Ceres. *Nature*, 528, 237–240.

Ness, R. G. and Emery, J. P. (2014) Thermal inertia estimates of four near-Earth asteroids from Spitzer Space Telescope spectral observations. Lunar and Planetary Science Conference, 45, 1430. www.hou.usra.edu/meetings/lpsc2014/pdf/1430.pdf.

Nesse, W. (2012) *Introduction to Optical Mineralogy (Fourth Edition)*. Oxford: Oxford University Press, 384 pp.

Nesvorný, D. (2015) Nesvorný HCM Asteroid Families V3.0. EAR-A-VARGBDET-5-NESVORNYFAM-V3.0. NASA Planetary Data System.

Nesvorný, D., Vokrouhlický, D., Morbidelli, A. and Bottke, W. F. (2009) Asteroidal source of L chondrite meteorites. *Icarus*, 200, 698–701.

Nesvorný, D., Brož, M. and Carruba, V. (2015) Identification and dynamical properties of asteroid families. In *Asteroids IV*, eds. Michel, P., DeMeo, F. E. and Bottke, W. F. Tucson: University of Arizona Press, pp. 297–321.

Neukum, G., Ivanov, B. A. and Hartmann, W. K. (2001) Cratering records in the early Solar System in relation to the lunar reference system. *Space Science Reviews*, 96, 55–86.

Newburn, R. L. Jr. and Yeomans, D. K. (1982) Halley's comet. *Annual Review of Earth and Planetary Sciences*, 10, 297–326.

Newburn, R. L. Jr. Duxbury, T. C., Hanner, M., Semenov, B. V., Hirst, E. E., Bhat, R. S., Bhaskaran, S., Wang, T.-C. M., Tsou, P., Brownlee, D. E., Cheuvront, A. R., Gingerich, D. E., Bollendonk, G. R., Vellinga, J. M., Parham, K. A. and Mumaw, S. J. (2003) Phase curve and albedo of asteroid 5535 Annefrank. *Journal of Geophysical Research*, 108, 3-1–3-7.

Newcott, W. R. (1997) Age of comets. *National Geographic*, 192, 94–109.

Nguyen, A. N. and Messenger, S. (2011) Presolar history recorded in extraterrestrial materials. *Elements*, 7, 17–22.

Nieto, M. M. (1972) *The Titius-Bode's Law of Planetary Distances*. New York: Pergamon Press. 186 pp.

Nishiizumi, K., Elmore, D. and Kubik, P. W. (1989) Update on terrestrial ages of Antarctic meteorites. *Earth and Planetary Science Letters*, 93, 299–313.

Nittler, L. R., Starr, R. D., Lim, L., *et al.* (2001) X-ray fluorescence measurements of the surface elemental composition of asteroid 433 Eros. *Meteoritics & Planetary Science*, 36, 1673–1695.

Nittler, L. R., McCoy, T. J., Clark, P. E., Murphy, M. E., Trombka, J. I. and Jarosewich, E. (2004) Bulk element compositions of meteorites: A guide for interpreting remote-sensing geochemical measurements of planets and asteroids. *Antarctic Meteorite Research*, 17, 231–251.

Noguchi, T., Nakamura, T., Kimura, M., *et al.* (2011) Incipient space weathering observed on the surface of Itokawa dust particles. *Science,* 333, 1121–1125.

Nozette, S. and Shoemaker, E. M. (1994) Clementine goes exploring. *Sky and Telescope*, 87, 38–39.

Nugent, C. R., Margot, J. L., Chesley, S. R. and Vokrouhlický, D. (2012a) detection of semimajor axis drifts in 54 near-Earth asteroids: New measurements of the Yarkovsky effect. *Astronomical Journal*, 144, 60.

Nugent, C. R., Mainzer, A., Masiero, J., Grav, T. and Bauer, J. (2012b) The Yarkovsky drift's influence on NEAs: Trends and predictions with NEOWISE measurements. *Astronomical Journal*, 144, 75.

Nyquist, L. E., Bogard, D. D., Shih, C.-Y., Greshake, A., Stöffler, D. and Eugster, O. (2001) Ages and geologic histories of Martian meteorites. *Space Science Reviews*, 96, 105–164.

Nyquist, L. E., Kleine, T., Shih, C.-Y. and Reese, Y. D. (2009) The distribution of short-lived radioisotopes in the early Solar System and the chronology of asteroid accretion, differentiation, and secondary mineralization. *Geochimica et Cosmochimica Acta*, 73, 5115–5136.

Oberc, P. (1996) Disintegration of dust aggregates as origin of the boundaries in Halley's coma: Derivation of the sublimation parameters. *Icarus*, 124, 195–208.

Oberst, J., Mottola, S., Di Martino, M., *et al.* (2001) A model for rotation and shape of asteroid 9969 Braille from ground-based observations and images obtained during the Deep Space 1 (DS1) flyby. *Icarus*, 153, 16–23.

O'Brien, D. P., Greenberg, R. and Richardson, J. E. (2006) Craters on asteroids: Reconciling diverse impact records with a common impacting population. *Icarus,* 183, 79–92.

Ockert-Bell, M. E., Clark, B. E., Shepard, M. K., *et al.* (2010) The composition of M-type asteroids: Synthesis of spectroscopic and radar observations. *Icarus*, 210, 674–692.

Okada, A., Keil, K. and Taylor, G. J. (1981) Unusual weathering products of oldhamite parentage in the Norton County enstatite achondrite. *Meteoritics*, 16, 141–152.

Okada, T., Shirai, K., Yamamoto, Y., *et al.* (2006) X-ray fluorescence spectrometry of asteroid Itokawa by Hayabusa. *Science*, 312, 1338–1341.

Olkin, C. B., Reuter, D., Lunsford, A., Binzel, R. P. and Stern, S. A. (2006) The New Horizons distant flyby of asteroid 2002 JF56. *Bulletin of the American Astronomical Society*, 38, 597.

Oort, J. H. (1950) The structure of the cloud of comets surrounding the Solar System and a hypothesis concerning its origin. *Bulletin of the Astronomical Institutes of the Netherlands*, 11, 91–110.

Opeil, C. P., Consolmagno, G. J. and Britt, D. T. (2010) The thermal conductivity of meteorites: New measurements and analysis. *Icarus*, 208, 449–454.

Opeil, C. P., Consolmagno, G. J., Safarik, D. J. and Britt, D. T. (2012) Stony meteorite thermal properties and their relationship with meteorite chemical and physical states. *Meteoritics & Planetary Science,* 47, 319–329.

Ortiz, J. L., Moreno, F., Molina, A., Sanz, P. S. and Gutiérrez, P. J. (2007) Possible patterns in the distribution of planetary formation regions. *Monthly Notices of the Royal Astronomical Society*, 379, 1222–1226.

Ortiz, J. L., Sicardy, B., Braga-Ribas, F., *et al.* (2012) Albedo and atmospheric constraints of dwarf planet Makemake from a stellar occultation. *Nature*, 491, 566–569.

Ortiz, J. L., Duffard, R., Pinilla-Alonso, N., *et al.* (2015) Possible ring material around centaur (2060) Chiron. *Astronomy & Astrophysics*, 576, A18.

Ostro, S. J. (1993) Planetary radar astronomy. *Reviews of Modern Physics*, 65, 1235–1279.

Ostro, S. J., Campbell, D. B. and Shapiro, I. I. (1985) Mainbelt asteroids: Dual-polarization radar observations. *Science*, 229, 442–446.

Ostro, S. J., Hudson, R. S., Jurgens, R. F., *et al.* (1995) Radar images of asteroid 4179 Toutatis. *Science*, 270, 80–83.

Ostrowski, D. R., Gietzen, K., Lacy, C. and Sears, D. W. G. (2010) An investigation of the presence and nature of phyllosilicates on the surfaces of C asteroids by an analysis of the continuum slopes in their near-infrared spectra. *Meteoritics & Planetary Science*, 45, 615–637.

Ostrowski, D. R., Lacy, C. H. S., Gietzen, K. M. and Sears, D. W. G. (2011) IRTF spectra for 17 asteroids from the C and X complexes: A discussion of continuum slopes and their relationships to C chondrites and phyllosilicates. *Icarus*, 212, 682–696.

Palmer, J. and Davenhall, A. C. (2001) *The CCD Photometric Calibration Cookbook.* Oxfordshire: Council for the Central Laboratory of the Research Councils, 63 pp.

Park, R. S., Konopliv, A. S., Asmar, S. W., *et al.* (2014) Gravity field expansion in ellipsoidal harmonic and polyhedral internal representations applied to Vesta. *Icarus*, 240, 118–132.

Park, R., Konopliv, A., Bills, B., Vaughan, A., Raymond, C. and Russell, C. (2015) Physical properties of Ceres from the Dawn mission. *American Astronomical Society, DPS meeting*, 47, 212.10.

Parker, A., Ivezić, Ž., Jurić, M., Lupton, R., Sekora, M. D. and Kowalski, A. (2008) The size distributions of asteroid families in the SDSS Moving Object Catalog 4. *Icarus*, 198, 138–155.

Parker, A. S, Buie, M. W., Grundy, W. M. and Noll, K. S., (2016) S/2015 (136472) 1. *Minor Planet Electronic Circular*, 2016-H46.

Pätzold, M., Andert, T. P., Asmar, S. W., *et al.* (2011) Asteroid 21 Lutetia: Low mass, high density. *Science*, 334, 491–492.

Pätzold, M., Andert, T., Hahn, M., *et al.* (2016) A homogeneous nucleus for comet 67P/ Churyumov– Gerasimenko from its gravity field. *Nature*, 530, 63–65.

Peplowski, P. N., Lawrence, D. J., Prettyman, T. H., *et al.* (2013) Compositional variability on the surface of 4 Vesta revealed through GRaND measurements of high-energy gamma rays. *Meteoritics & Planetary Science*, 48, 2252–2270.

Peplowski, P. N., Bazell, D., Evans, L. G., Goldsten, J. O., Lawrence, D. J. and Nittler, L. R. (2015) Hydrogen and major element concentrations on asteroid 433 Eros: Evidence for an L and LL chondrite-like surface composition. *Meteoritics & Planetary Science*, 50, 353–367.

Perna, D., Dotto, E., Ieva, S., *et al.* (2016) Grasping the nature of potentially hazardous asteroids. *Astronomical Journal*, 151, 11.

Pesonen, L. J., Terho, M. and Kukkonen, I. (1993) Physical properties of 368 meteorites: Implications for meteorite magnetism and planetary geophysics. *Proceedings of the NIPR Symposium*, 6, 401–416.

Pieters, C. M., Taylor, L. A., Noble, S. K., *et al.* (2000) Space weathering on airless bodies: Resolving a mystery with lunar samples. *Meteoritics & Planetary Science*, 35, 1101–1107.

Pilger, C., Ceranna, L., Ross, J. O., Le Pinchon, A., Mialle, P. and Garcés, M. A. (2015) CBT infrasound network performance to detect the 2013 Russian fireball event. *Geophysical Research Letters*, 42, 2523–2531.

Pizzarello, S. and Shock, E. (2010) The organic composition of carbonaceous meteorites: The evolutionary story ahead of biochemistry. *Cold Spring Harbor Perspectives in Biology*, 2, a002105.

Pizzarello, S., Huang, Y., Becker, L., *et al.* (2001) The organic content of the Tagish Lake meteorite. *Science*, 293, 2236–2239.

Pizzarello, S., Cooper, G. W. and Flynn, G. J. (2006) The nature and distribution of the organic material in carbonaceous chondrites and interplanetary dust particles. In *Meteorites and the Early Solar System,* eds. Lauretta, D. S. and McSween, H. Y. Jr. Tucson: University of Arizona Press, pp. 625–651.

Plotkin, H., Clarke, R. S., McCoy, T. J. and Corrigan, C. M. (2012) The Old Woman, California, IIAB iron meteorite. *Meteoritics & Planetary Science*, 47, 929–946.

Polishook, D. (2012) Lightcurves and spin periods from the Wise Observatory – October 2011. *Minor Planet Bulletin*, 39, 88–89.

Polishook, D., Moskovitz, N., Binzel, R. P., *et al.* (2016) A 2 km-size asteroid challenging the rubble-pile spin barrier – A case for cohesion. *Icarus*, 267, 243–254.

Popova, O. P., Jenniskens, P., Emel'yanenko, V., *et al.* (2013) Chelyabinsk airburst, damage assessment, meteorite recovery, and characterization. *Science*, 342, 1069–1073.

Pravec, P., Harris, A. W. and Michałowski, T. (2002) Asteroid rotations. In *Asteroids III*, eds. Bottke, W. F. Jr., Cellino, A., Paolicchi, P. and Binzel, R. P. Tucson: University of Arizona Press, pp. 113–122.

Prettyman, T. H. (2007) Remote chemical sensing using nuclear spectroscopy. In *Encyclopedia of the Solar System (Second Edition)*, eds. McFadden, L.-A., Weissman P. R. and Johnson, T. V. San Diego: Elsevier, pp. 765–786.

Prettyman, T. H., Feldman, W. C., McSween, H. Y., *et al.* (2011) Dawn's gamma ray and neutron detector. *Space Science Reviews*, 163, 371–459.

Prettyman, T. H., Mittlefehldt, D. W., Yamashita, N., *et al.* (2012) Elemental mapping by Dawn reveals exogenic H in Vesta's regolith. *Science*, 338, 242–246.

Prettyman, T. H., Yamashita, N., Castillo-Rogez, J. C., *et al.* (2016) Elemental composition of Ceres by Dawn's gamma ray and neutron detector. Lunar and Planetary Science Conference, 47, 2228. www.hou.usra.edu/meetings/lpsc2016/pdf/2228.pdf.

Prinz, M., Keil, K., Hlava, P. F., Berkley, J. L., Gomes, C. B. and Curvello, W. S. (1977) Studies of Brazilian meteorites., III – Origin and history of the Angra Dos Reis achondrite. *Earth and Planetary Science Letters*, 35, 317–330.

Prior, G. (1916) On the genetic relationship and classification of meteorites. *Mineralogical Magazine*, 18, 26–44.

Prior, G. T. (1920) The classification of meteorites. *Mineralogical Magazine*, 19, 51–63.

Probst, L. W., Desch, S. J. and Thirumalai, A. (2015) The internal structure of Haumea. Lunar and Planetary Science Conference, 46, 2183. www.hou.usra.edu/meetings/lpsc2015/pdf/2183.pdf.

Puckett, A. W., Rector, T. A., Baalke, R. and Ajiki, O. (2016) OrbitMaster: An online tool for investigating Solar System dynamics and visualizing orbital uncertainties in the undergraduate classroom. *American Astronomical Society Meeting*, 227, 328.09.

Putnis, A. (1992) *An Introduction to Mineral Sciences.* Cambridge: Cambridge University Press, 480 pp.

Qin, L., Dauphas, N., Wadhwa, M., Masarik, J. and Janney, P. E. (2008) Rapid accretion and differentiation of iron meteorite parent bodies inferred from ^{182}Hf-^{182}W chronometry and thermal modeling. *Earth and Planetary Science Letters*, 273, 94–104.

Rabinowitz, D., Tourtellotte, S., Brown, M. and Trujillo, C. (2005) Photometric observations of a very bright TNO with an extraordinary lightcurve. *Bulletin of the American Astronomical Society*, 37, 746.

Rabinowitz, D. L., Barkume, K., Brown, M. E., *et al.* (2006) Photometric observations constraining the size, shape, and albedo of 2003 EL61, a rapidly rotating, Pluto-sized object in the Kuiper belt. *Astrophysical Journal*, 639, 1238–1251.

Ragozzine, D. and Brown, M. E. (2009) Orbits and masses of the satellites of the dwarf planet Haumea (2003 EL61). *Astronomical Journal*, 137, 4766–4776.

Rayman, M. D., Varghese, P., Lehman, D. H. and Livesay, L. L. (2000) Results from the Deep Space 1 technology validation mission. *Acta Astronautica*, 47, 475–487.

Raymond, S. N., O'Brien, D. P., Morbidelli, A. and Kaib, N. A. (2009) Building the terrestrial planets: Constrained accretion in the inner Solar System. *Icarus*, 203, 644–662.

Reddy, V., Emery, J. P., Gaffey, M. J., Bottke, W. F., Cramer, A. and Kelley, M. S. (2009) Composition of 298 Baptistina: Implications for the K/T impactor link. *Meteoritics & Planetary Science,* 44, 1917–1927.

Reddy, V., Gaffey, M. J., Kelley, M. S., Nathues, A., Li, J.-Y. and Yarbrough, R. (2010) Compositional heterogeneity of asteroid 4 Vesta's southern hemisphere: Implications for the Dawn mission. *Icarus*, 210, 693–706.

Reddy, V., Carvano, J. M., Lazzaro, D., *et al.* (2011) Mineralogical characterization of Baptistina asteroid family: Implications for K/T impactor source. *Icarus,* 216, 184–197.

Reddy, V., Nathues, A., Le Corre, L., *et al.* (2012a) Color and albedo heterogeneity of Vesta from Dawn. *Science*, 336, 700–704.

Reddy, V., Le Corre, L., O'Brien, D. P., *et al.* (2012b) Delivery of dark material to Vesta via carbonaceous chondritic impacts. *Icarus*, 221, 544–559.

Reddy, V., Sanchez, J. A., Nathues, A., *et al.* (2012c) Photometric, spectral phase and temperature effects on 4 Vesta and HED meteorites: Implications for the Dawn mission. *Icarus*, 217, 153–168.

Reddy, V., Li, J.-Y., Le Corre, L., *et al.* (2013) Comparing Dawn, Hubble Space Telescope, and ground-based interpretations of (4) Vesta. *Icarus*, 226, 1103–1114.

Reddy, V., Gary, B. L., Sanchez, J. A., *et al.* (2015) The physical characterization of the potentially hazardous asteroid 2004 BL86: A fragment of a differentiated asteroid. *Astrophysical Journal*, 811, 65.

Reichhardt, T. (1998) Asteroid watchers debate false alarm. *Nature*, 392, 215.

Renne, P. R., Deino, A. L., Hilgen, F. J., *et al.* (2013) Time scales of critical events around the Cretaceous-Paleogene boundary. *Science*, 339, 684–687.

Reuter, D. C., Stern, S. A., Scherrer, J., *et al.* (2008) Ralph: A visible/infrared imager for the New Horizons Pluto/Kuiper belt mission. *Space Science Reviews*, 140, 129–154.

Richardson, J. E. and Bowling, T. J. (2014) Investigating the combined effects of shape, density, and rotation on small body surface slopes and erosion rates. *Icarus*, 234, 53–65.

Richardson, J. E. and Melosh, H. J. (2013) An examination of the Deep Impact collision site on comet Tempel 1 via Stardust-NExT: Placing further constraints on cometary surface properties. *Icarus*, 222, 492–501.

Richardson, J. E., Melosh, H. J. and Greenberg, R. (2004) Impact-induced seismic activity on asteroid 433 Eros: A surface modification process. *Science*, 306, 1526–1529.

Richter, S, Eykins, R., Kühn, H., Aregbe, Y., Verbruggen, A. and Weyer, S. (2010) New average values for the $n(^{238}U)/n(^{235}U)$ isotope ratios of natural uranium standards. *International Journal of Mass Spectrometry*, 295, 94–97.

Rieke, G. H. (2007) Infrared detector arrays for astronomy. *Annual Review of Astronomy and Astrophysics*, 45, 77–115.

Righter, K. (2016) Curator's comments. *Antarctic Meteorite Newsletter*, 39, 1.

Rivkin, A. S. (2012) The fraction of hydrated C-complex asteroids in the asteroid belt from SDSS data. *Icarus*, 221, 744–752.

Rivkin, A. S. and Emery, J. P. (2010) Detection of ice and organics on an asteroidal surface. *Nature*, 464, 1322–1323.

Rivkin, A. S., Howell, E. S., Britt, D. T., Lebofsky, L. A., Nolan, M. C. and Branston, D. D. (1995) Three-micron spectrometric survey of M-and E-class asteroids. *Icarus*, 117, 90–100.

Rivkin, A. S., Howell, E. S., Lebofsky, L. A., Clark, B. E. and Britt, D. T. (2000) The nature of M-class asteroids from 3-micron observations. *Icarus*, 145, 351–368.

Rivkin, A. S., Davies, J. K., Clark, B. E., Trilling, D. E. and Brown, R. H. (2001) Aqueous alteration on S asteroid 6 Hebe? Lunar and Planetary Science Conference, XXXII, 1723. www.lpi.usra.edu/meetings/lpsc2001/pdf/1723.pdf.

Rivkin, A. S., Howell, E. S., Vilas, F. and Lebofsky, L. A. (2002) Hydrated minerals on asteroids: The astronomical record. In *Asteroids III*, eds. Bottke, W. F. Jr., Cellino, A., Paolicchi, P. and Binzel, R. P. Tucson: University of Arizona Press, pp. 235–253.

Rivkin, A. S., Trilling, D. E., Thomas, C. A., DeMeo, F., Spahr, T. B. and Binzel, R. P. (2007) Composition of the L5 Mars Trojans: Neighbors, not siblings. *Icarus*, 192, 434–441.

Rivkin, A. S., Li, J.-Y., Milliken, R. E., *et al.* (2011a) The surface composition of Ceres. *Space Science Reviews*, 163, 95–116.

Rivkin, A. S., Thomas, C. A., Trilling, D. E., Enga, M.-t. and Grier, J. A. (2011b) Ordinary chondrite-like colors in small Koronis family members. *Icarus*, 211, 1294–1297.

Rivkin, A. S., Thomas, C. A., Howell, E. S. and Emery, J. P. (2015) The Ch-class asteroids: Connecting a visible taxonomic class to a 3 μm band shape. *Astronomical Journal*, 150, 198.

Robbins, S. J., Antonenko, I., Kirchoff, M. R., *et al.* (2014) The variability of crater identification among expert and community crater analysts. *Icarus,* 234, 109–131.

Robertson, D. S., Lewis, W. M., Sheehan, P. M. and Toon, O. B. (2013) K-Pg extinction: Reevaluation of the heat-fire hypothesis. *Journal of Geophysical Research: Biogeosciences*, 118, 329–336.

Robinson, M. S., Thomas, P. C., Veverka, J., Murchie, S. and Carcich, B. (2001) The nature of ponded deposits on Eros. *Nature*, 413, 396–400.

Rogalski, A. (2002) Infrared detectors: An overview. *Infrared Physics & Technology*, 43, 187–210.

Roig, F. and Gil-Hutton, R. (2006) Selecting candidate V-type asteroids from the analysis of the Sloan Digital Sky Survey colors. *Icarus*, 183, 411–419.

Roig, F., Nesvorný, D., Gil-Hutton, R. and Lazzaro, D. (2008a) V-type asteroids in the middle main belt. *Icarus*, 194, 125–136.

Roig, F., Ribeiro, A. O. and Gil-Hutton, R. (2008b) Taxonomy of asteroid families among the Jupiter Trojans: Comparison between spectroscopic data and the Sloan Digital Sky Survey colors. *Astronomy & Astrophysics*, 483, 911–931.

Rose, A. and Weimer, P. K. (1989) Physical limits to the performance of imaging systems. *Physics Today*, 42, 24–32.

Roush, T. L. and Singer, R. B. (1987) Possible temperature variation effects on the interpretation of spatially resolved reflectance observations of asteroid surfaces. *Icarus*, 69, 571–574.

Rozitis, B., MacLennan, E. and Emery, J. P. (2014) Cohesive forces prevent the rotational breakup of rubble-pile asteroid (29075) 1950 DA. *Nature*, 512, 174–176

Rubin, A. E. (1997a) Mineralogy of meteorite groups. *Meteoritics*, 32, 231–247.

(1997b) Mineralogy of meteorite groups: An update. *Meteoritics*, 32, 733–734.

Rubin, A. E., Ulff-Moller, F., Wasson, J. T. and Carlson, W. D. (2001) The Portales Valley meteorite breccia: evidence for impact-induced melting and metamorphism of an ordinary chondrite. *Geochimica et Cosmochimica Acta*, 65, 323–342.

Rubin, A. E., Zolensky, M. E. and Bodnar, R. J. (2002) The halite-bearing Zag and Monahans (1998) meteorite breccias: Shock metamorphism, thermal metamorphism and aqueous alteration on the H-chondrite parent body. *Meteoritics & Planetary Science*, 37, 125–141.

Rubin, M., Altwegg, K., Balsiger, H., *et al.* (2015) Molecular nitrogen in comet 67P/Churyumov-Gerasimenko indicates a low formation temperature. *Science*, 348, 232–235.

Rubincam, D. P. (1987) LAGEOS orbit decay due to infrared radiation from Earth. *Journal of Geophysical Research*, 92, 1287–1294.

(1990) Drag on the LAGEOS satellite. *Journal of Geophysical Research*, 95, 4881–4886.

(1993) The LAGEOS along-track acceleration: A review. In *Relativistic Gravitational Experiments in Space*, eds. Demianski, M. and Everitt, C. W. F. Singapore: World Scientific Publishing, pp. 195–209.

(1998) Does sunlight change the spin of small asteroids? *Bulletin of the American Astronomical Society*, 30, 1035.

(2000) Radiative spin-up and spin-down of small asteroids. *Icarus*, 148, 2–11.

Rumpf, C., Lewis, H. G. and Atkinson, P. M. (2016) The global impact distribution of near-Earth objects. *Icarus*, 265, 209–217.

Ruprecht, J. D., Bosh, A. S., Person, M. J., *et al.* (2015) 29 November 2011 occultation by 2060 Chiron: Symmetric jet-like features. *Icarus*, 252, 271–276.

Russell, C. T., Raymond, C. A., Coradini, A., *et al.* (2012) Dawn at Vesta: Testing the protoplanetary paradigm. *Science*, 336, 684–686.

Russell, C. T., Raymond, C. A., Nathues, A., *et al.* (2015) Dawn explores Ceres: Results from the survey orbit. http://nesf2015.arc.nasa.gov/sites/default/files/downloads/pdf/05.pdf.

Russell, S. and Grady, M. M. (2006) A history of the meteorite collection at the Natural History Museum, London. In *The History of Meteoritics and Key Meteorite Collections: Fireballs, Falls and Finds, Geological Society Special Publication, no. 256*, eds. McCall, G. J. H., Bowden, A. J. and Bowden, A. J. Bath: Geological Society Publishing House, pp. 153–162.

Ruzicka, A. (2014) Silicate-bearing iron meteorites and their implications for the evolution of asteroidal parent bodies. *Chemie der Erde*, 74, 3–48.

Sagan, C. and Khare, B. (1979) Tholins: Organic chemistry of interstellar grains and gas. *Nature*, 277, 102–107.

Saito, J., Miyamoto, H., Nakamura, R., *et al.* (2006) Detailed images of asteroid 25143 Itokawa from Hayabusa. *Science*, 312, 1341–1344.

Salisbury, J. W. and Hunt, G. R. (1974) Meteorite spectra and weathering. *Journal of Geophysical Research*, 79, 4439–4441.

Salisbury, J. W., Hunt, G. R. and Lenhoff, C. J. (1975) Visible and near-infrared spectra: X. Stony meteorites. *Modern Geology*, 5, 115–126.

Sanchez, J. A., Reddy, V., Nathues, A., Cloutis, E. A., Mann, P. and Hiesinger, H. (2012) Phase reddening on near-Earth asteroids: Implications for mineralogical analysis, space weathering and taxonomic classification. *Icarus*, 220, 36–50.

Sasaki, S., Nakamura, K., Hamabe, Y., Kurahashi, E. and Hiroi, T. (2001) Production of iron nanoparticles by laser irradiation in a simulation of lunar-like space weathering. *Nature*, 410, 555–557.

Sato, K., Miyamoto, M. and Zolensky, M. E. (1997) Absorption bands near 3 micrometers in diffuse reflectance spectra of carbonaceous chondrites: Comparison with asteroids. *Meteoritics*, 32, 503–507.

Sawyer, S. R., Barker, E. S., Cochran, A. L. and Cochran, W. D. (1987) Spectrophotometry of Pluto-Charon mutual events: Individual spectra of Pluto and Charon. *Science*, 238, 1560–1563.

Schade, U. and Wäsch, R. (1999) NIR reflectance spectroscopy of mafic minerals in the temperature range between 80 and 473 K. *Advances in Space Research*, 23, 1253–1256.

Schaefer, M. W., Schaefer, B. E., Rabinowitz, D. L. and Tourtellotte, S. W. (2010) Phase curves of nine Trojan asteroids over a wide range of phase angles. *Icarus*, 207, 699–713.

Schechner, S. (1999) *Comets, Popular Culture, and the Birth of Modern Cosmology*. Princeton: Princeton University Press, 384 pp.

Scheeres, D. J., Ostro, S. J., Hudson, R. S., DeJong, E. M. and Suzuki, S. (1998) Dynamics of orbits close to asteroid 4179 Toutatis. *Icarus*, 132, 53–79.

Schmadel, L. (2003) *Dictionary of Minor Planet Names (5th Revised and Enlarged Edition)*. Berlin: Springer-Verlag, 1008 pp.

Schmedemann, N., Kneissl, T., Ivanov, B. A., *et al.* (2014) The cratering record, chronology and surface ages of (4) Vesta in comparison to smaller asteroids and the ages of HED meteorites. *Planetary and Space Science*, 103, 104–130.

Schmidt, B. E., Thomas, P. C., Bauer, J. M., *et al.* (2009) The shape and surface variation of 2 Pallas from the Hubble Space Telescope. *Science*, 326, 275–278.

Schmitt, D. G. (2002) The law of ownership and control of meteorites. *Meteoritics & Planetary Science*, 37 (Supplement), B5–B11.

Schmitt-Kopplin, P., Gabelica, Z., Gougeon, R. D., *et al.* (2010) High molecular diversity of extraterrestrial organic matter in Murchison meteorite revealed 40 years after its fall. *Proceedings of the National Academy of Sciences*, 107, 2763–2768.

Schmitz, B., Tassinari, M. and Peucker-Ehrenbrink, B. (2001) A rain of ordinary chondritic meteorites in the early Ordovician. *Earth and Planetary Science Letters*, 194, 1–15.

Schmitz, B., Heck, P. R., Alwmark, C., *et al.* (2011) Determining the impactor of the Ordovician Lockne crater: Oxygen and neon isotopes in chromite versus sedimentary PGE signatures. *Earth and Planetary Science Letters*, 306, 149–155.

Schoene, B. (2014) U–Th–Pb geochronology. *Treatise on Geochemistry (Second Edition), Volume 4: The Crust*, eds. Holland, H. D. and Turekian, K. K. Amsterdam: Elsevier, pp. 341–378.

Score, R. and Mason, B. (1983) ALHA81005. *Antarctic Meteorite Newsletter*, 6, 3.

Scott, E. R. D. (1977) Pallasites – Metal composition, classification and relationships with iron meteorites. *Geochimica et Cosmochimica Acta*, 41, 349–360.

Scott, E. R. D. and Krot, A. N. (2005) Chondritic meteorites and the high-temperature nebular origins of their components. In *Chondrites and the Protoplanetary Disk, ASP Conference Series, Volume 341*, eds. Krot, A. N., Scott, E. R. D. and Reipurth, B. San Francisco: Astronomical Society of the Pacific, pp. 15–53.

(2014) Chondrites and their components. In *Treatise on Geochemistry (Second Edition), Volume 1: Meteorites and Cosmochemical Processes*, eds. Holland, H. D. and Turekian, K. K. Amsterdam: Elsevier, pp. 65–137.

Scott, E. R. D., Haack, H. and Love, S. J. (2001) Formation of mesosiderites by fragmentation and reaccretion of a large differentiated asteroid. *Meteoritics & Planetary Science*, 36, 869–891.

Scott, E. R. D., Greenwood, R. C., Franchi, I. A. and Sanders, I. A. (2009) Oxygen isotopic constraints on the origin and parent bodies of eucrites, diogenites, and howardites. *Geochimica et Cosmochimica Acta,* 73, 5835–5853.

Scully, J. E. C., Russell, C. T., Yin, A., *et al.* (2015) Geomorphological evidence for transient water flow on Vesta. *Earth and Planetary Science Letters*, 411, 151–163.

Seares, F. H. (1930) Address of the retiring president of the society in awarding the Bruce Medal to professor Max Wolf. *Publications of the Astronomical Society of the Pacific*, 42, 5–22.

Sears, D. W. G. (2004) *The Origin of Chondrules and Chondrites*. Cambridge: Cambridge University Press, 222 pp.

Sekanina, Z. (1991) Cometary activity, discrete outgassing areas, and dust-jet formation. In *Comets in the Post-Halley Era, Volume 2*, eds. Newburn, R. L., Neugebauer, M. and Rahe, J. H. Dordrecht: Kluwer Academic Publishers. pp. 769–823.

Sephton, M. A. (2002) Organic compounds in carbonaceous meteorites. *Natural Product Reports*, 19, 292–311.

Sephton, M. A., Love, G. D., Watson, J. S., *et al.* (2004) Hydropyrolysis of insoluble carbonaceous matter in the Murchison meteorite: new insights into its macromolecular structure. *Geochimica et Cosmochimica Acta*, 68, 1385–1393.

Sharp, T. G. and DeCarli, P. S. (2006) Shock effects in meteorites. In *Meteorites and the Early Solar System*, eds. Lauretta, D. S. and McSween, H. Y. Jr. Tucson: University of Arizona Press, pp. 653–677.

Shepard, M. K. (2015) *Asteroids: Relics of Ancient Time*. Cambridge: Cambridge University Press, 368 pp.

Shepard, M. K., Clark, B. E., Ockert-Bell, M., *et al.* (2010) A radar survey of M- and X-class asteroids II. Summary and synthesis. *Icarus*, 208, 221–237.

Shepard, M. K., Harris, A. W., Taylor, P. A., *et al.* (2011) Radar observations of asteroids 64 Angelina and 69 Hesperia. *Icarus*, 215, 547–551.

Shepard, M. K., Taylor, P. A., Nolan, M. C., *et al.* (2015) A radar survey of M- and X-class asteroids. III. Insights into their composition, hydration state, & structure. *Icarus*, 245, 38–55.

Sheppard, S. S. and Trujillo, C. A. (2010) The size distribution of the Neptune Trojans and the missing intermediate-sized planetesimals. *Astrophysical Journal Letters*, 723, L233–L237.

Shestopalov, D. I., Golubeva, L. F., McFadden, L. A., Fornasier, S. and Taran, M. N. (2010) Titanium-bearing pyroxenes of some E asteroids: Coexisting of igneous and hydrated rocks. *Planetary and Space Science*, 58, 1400–1403.

Shkuratov, Y., Kaydash, V., Korokhin, V., *et al.* (2012) A critical assessment of the Hapke photometric model. *Journal of Quantitative Spectroscopy & Radiative Transfer*, 113, 2431–2456.

Shu, F. H., Shang, H. and Lee, T. (1996) Toward an astrophysical theory of chondrites. *Science*, 271, 1545–1552.

Shu, F. H., Shang, H., Glassgold, A. E. and Lee, T. (1997) X-rays and fluctuating X-winds from protostars. *Science*, 277, 1475–1479.

Shukolyukov, A. and Lugmair, G. W. (1998) Isotopic evidence for the Cretaceous-Tertiary impactor and its type. *Science,* 282, 927–930.

Sicardy, B., Ortiz, J. L., Assafin, M., *et al.* (2011) A Pluto-like radius and a high albedo for the dwarf planet Eris from an occultation. *Nature*, 478, 493–496.

Sierks, H., Lamy, P., Barbieri, C., *et al.* (2011) Images of asteroid 21 Lutetia: A remnant planetesimal from the early Solar System. *Science*, 334, 487–490.

Sierks, H., Barbieri, C., Lamy, P. L., *et al.* (2015) On the nucleus structure and activity of comet 67P/Churyumov-Gerasimenko. *Science*, 347, aaa104-1–aaa1044-5.

Singer, R. B. and Roush, T. L. (1985) Effects of temperature on remotely sensed mineral absorption features. *Journal of Geophysical Research*, 90, 12434–12444.

Sklute, E. C. (2014) On the subject of analyzing iron and sulfur bearing minerals from three extreme environments: Geological carbon sequestration, acid mine drainage, and Mars. PhD thesis, Stony Brook University, Stony Brook, New York, 415 pp.

Slade, M. A., Benner, L. A. M. and Silva, A. (2011) Goldstone Solar System Radar Observatory: Earth-based planetary mission support and unique science results. *Proceedings of the IEEE*, 99, 757–769.

Slivan, S. M. (2002) Spin vector alignment of Koronis family asteroids. *Nature*, 419, 49–51.

Slivan, S. M., Binzel, R. P., Kaasalainen, M., *et al.* (2009) Spin vectors in the Koronis family. II. Additional clustered spins, and one stray. *Icarus*, 200, 514–530.

Smith, R. C. (1995) *Observational Astrophysics*. Cambridge: Cambridge University Press, 468 pp.

Soter, S. (2006) What is a planet? *Astronomical Journal*, 132, 2513–2519.

Southwood, D. J. (1994) Recent magnetic field results from the Galileo and Ulysses spacecraft. *Philosophical Transactions: Physical Sciences and Engineering,* 349, 261–270.

Stansberry, J., Grundy, W., Brown, M., *et al.* (2008) Physical properties of Kuiper belt and centaur objects: Constraints from Spitzer Space Telescope. In *The Solar System Beyond Neptune*, eds. Barucci, M. A., Boehnhardt, H., Cruikshank, D. P. and Morbidelli, A. Tucson: University of Arizona Press, pp. 161–179.

Stephens, D. C. and Noll, K. S. (2006) Detection of six trans-Neptunian binaries with NICMOS: A high fraction of binaries in the cold classical disk. *Astronomical Journal*, 131, 1142–1148.

Stephenson, F. R., Yau, K. K. C. and Hunger, H. (1985) Records of Halley's comet on Bablyonian tablets. *Nature*, 314, 587–592.

Stern, S. A., Weaver, H. A., Steffl, A. J., *et al.* (2006) A giant impact origin for Pluto's small moons and satellite multiplicity in the Kuiper belt. *Nature*, 439, 946–948.

Stern, S. A., Bagenal, F., Ennico, K., *et al.* (2015) The Pluto system: Initial results from its exploration by New Horizons. *Science*, 350, aad1815-1–aad1815-8.

Stöffler, D., Keil, K. and Scott, E. R. D. (1991) Shock metamorphism of ordinary chondrites. *Geochimica et Cosmochimica Acta*, 55, 3845–3867.

Stooke, P. J. and Ford, H. A. (2001) Gaspra: Revised crater counts. Lunar and Planetary Science Conference, XXXII, 1073. www.lpi.usra.edu/meetings/lpsc2001/pdf/1073.pdf.

Storrs, A., Weiss, B., Zellner, B., *et al.* (1999) Imaging observations of asteroids with Hubble Space Telescope. *Icarus*, 137, 260–268.

Strom, R. G. and Neukum, G. (1988) The cratering record on Mercury and the origin of impacting objects. In *Mercury*, eds. Vilas, F., Chapman, C. R. and Shapley, M. S. Tucson: University of Arizona Press, pp. 336–373.

Strom, R. G., Malhotra, R., Ito, T., Yoshida, F. and Kring, D. A. (2005) The origin of planetary impactors in the inner Solar System. *Science*, 309, 1847–1850.

Strom, R. G., Malhotra, R., Xiao, Z.-Y., Ito, T., Yoshida, F. and Ostrach, L. R. (2015) The inner Solar System cratering record and the evolution of impactor populations. *Research in Astronomy and Astrophysics*, 15, 407-434.

Sunshine, J. M., Pieters, C. M. and Pratt, S. F. (1990) Deconvolution of mineral absorption bands: An improved approach. *Journal of Geophysical Research*, 95, 6955–6966.

Sunshine, J. M., Bus, S. J., McCoy, T. J., Burbine, T. H., Corrigan, C. M. and Binzel, R. P. (2004) High-calcium pyroxene as an indicator of igneous differentiation in asteroids and meteorites. *Meteoritics & Planetary Science*, 39, 1343–1357.

Sunshine, J. M., Bus, S. J., Corrigan, C. M., McCoy, T. J. and Burbine, T. H. (2007) Olivine-dominated asteroids and meteorites: Distinguishing nebular and igneous histories. *Meteoritics & Planetary Science*, 42, 155–170.

Sunshine, J. M., Connolly, H. C. Jr., McCoy, T. J., Bus, S. J. and La Croix, L. M. (2008) Ancient asteroids enriched in refractory inclusions. *Science*, 320, 514–517.

Swamy, K. K. S. (2010) *Physics of Comets (Third Edition)*. Singapore: World Scientific Publishing Co. Pte. Ltd., 460 pp.

Swisher, C. C., III, Grajales-Nishimura, J. M., Montanari, A., *et al.* (1992) Coeval ^{40}Ar/^{39}Ar ages of 65.0 million years ago from Chicxulub crater melt rock and Cretaceous-Tertiary boundary tektites. *Science*, 257, 954–958.

Tachibana, S., Abe, M., Arakawa, M., *et al.* (2014) Hayabusa2: Scientific importance of samples returned from C-type near-Earth asteroid (162173) 1999 JU3. *Geochemical Journal*, 48, 571–587.

Tait, A. W., Tomkins, A. G., Godel, B. M., Wilson, S. A. and Hasalova, P. (2014) Investigation of the H7 ordinary chondrite, Watson 012: Implications for recognition and classification of Type 7 meteorites. *Geochimica et Cosmochimica Acta*, 134, 175–196.

Takir, D., Emery, J. P., McSween, H. Y., Hibbitts, C. A., Clark, R. N., Pearson, N. and Wang, A. (2013) Nature and degree of aqueous alteration in CM and CI carbonaceous chondrites. *Meteoritics & Planetary Science*, 48, 1618–1637.

Taylor, P. A., Margot, J.-L., Vokrouhlický, D., *et al.* (2007) Spin rate of asteroid (54509) 2000 PH5 increasing due to the YORP effect. *Science*, 316, 274–277.

Tedesco, E. F., Tholen, D. J. and Zellner, B. (1982) The Eight-Color Asteroid Survey: Standard stars. *Astronomical Journal*, 87, 1585–1592.

Tedesco, E. F., Noah, P. V., Noah, M. and Price, S. D. (2002) The supplemental IRAS Minor Planet Survey. *Astronomical Journal*, 123, 1056–1085.

Teets, D. and Whitehead, K. (1999) The discovery of Ceres: How Gauss became famous. *Mathematics Magazine*, 72, 83–93.

Tegler, S. C. (2014) Kuiper belt objects: Physical studies. In *Encyclopedia of the Solar System (Third Edition)*, eds. Spohn, T., Breuer, D., Weissman P. R. and Johnson, T. V. Amsterdam: Elsevier, pp. 941–955.

Tegler, S. C. and Romanishin, W. (2000) Extremely red Kuiper-belt objects in near-circular orbits beyond 40 AU. *Nature*, 407, 979–981.

Tegler, S. C., Grundy, W. M., Romanishin, W., Consolmagno, G. J., Mogren, K. and Vilas, F. (2007) Optical spectroscopy of the large Kuiper belt objects 136472 (2005 FY9) and 136108 (2003 EL61). *Astronomical Journal*, 133, 526–530.

Thangjam, G., Reddy, V., Le Corre, L., *et al.* (2013) Lithologic mapping of HED terrains on Vesta using Dawn Framing Camera color data. *Meteoritics & Planetary Science*, 48, 2199–2210.

Thangjam, G., Nathues, A., Mengel, K., *et al.* (2014) Olivine-rich exposures at Bellicia and Arruntia craters on (4) Vesta from Dawn FC. *Meteoritics & Planetary Science*, 49, 1831–1850.

Tholen, D. J. (1984) Asteroid taxonomy from cluster analysis of photometry. PhD thesis, University of Arizona, Tucson, Arizona, 150 pp.

Thomas, C. A., Trilling, D. E. and Rivkin, A. S. (2012) Space weathering of small Koronis family asteroids in the SDSS Moving Object Catalog. *Icarus*, 219, 505–507.

Thomas, N., Sierks, H., Barbieri, C., *et al.* (2015) The morphological diversity of comet 67P/Churyumov-Gerasimenko. *Science*, 347, aaa0440-1–aaa0440-6.

Thomas, P. C., Veverka, J., Simonelli, D., *et al.* (1994) The shape of Gaspra. *Icarus*, 107, 23–36.

Thomas, P. C., Binzel, R. P., Gaffey, M. J., Zellner, B. H., Storrs, A. D. and Wells, E. (1997) Vesta: Spin pole, size, and shape from HST images. *Icarus*, 128, 88–94.

Thomas, P. C., Joseph, J., Carcich, B., *et al.* (2002) Eros: Shape, topography, and slope processes. *Icarus*, 155, 18–37.

Thomas, P. C., Parker, J. Wm., McFadden, L. A., *et al.* (2005) Differentiation of the asteroid Ceres as revealed by its shape. *Nature*, 437, 224–226.

Thompson, M. S., Christoffersen, R., Zega, T. J. and Keller, L. P. (2014) Microchemical and structural evidence for space weathering in soils from asteroid Itokawa. *Earth, Planets and Space*, 66, 89.

Tissot, F. L. H., Dauphas, N. and Grossman, L. (2016) Origin of uranium isotope variations in early solar nebula condensates. *Science Advances*, 2, e1501400.

Tombaugh, C. W. and Moore, P. (1980) *Out of the Darkness: The Planet Pluto*. Guilford: Stackhole Books, 221 pp.

Tremain, A. H., Gleason, J. D. and Bogard, D. D. (2000) The SNC meteorites are from Mars. *Planetary and Space Science*, 48, 1213–1230.

Trigo-Rodriguez, J. M., Llorca, J., Borovička, J. and Fabregat, J. (2003) Chemical abundances determined from meteor spectra: I Ratios of the main chemical elements. *Meteoritics & Planetary Science*, 38, 1283–1294.

Trinquier, A., Birck, J.-L. and Allègre, J. C. (2006) The nature of the K/T impactor. A ^{54}Cr reappraisal. *Earth and Planetary Science Letters*, 241, 780–788.

Trinquier, A., Birck, J.-L., Allègre, C. J., Göpel, C. and Ulfbeck, D. (2008) ^{53}Mn–^{53}Cr systematics of the early Solar System revisited. *Geochimica et Cosmochimica Acta*, 72, 5146–5163.

Trombka, J. I., Squyres, S. W., Brückner, J., *et al.* (2000) The elemental composition of asteroid 433 Eros: Results of the NEAR Shoemaker X-ray spectrometer. *Science*, 289, 2101–2105.

Trombka, J. I., Nittler, L. R., Starr, R. D., *et al.* (2001) The NEAR-Shoemaker X-ray/gamma-ray spectrometer experiment: Overview and lessons learned. *Meteoritics & Planetary Science*, 36, 1605–1616.

Trujillo, C. A. and Sheppard, S. S. (2014) A Sedna-like body with a perihelion of 80 astronomical units. *Nature*, 507, 471–474.

Trujillo, C. A., Brown, M. E., Rabinowitz, D. L. and Geballe, T. R. (2005) Near-infrared surface properties of the two intrinsically brightest minor planets: (90377) Sedna and (90482) Orcus. *Astrophysical Journal*, 627, 1057–1065.

Trujillo, C. A., Brown, M. E., Barkume, K. M., Schaller, E. L. and Rabinowitz, D. L. (2007) The surface of 2003 EL$_{61}$ in the near-infrared. *Astrophysical Journal*, 655, 1172–1178.

Tsiganis, K., Gomes, R., Morbidelli, A. and Levison, H. F. (2005). Origin of the orbital architecture of the giant planets of the Solar System. *Nature*, 435, 459–461.

Tubiana, C., Duffard, R., Barrera, L. and Boehnhardt, H. (2007) Photometric and spectroscopic observations of (132524) 2002 JF_{56}: Fly-by target of the New Horizons mission. *Astronomy & Astrophysics*, 463, 1197–1199.

Urey, H. C. (1956) Diamonds, meteorites, and the origin of the Solar System. *Astrophysical Journal*, 124, 623–637.

van Flandern, T. C., Tedesco, E. F. and Binzel, R. P. (1979) Satellites of asteroids. In *Asteroids*. Tucson: University of Arizona Press, pp. 443–465.

Van Schmus, W. R. and Wood, J. A. (1967) A chemical-petrologic classification for the chondritic meteorites. *Geochimica et Cosmochimica Acta*, 31, 747–765.

Vaughan, W. M. and Head, J. W. (2014) Criteria for identifying Mercurian meteorites. Lunar and Planetary Science Conference, 45, 2013. www.hou.usra.edu/meetings/lpsc2014/pdf/2013.pdf.

Veeder, G. J., Hanner, M. S., Matson, D. L., Tedesco, E. F., Lebofsky, L. A. and Tokunaga, A. T. (1989) Radiometry of near-Earth asteroids. *Astronomical Journal*, 97, 1211–1219.

Vereš, P., Jedicke, R., Fitzsimmons, A., *et al.* (2015) Absolute magnitudes and slope parameters for 250,000 asteroids observed by Pan-STARRS PS1 – preliminary results. *Icarus*, 261, 34–47.

Vernazza, P., Mothé-Diniz, T., Barucci, M. A., *et al.* (2005) Analysis of near-IR spectra of 1 Ceres and 4 Vesta, targets of the Dawn mission. *Astronomy & Astrophysics*, 436, 1113–1121.

Vernazza, P., Birlan, M., Rossi, A., *et al.* (2006a) Physical characterization of the Karin family. *Astronomy & Astrophysics*, 460, 945–951.

Vernazza, P., Brunetto, R., Strazzulla, G., *et al.* (2006b) Asteroid colors: A novel tool for magnetic field detection? The case of Vesta. *Astronomy & Astrophysics*, 451, L43–L46.

Vernazza, P., Binzel, R. P., Thomas, C. A., *et al.* (2008) Compositional differences between meteorites and near-Earth asteroids. *Nature*, 454, 858–860.

Vernazza, P., Binzel, R. P., Rossi, A., Fulchignoni, M. and Birlan, M. (2009) Solar wind as the origin of rapid reddening of asteroid surfaces. *Nature*, 458, 993–995.

Vernazza, P., Lamy, P., Groussin, O., *et al.* (2012) Asteroid (21) Lutetia as a remnant of Earth's precursor planetesimals. *Icarus*, 216, 650–659.

Vernazza, P., Zanda, B., Binzel, R. P., *et al.* (2014) Multiple and fast: The accretion of ordinary chondrite parent bodies. *Astrophysical Journal*, 791, 120.

Veverka, J., Thomas, P., Harch, A., *et al.* (1997) NEAR's flyby of 253 Mathilde: Images of a C asteroid. *Science*, 278, 2109–2114.

Veverka, J., Thomas, P., Harch, A., *et al.* (1999) NEAR encounter with asteroid 253 Mathilde: Overview. *Icarus*, 140, 3–16.

Veverka, J., Farquhar, B., Robinson, M., *et al.* (2001) The landing of the NEAR-Shoemaker spacecraft on asteroid 433 Eros. *Nature*, 413, 390–393.

Viateau, B. and Rapaport, M. (2001) Mass and density of asteroids (4) Vesta and (11) Parthenope. *Astronomy & Astrophysics*, 370, 602–609.

Vilas, F. (1994) A cheaper, faster, better way to detect water of hydration on Solar System bodies. *Icarus*, 111, 456–467.

(2008) Spectral characteristics of Hayabusa 2 near-Earth asteroid targets 162173 1999 JU3 and 2001 QC34. *Astronomical Journal*, 135, 1101–1105.

Vilas, F. and Gaffey, M. J. (1989) Phyllosilicate absorption features in main-belt and outer-belt asteroid reflectance spectra. *Science*, 246, 790–792.

Vilas, F. and Hendrix, A. R. (2015) The UV/blue effects of space weathering manifested in S-complex asteroids. I. Quantifying change with asteroid age. *Astronomical Journal*, 150, 64.

Vilas, F. and Smith, B. A. (1985) Reflectance spectrophotometry (about 0.5–1.0 micron) of outer-belt asteroids: Implications for primitive, organic Solar System material. *Icarus*, 64, 503–516.

Vilas, F., Cochran, A. L. and Jarvis, K. S. (2000) Vesta and the Vestoids: A new rock group? *Icarus*, 147, 119–128.

Vokrouhlický, D. and Farinella, P. (2000) Efficient delivery of meteorites to the Earth from a wide range of asteroid parent bodies. *Nature*, 407, 606–608.

Vokrouhlický, D., Milani, A. and Chesley, S. R. (2000) Yarkovsky effect on small near-Earth asteroids: Mathematical formulation and examples. *Icarus*, 148, 118–138.

Vokrouhlický, D., Nesvorný, D. and Bottke, W. F. (2003) The vector alignments of asteroid spins by thermal torques. *Nature*, 425, 147–151.

Vokrouhlický, D., Bottke, W. F., Chesley, S. R., Scheeres, D. J. and Statler, T. S. (2015) The Yarkovsky and YORP effects. In *Asteroids IV*, eds. Michel, P., DeMeo, F. E. and Bottke, W. F. Tucson: University of Arizona Press, pp. 509–531.

Vokrouhlický, D., Ďurech, J., Pravec, P., *et al.* (2016) The Schulhof family: Solving the age puzzle. *Astronomical Journal*, 151, 56.

Wall, M. (2014) Asteroid zoo asks public to find dangerous space rocks. www.space .com/26349-asteroid-zoo-zooniverse-planetary-resources.html.

Walsh, K. J., Morbidelli, A., Raymond, S. N., O'Brien, D. P. and Mandell, A. M. (2011) A low mass for Mars from Jupiter's early gas-driven migration. *Nature*, 475, 206–209.

Wang, Z., Kivelson, M. G., Joy, S., Khurana, K. K., Polanskey, C., Southwood, D. J. and Walker, R. J. (1995) Solar wind interaction with small bodies: 1: Whistler wing signatures near to Gaspra and Ida. *Advances in Space Research*, 16, 447–457.

Warner, B. D. (2006) *A Practical Guide to Lightcurve Photometry and Analysis*. New York: Springer Science, 298 pp.

(2011) *MPO Canopus and PhotoRed*. Colorado Springs: Bdw Publishing, 286 pp.

Warner, B. D., Harris, A. W. and Pravec, P. (2009) The asteroid lightcurve database. *Icarus*, 202, 134–146.

Wasserburg, G. J., Wimpenny, J. and Yin, Q.-z. (2012) Mg isotopic heterogeneity, Al-Mg isochrons, and canonical ^{26}Al/^{27}Al in the early Solar System. *Meteoritics & Planetary Science*, 47, 1980–1997.

Wasson, J. T. (1995) Sampling the asteroid belt: How biases make it difficult to establish meteorite-asteroid connections. *Meteoritics*, 30, 595.

Wasson, J. T. and Wang, J. (1986) A nonmagmatic origin of group-IIE iron meteorites. *Geochimica et Cosmochimica Acta*, 50, 725–732.

Watson, T. (2015) Falling junk has scientific value. *Nature*, 526: 621–622.

(2016) Falling space debris traced to 1998 lunar mission. *Nature*, 10.1038/nature .2016.19162.

Watters, T. R. and Prinz, M. (1979) Aubrites: Their origin and relationship to enstatite chondrites. *Proceedings of the Tenth Lunar and Planetary Science Conference*, 1073–1093.

Weaver, H. A., Feldman, P. D., A'Hearn, M. F., *et al.* (1997) The activity and size of the nucleus of comet Hale-Bopp (C/1995 O1). *Science*, 275, 1900–1904.

Weaver, H. A., Buie, M. W., Buratti, B., *et al.* (2016) The small satellites of Pluto as observed by New Horizons. *Science*, 351, aae0030-1–aae0030-5.

Weisberg, M. K., Prinz, M., Clayton, R. N., *et al.* (1996) The K (Kakangari) chondrite grouplet. *Geochimica et Cosmochimica Acta*, 60, 4253–4263.

Weisberg, M. K., Prinz, M., Clayton, R. N. and Mayeda, T. K. (1997) CV3 chondrites: Three subgroups, not two. *Meteoritics & Planetary Science*, 32, A138–A139.

Weisberg, M. K., Prinz, M., Clayton, R. N., *et al.* (2001) A new metal-rich chondrite grouplet. *Meteoritics & Planetary Science*, 36, 401–418.

Weiss, B. P., Bryson J. F. J., Harrison, R. J., *et al.* (2016) A core dynamo on an iron meteorite parent body and the magnetism of metallic asteroids. Lunar and Planetary Science Conference, 47, 1661. www.hou.usra.edu/meetings/lpsc2016/pdf/1661.pdf.

Weiss, R., Wünnemann, K. and Bahlburg, H. (2006) Numerical modelling of generation, propagation and run-up of tsunamis caused by oceanic impacts: Model strategy and technical solutions. *Geophysical Journal International*, 167, 77–88.

Welten, K. C., Alderliesten, C., van der Borg, K., Lindner, L., Loeken, T. and Schultz, L. (1997a) Lewis Cliff 86360: An Antarctic L-chondrite with a terrestrial age of 2.35 million years. *Meteoritics & Planetary Science*, 32, 775–780.

Welten, K. C., Lindner, L., van der Borg, K., Loeken, T., Scherer, P. and Schultz, L. (1997b) Cosmic-ray exposure ages of diogenites and the recent collisional history of the howardite, eucrite and diogenite parent body/bodies. *Meteoritics & Planetary Science*, 32, 891–902.

Welten, K. C., Nishiizumi, K., Finkel, R. C., *et al.* (2004) Exposure history and terrestrial ages of ordinary chondrites from the Dar al Gani region, Libya. *Meteoritics & Planetary Science*, 39, 481–498.

Welten, K. C., Folco, L., Nishiizumi, K., *et al.* (2008) Meteoritic and bedrock constraints on the glacial history of Frontier Mountain in northern Victoria Land, Antarctica. *Earth and Planetary Science Letters*, 270, 308–315.

Werner, S. C. (2005) Major aspects of the chronostratigraphy and geologic evolutionary history of Mars. PhD thesis, Free University of Berlin, Berlin, Germany, 252 pp.

Wetherill, G. W. (1987) Dynamical relations between asteroids, meteorites and Apollo-Amor objects. *Philosophical Transactions of the Royal Society A*, 323, 323–337.

Whipple, F. L. (1950) A comet model. I. The acceleration of comet Encke. *Astrophysical Journal*, 111, 375–394.

    (1989) Comets in the space age. *Astrophysical Journal*, 341, 1–15.

Williams, D. A., Jaumann, R., McSween, H. Y. Jr., *et al.* (2014) The chronostratigraphy of protoplanet Vesta. *Icarus*, 244, 158–165.

Wisdom, J. (1983) Chaotic behavior and the origin of the 3:1 Kirkwood gap. *Icarus*, 56, 51–74.

Wlotzka, F. (1993) A weathering scale for the ordinary chondrites. *Meteoritics*, 28, 460.

Wood, C. A. and Ashwal, L. D. (1981) SNC meteorites: Igneous rocks from Mars? *Proceedings of the Twelfth Lunar Planetary Science Conference*, 1359–1375.

Wood, H. J. and Kuiper, G. P. (1963) Photometric studies of asteroids. *Astrophysical Journal*, 137, 1279–1285.

Wright, J. T., Marcy, G. W., Howard, A. W., Johnson J. A., Morton, T. D. and Fischer, D. A. (2012) The frequency of hot Jupiters orbiting nearby solar-type stars. *Astrophysical Journal*, 753, 160.

Xu, S., Binzel, R. P., Burbine, T. H. and Bus, S. J. (1995) Small Main-belt Asteroid Spectroscopic Survey: Initial results. *Icarus*, 115, 1–35.

Yamaguchi, A., Clayton, R. N., Mayeda, T. K., *et al.* (2002) A new source of basaltic meteorites inferred from Northwest Africa 011. *Science*, 296, 334–336.

Yang, B. and Jewitt, D. (2010) Identification of magnetite in B-type asteroids. *Astronomical Journal*, 140, 692–698.

Yang, J., Goldstein, J. I. and Scott, E. R. D. (2010) Main-group pallasites: Thermal history, relationship to IIIAB irons, and origin. *Geochimica et Cosmochimica Acta*, 74, 4471–4492.

Yano, H., Kubota, T., Miyamoto, H., *et al.* (2006) Touchdown of the Hayabusa spacecraft at the Muses Sea on Itokawa. *Science*, 312, 1350–1353.

Yau, K., Yeomans, D. and Weissman, P. (1994) The past and future motion of Comet P/Swift–Tuttle. *Monthly Notices of the Royal Astronomical Society*, 266, 305–316.

Yeomans, D. K., Barriot, J.-P., Dunham, D. W., *et al.* (1997) Estimating the mass of asteroid 253 Mathilde from tracking data during the NEAR flyby. *Science*, 278, 2106–2109.

Yeomans, D. K., Antreasian, P. G., Barriot, J.-P., *et al.* (2000) Radio science results during the NEAR-Shoemaker spacecraft rendezvous with Eros. *Science*, 289, 2085–2088.

Yoshida, F. and Nakamura, T. (2005) Size distribution of faint Jovian L4 Trojan asteroids. *Astronomical Journal*, 130, 2900–2911.

Young, E. D. and Russell, S. S. (1998) Oxygen reservoirs in the early solar nebula inferred from an Allende CAI. *Science*, 282, 452–455.

Young, E. D., Kohl, I. E., Warren, P. H., Rubie, D. C., Jacobsen, S. A. and Morbidelli, A. (2016) Oxygen isotopic evidence for vigorous mixing during the Moon-forming giant impact. *Science*, 351, 493–496.

Yurimoto, H., Abe, K.-i., Abe, M., *et al.* (2011) Oxygen isotopic compositions of asteroidal materials returned from Itokawa by the Hayabusa mission. *Science,* 333, 1116–1119.

Zambon, F., De Sanctis, M. C., Schröder, S., *et al.* (2014) Spectral analysis of the bright materials on the asteroid Vesta. *Icarus*, 240, 73–85.

Zappalà, V., Cellino, A., Farinella, P. and Knežević, Z. (1990) Asteroid families. I. Identification by hierarchical clustering and reliability assessment. *Astronomical Journal*, 100, 2030–2046.

Zappalà, V., Cellino, A., Farinella, P. and Milani, A. (1994) Asteroid families. 2: Extension to unnumbered multiopposition asteroids. *Astronomical Journal*, 107, 772–801.

Zappalà, V., Bendjoya, Ph., Cellino, A., Farinella, P. and Froeschlé, C. (1995) Asteroid families: Search of a 12,487-asteroid sample using two different clustering techniques. *Icarus*, 116, 291–314.

Zeller, E. J. and Ronca, L. B. (1967) Space weathering of lunar and asteroidal surfaces. *Icarus*, 7, 372–379.

Zellner, B. (1973) Polarimetric albedos of asteroids. *Bulletin of the American Astronomical Society*, 5, 388.

(1975) 44 Nysa: An iron-depleted asteroid. *Astrophysical Journal*, 198, L45–L47.

Zellner, B. and Gradie, J. (1976) Minor planets and related objects. XX. Polarimetric evidence for the albedos and compositions of 94 asteroids. *Astronomical Journal*, 81, 262–280.

Zellner, B., Leake, M., Williams, J. G. and Morrison, D. (1977) The E asteroids and the origin of the enstatite achondrites. *Geochimica Cosmochimica Acta*, 41, 1759–1767.

Zellner, B., Tholen, D. J. and Tedesco, E. F. (1985) The eight-color asteroid survey: Results for 589 minor planets. *Icarus*, 61, 355–416.

Zielenbach, W. (2011) Mass determination studies of 104 large asteroids. *Astronomical Journal*, 142, 120.

Zolensky, M. E., Weisberg, M. K., Buchanan, P. C. and Mittlefehldt, D. W. (1996) Mineralogy of carbonaceous chondrite clasts in HED achondrites and the Moon. *Meteoritics & Planetary Science*, 31, 518–537.

Zolensky, M. E., Bodnar, R. J., Gibson, E. K. Jr., *et al.* (1999) Asteroidal water within fluid inclusion-bearing halite in an H5 chondrite, Monahans (1998). *Science*, 285, 1377–1379.

Zolensky, M., Herrin, J., Mikouchi, T., *et al.* (2010) Mineralogy and petrography of the Almahata Sitta ureilite. *Meteoritics & Planetary Science*, 45, 1618–1637.

Zolensky, M., Mikouchi, T., Fries, M., *et al.* (2014) Mineralogy and petrography of C asteroid regolith: The Sutter's Mill CM meteorite. *Meteoritics & Planetary Science*, 49, 1997–2016.

Zou, X., Li, C., Liu, J., Wang, W., Li, H. and Ping, J. (2014) The preliminary analysis of the 4179 Toutatis snapshots of the Chang'E-2 flyby. *Icarus*, 229, 348–354.

Index